Robot Technology Fundamentals

Robot Technology Fundamentals

James G. Keramas

Professor of Engineering Technology
University of Massachusetts (ret.)

Delmar Publishers

an International Thomson Publishing company

Albany • Bonn • Boston • Cincinnati • Detroit • London • Madrid
Melbourne • Mexico City • New York • Pacific Grove • Paris • San Francisco
Singapore • Tokyo • Toronto • Washington

NOTICE TO THE READER

Publisher does not warrant or guarantee any of the products described herein or perform any independent analysis in connection with any of the product information contained herein. Publisher does not assume, and expressly disclaims, any obligation to obtain and include information other than that provided to it by the manufacturer.

The reader is expressly warned to consider and adopt all safety precautions that might be indicated by the activities herein and to avoid all potential hazards. By following the instructions contained herein, the reader willingly assumes all risks in connection with such instructions. The Publisher makes no representation or warranties of any kind, including but not limited to, the warranties of fitness for particular purpose or merchantability, nor are any such representations implied with respect to the material set forth herein, and the publisher takes no responsibility with respect to such material. The publisher shall not be liable for any special, consequential, or exemplary damages resulting, in whole or part, from the readers' use of, or reliance upon, this material.

Cover Design: Nicole Reamer

Delmar Staff

Publisher: Alar Elkin
Acquisitions Editor: Thomas Schin
Production Manager: Larry Main

Art Director: Nicole Reamer
Editorial Assistant: Fionnuala McAvey

COPYRIGHT © 1999
Delmar is a division of Thomson Learning. The Thomson Learning logo is a registered trademark used herein under license.

Printed in the United States of America
   6 7 8 9 10 XXX 07 06 05

For more information, contact Delmar, 3 Columbia Circle, PO Box 15015, Albany, NY 12212-0515; or find us on the World Wide Web at http://www.delmar.com

International Division List

Japan:
Thomson Learning
Palaceside Building 5F
1-1-1 Hitotsubashi, Chiyoda-ku
Tokyo 100 0003 Japan
Tel: 813 5218 6544
Fax: 813 5218 6551

Australia/New Zealand:
Nelson/Thomson Learning
102 Dodds Street
South Melbourne, Victoria 3205
Australia
Tel: 61 39 685 4111
Fax: 61 39 685 4199

UK/Europe/Middle East:
Thomson Learning
Berkshire House
168-173 High Holborn
London
WC1V 7AA United Kingdom
Tel: 44 171 497 1422
Fax: 44 171 497 1426

Latin America:
Thomson Learning
Seneca, 53
Colonia Polanco
11560 Mexico D.F. Mexico
Tel: 525-281-2906
Fax: 525-281-2656

Canada:
Nelson/Thomson Learning
1120 Birchmount Road
Scarborough, Ontario
Canada M1K 5G4
Tel: 416-752-9100
Fax: 416-752-8102

Asia:
Thomson Learning
60 Albert Street, #15-01
Albert Complex
Singapore 189969
Tel: 65 336 6411
Fax: 65 336 7411

All rights reserved Thomson Learning 1999. The text of this publication, or any part thereof, may not be reproduced or transmitted in any form or by any means, electronics or mechanical, including photocopying, recording, storage in an information retrieval system, or otherwise, without prior permission of the publisher.

You can request permission to use material from this text through the following phone and fax numbers.
Phone: 1-800-730-2214; Fax 1-800-730-2215; or visit our Web site at http://www.thomsonrights.com

Library of Congress Cataloging-in-Publication Data
Keramas, James G.
    Robot technology fundamentals/James G. Keramas.
      p. cm.
    Includes bibliographical references and index.
    ISBN 0-8273-8236-7
     1. Robotics. 2. Robots, Industrial. I. Title.
TJ211.K47   1998   98-9882
670.42'72—dc21   CIP

*To Philip and Colin
and in memory of my parents
George and Irene Keramas
and my brother Anthony*

Contents

Preface	**xiii**
1. Introduction	**1**
Objectives	1
Automation and Robots	1
Brief History	3
The Technology of Robots	7
Economic and Social Issues	12
Present and Future Applications	15
Summary	17
Review Questions	18
Problems	19
References	20
2. Robot Technology	**21**
Objectives	21
Fundamentals	21
General Characteristics	21
Basic Components	23
Robot Anatomy	41
Robot Generations	43
Robot Selection	45
Summary	47
Review Questions	47
Problems	48
References	48

3. Robot Classification — 51

- Objectives — 51
- Classification — 51
- Arm Geometry — 51
- Degrees of Freedom — 64
- Power Sources — 65
- Types of Motion — 72
- Path Control — 73
- Intelligence Level — 78
- Summary — 80
- Review Questions — 81
- Problems — 81
- References — 82

4. Robot System Analysis — 83

- Objectives — 83
- Robot Operation — 83
- Hierarchical Control Structure — 90
- Line Tracking — 93
- Dynamic Properties of Robots — 94
- Modular Robot Components — 104
- Summary — 107
- Review Questions — 107
- Problems — 108
- References — 109

5. Robot End Effectors — 111

- Objectives — 111
- Types of End Effectors — 111
- Mechanical Grippers — 114
- Gripper Force Analysis — 117
- Other Types of Grippers — 124
- Special-Purpose Grippers — 131
- Gripper Selection and Design — 133
- Process Tooling — 134

Compliance	136
Summary	147
Review Questions	148
Problems	149
References	150

6. Sensors — 153

Objectives	153
Robot Sensors	153
Sensor Classification	154
Microswitches	160
Solid-State Switches	170
Proximity Sensors	172
Photoelectric Sensors	173
Rotary Position Sensors	176
Usage and Selection of Sensors	183
Signal Processing	184
Sensors and Control Integration	186
Summary	187
Review Questions	188
Problems	189
References	190

7. Vision — 193

Objectives	193
Visual Sensing	193
Machine Vision	195
Machine Vision Applications	206
Other Optical Methods	207
Summary	213
Review Questions	215
Problems	215
References	219

8. Control Systems — 221

Objectives — 221
Control System Correlation — 221
Control System Requirements — 222
Programmable Logic Controller — 224
PLC Programming Terminals — 239
Proportional-Integral-Derivative — 246
Computer Numerical Control — 248
Microprocessor Unit — 249
Universal Robot Controller — 254
Interfacing — 254
Workcell Control — 256
Summary — 259
Review Questions — 260
Problems — 261
References — 266

9. Programming — 267

Objectives — 267
Robot Programming — 267
Programming Methods — 270
Programming Languages — 274
Levels of Robot Programming — 280
Space Position Programming — 288
Motion Interpolation — 292
Program Statements — 294
Sample Programs — 296
Summary — 302
Review Questions — 303
Problems — 304
References — 306

10. Artificial Intelligence — 309

Objectives — 309
Intelligent Systems — 309

Elements of Artificial Intelligence	309
System Architecture	316
Applications of Advanced Robots	321
Fuzzy Logic Controls	325
Advanced Concepts and Procedures	325
Future Developments	327
Impact on Employment	328
Summary	328
Review Questions	330
Problems	330
References	331

11. Safety — 333

Objectives	333
Robot Safety	333
Safety Standards	334
System Reliability	336
Human Factor Issues	337
Safety Sensors and Monitoring	338
Safeguarding	340
Training	342
Safety Guidelines	343
Definitions	344
Summary	346
Review Questions	347
Problems	348
References	348

12. Industrial Applications — 351

Objectives	351
Automation in Manufacturing	351
Robot Applications	352
Material-Handling Applications	353
Processing Operations	357
Assembly Operations	359

Inspection Operations	362
Evaluating the Potential of a Robot Application	363
Future Applications	364
Challenge for the Future	365
Innovations	367
Case Studies	368
Summary	375
Review Questions	376
Problems	377
References	378

Glossary of Selected Terms 381

Robot Manufacturers 403

Index 405

Preface

In the last three decades, the robotics field benefited considerably from the advancement of microelectronics, computer science, and improved design of electrical, electromechanical, and hydromechanical servo systems.

Over the past decade, highly selective applications for robots resulted in so-called "islands" of automation. With the development of more sophisticated automation concepts, such as computer-integrated manufacturing (CIM), users learned that industrial operations are usually best automated through the integration of robots with machines into what is often referred to as a "work cell." In these configurations, the robots, along with machines that they serve, are treated as a "unified system." This integration, which causes the need for knowledge about robotics, has become very important in flexible automation today.

Robots can be used in any industry providing work and services, and can also be adapted easily to numerous job functions with uncanny skill and unmatched endurance. Just as computing technology is being absorbed into systems as a result of integration and thus losing its former separate identity, so it is predicted that the robot will ultimately be regarded as just one more part of an automated complex.

This text provides a comprehensive approach to learning the technical aspects of industrial robots. The material in the text not only emphasizes current technology, but also looks into the future where it describes the direction of robot technology in the years ahead.

The text is written for both the industrial and technical reader as well as the student in two- and four-year colleges or universities. To support classroom instruction, a practical approach is used, with examples, questions, and problems designed to provide the student with a sense of relevancy. Principles and techniques are introduced by concrete examples rather than by abstract rhetoric. Simple language is used, and excess talkativeness is avoided.

To aid the student in understanding the purpose of studying robot systems, a list of objectives is given at the beginning of each chapter. At the end of each chapter is a summary of the chapter's content for review, together with questions and problems as well as a list of reference material for additional reading.

The text can be used for a one-quarter or one-semester course to introduce robot technology fundamentals. Chapter 1 introduces industrial robots along with their history and present and future applications. Chapter 2 deals with the fundamentals of robot technology, including general characteristics, basic components, anatomy and classes of robots, methods of path control, a brief description of robot's generations, and factors that influence their selection. Chapter 3 presents the topic of robot classification, which includes manipulator arm geometry, degrees of freedom, types of power sources, types of motion, and path control. Chapter 4 describes the robot operation, its control structure and analysis, line tracking, its dynamic properties, and modular components.

The types of end effectors are described in Chapter 5. The material includes mechanical grippers, gripper force analysis, other types and special-purpose grippers, gripper selection and design, process tooling, and compliance. Chapter 6 describes robot sensors, including sensor classification, microswitches, solid-state switches, proximity sensors, photoelectric sensors, rotary position sensors, usage and selection of sensors, signal processing, and sensor and control integration. Chapter 7 describes robot vision, visual sensing, machine vision, machine vision applications, and other optical methods available.

Robot control systems are discussed in Chapter 8. Control systems correlation and requirements are included, together with Programmable Logic Controllers (PLCs), PLC programming terminals, Proportional Integral Derivative (PID) control, Computer Numerical Control (CNC), Microprocessor Unit (MCU), Universal Robot Controller (URC), interfacing, and workcell controls. An overview of robot programming is included in Chapter 9 with current programming methods, programming languages, levels of robot programming, space position programming, motion interpolation, program statements, and sample programs. Chapter 10 introduces Artificial Intelligence (AI) in robotics systems, including elements of artificial intelligence, system architecture, applications of advanced robots, fuzzy logic controls, advanced concepts and procedures, future developments, and the impact on employment. Chapter 11 deals with safety, including robot safety, safety standards, safety reliability, human factors, safety sensors and monitoring, safeguarding, training, safety guidelines, and definition of terms. Chapter 12, the concluding chapter, brings together many of the impact technological ideas presented in the preceding chapters. Industrial applications of robots in an automated manufacturing environment, material handling, processing operations, assembly and inspection, methods for evaluating potential robot applications, future applications, and the challenges to come are discussed extensively in this chapter, including case studies.

A glossary is also provided at the end of the book with terms and definitions used throughout the text. All new words and terms important enough to be added to the students' technical vocabulary first appear **boldface** in the text and are located in this glossary.

Chapters 5 through 12 are independent of each other and can be read or used in any order. In a semester system, Chapters 1 through 12 would support a good introductory course in robotics. If, however, the course is taught in one quarter, some chapters should be delayed for a later course or second quarter.

An instructor's manual, which includes answers to all end-of-chapter questions and problems, is combined with this text.

I would like to thank the many organizations and institutions that have contributed material and have been supportive and helpful. A very special thanks to

John Fisher of Delmar Publishers for suggestions and motivation throughout the text; to Dr. Richard Polanin of Illinois Central College for reviewing the manuscript and for his valuable assistance in major issues. Thanks also to students, instructors, and professors in automated manufacturing technology programs across the country and abroad who have provided many suggestions during my survey for this book.

About the Author

James G. Keramas, a professor of engineering technology at the University of Massachusetts and at MIT, received his B.S. and M.S. in mechanical engineering from Athens Polytechnic Institute, Athens, Greece, and his doctorate from the University of Massachusetts, Amherst.

Dr. Keramas' experience as a professor in engineering technology is coupled with his industrial practice as a project leader, director of research, inventor, consultant, and entrepreneur. As a full faculty member at the University of Massachusetts-Lowell, he taught courses in mechanical engineering, electrical engineering, and industrial technology (both graduate and undergraduate levels) where he also received the Teacher of the Year Award in 1991. His research interest is in automated manufacturing, robotics, artificial intelligence, CAD/CAM, and computer-integrated manufacturing.

Professor Keramas has taught robotics and computer-integrated manufacturing courses at MIT. He has authored many technical journal articles, and a text entitled *Curriculum Development for Robotics and Automated Systems* (1991), and is a review board member for the *Journal of Industrial Technology*. He has given numerous presentations worldwide on issues related to automated manufacturing, robotics and the utilization of high technology. He has received numerous honors and awards for his work.

Dr. Keramas holds 23 patents in the United States and Canada for inventions in the automated manufacturing field and is an expert product liability witness listed in the Harvard Lawyer's Trial Book. He is a regular consultant for the Office of Technology Innovations of the U.S. Department of Commerce and the National Institute of Standards and Technology, and has provided consulting services to many corporations in the U.S. and abroad.

Dr. Keramas is an advisory committee member for the Pan European Network, creating joint education/industry alliances for technology transfer and training, and a member of the Ed/Tech Group at MIT's Lincoln Laboratory, working on curriculum development for The High-Tech Workforce of Tomorrow.

He has been a Registered Professional Engineer in Massachusetts since 1967 and is a member of many professional associations. He is listed in several issues of Marquis Biographies, including *Who's Who in Science and Engineering* (1994–1995) and *Who's Who in the World* (1998–1999).

CHAPTER 1

Introduction

1.0 OBJECTIVES

After studying this chapter, the reader should:

1. Be acquainted with automation and robots
2. Know some of the history of robots
3. Understand the technology of robots
4. Recognize the economic and social issues associated with robots
5. Have some familiarity with present and future applications of industrial robots

1.1 AUTOMATION AND ROBOTS

The field of **robotics** has its origins in science fiction. The term **robot** was derived from the English translation of a fantasy play written in Czechoslovakia around 1921. It took another forty years before the modern technology of industrial robotics began. Today, robots are highly automated mechanical manipulators controlled by computers.

Automation and robots are two closely related technologies. Both are connected with the use and control of production operations. In an industrial context, we can define automation as a technology that is concerned with the use of mechanical, electrical/electronic, and computer-based systems to control production processes. Examples of this technology include transfer lines, mechanized assembly machines, feedback control systems, numerically controlled machine tools, and robots. Accordingly, robots are mechanical devices that assist industrial automation. There are three types of industrial automation: fixed automation, programmable automation, and flexible automation.

Fixed automation is used when the volume of production is very high and it is, therefore, appropriate to design specialized equipment to process products at high rates and low cost. A good example of fixed automation can be found in the automobile industry, where highly integrated transfer lines are used to perform machining operations on engine and transmission components. The economics of fixed automation is such that the cost of the special equipment can be divided over a large number of units produced, so that the resulting unit costs can be lower relative to alternative methods of production. The risk encountered with fixed automation is that the initial investment cost is high and if the volume of production turns out to be lower than anticipated, then the unit costs become greater.

Another problem with fixed automation is that the equipment is specially designed to produce only one product, and after that product's life cycle is finished, the equipment is likely to become obsolete. Therefore, for products with short life cycles, fixed automation is not economical.

Programmable automation is used when the volume of production is relatively low and there is a variety of products to be made. In this case, the production equipment is designed to be adaptable to variations in a product configuration. This adaptability feature is accomplished by operating the equipment under the control of a "program" of instructions that has been prepared especially for the given product. The program is read into the production equipment, and the equipment performs the particular sequence of operations to make that product. In terms of economics, the cost of the programmable equipment can be spread over a large number of products even though the products are different. Because of the programming feature, and the resulting adaptability of the equipment, many different and unique products can be processed economically in small batches. There is a third category between fixed automation and programmable automation, which is called flexible automation.

Flexible automation has only developed within the past twenty-five or thirty years. This type of automation is most suitable for the mid-volume production range. Flexible automation possesses some of the features of both fixed and programmable automation. Other terms used for flexible automation include **Flexible Manufacturing Systems (FMS)** and **Computer-Integrated Manufacturing (CIM)**. Flexible automation typically consists of a series of workstations that are interconnected by material-handling and storage equipment to process different product configurations at the same time on the same manufacturing system. A central computer is used to control the various activities that occur in the system, routing the various parts to the appropriate stations and controlling the programmed operations at the different stations. The three types of industrial automation and manual labor are illustrated in Figure 1.1.1.

One of the features that distinguish programmable automation from flexible automation is that with programmable automation the products are made in batches. When one batch is completed, the equipment is reprogrammed to process the next batch.

With flexible automation, different products can be made at the same time on the same system. This feature allows a level of versatility that is not available in pure programmable automation, as we have defined it. This means that products can be produced on a flexible system in batches, if desirable, or that several products can be mixed on the same system. The computational power of the control computer is what makes this versatility possible.

Of the three types of automation, robots coincide most closely with programmable automation. The "official" definition of an industrial robot is provided by the **Robotics Industries Association (RIA)** as follows:

> An industrial robot is a reprogrammable, multifunctional manipulator designed to move materials, parts, tools, or special devices through variable programmed motions for the performance of a variety of tasks.

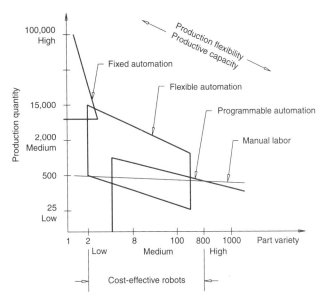

Figure 1.1.1 The three types of production automation and manual labor as a function of production volume and product variety relative to cost-effective robots

Though the robots themselves are examples of programmable automation, they are sometimes used in flexible automation and even fixed automation systems. These systems consist of several machines and/or robots working together, and are typically controlled by a computer or a programmable controller. Such a system might appropriately be considered a high-production fixed or flexible automation system.

Today, the human analogy of an industrial robot is very limited. Robots do not look or behave like humans. Instead, they are one- or multiarmed machines that almost always operate from a fixed location on the factory floor. Future robots are likely to have greater sensor capabilities, more intelligence, higher levels of manual dexterity, but a limited degree of mobility as compared to humans. There is no denying that the technology of robots is moving in a direction to provide us with more and more capabilities similar to those of humans.

1.2 BRIEF HISTORY

The word *robot* was introduced in 1921 by the Czech playwright Karel Capek, in his play *R.U.R. (Rossum's Universal Robots)*, and is derived from the Czech word *robota*, meaning "forced labor." The story concerns a brilliant scientist named Rossum and his son, who developed a chemical substance similar to protoplasm to manufacture robots. Their plan was that the robots would serve humankind obediently and do all physical labor. Finally, after improvements and eliminating unnecessary parts, they develop a "perfect" robot, which eventually goes out of control and attacks humans.

Although Capek introduced the word *robot* to the world, the term *robotics* was coined by Isaac Asimov in his science fiction story "Runaround," first published in the March 1942 issue of *Astounding*, where he portrayed robots not in a negative manner but built with safety measures in mind to assist human beings. Asimov established in his story the three fundamental laws of robotics as follows:

1. A robot may not injure a human being or, through inaction, allow a human being to come to harm.
2. A robot must obey the orders given it by human beings, except where such orders would conflict with the first law.
3. A robot must protect its own existence as long as such protection does not conflict with the first and second laws.

In a broader sense, Capek's term *robot* meant a manipulator that was activated directly by an operator or other mechanical or electrical means. More generally, an industrial robot has been described by the **International Standards Organization (ISO)** as follows:

> A machine formed by a mechanism, including several degrees of freedom, often having the appearance of one or several arms ending in a wrist capable of holding a tool, a workpiece, or an inspection device. In particular, its control unit must use a memorizing device and it may sometimes use sensing or adaptation appliances to take into account environment and circumstances. These multipurpose machines are generally designed to carry out a repetitive function and can be adapted to other functions.

According to Miller (1987), robots were introduced to industry in the early 1960s. Initially, robots sold for an average of $25,000 with a life expectancy of about eight years, cost approximately $4.00 per hour to operate, and had to compete for jobs with human workers earning slightly more per hour than the robot hourly operating cost. Robots, originally, were used in hazardous operations, such as handling toxic and radioactive materials, and loading and unloading hot workpieces from furnaces and handling them in foundries. Some rule-of-thumb applications for robots are the four *D*s (*d*ull, *d*irty, *d*angerous, and *d*ifficult, including demeaning but necessary tasks) and the four *H*s (*h*ot, *h*eavy, *h*azardous, and *h*umble).

By 1970 approximately two hundred robots were in use in U.S. manufacturing facilities. The jobs to which robots were assigned during that decade were primarily hazardous, strenuous, or repetitious and required the robot to respond only to simple input commands. Control and feedback technology at this evolutionary point remained relatively basic, limiting robots to jobs requiring a lot of "brawn" but very little "brain."

During the 1970s, with nationally declining productivity and increasing labor rates, a significant increase in robot usage began. Many improvements in controls increased the flexibility and capabilities of robots. The first robots had been introduced in the automotive industry. Ten years later, the same industry was contributing most to the growth of robotics by its widespread acceptance. The average prices of robots increased to approximately $45,000, life expectancy remained at about eight years, operating costs rose to approximately $5.00 per hour, and the average direct labor cost in the automotive industry was twice the hourly operating cost of an industrial robot.

In 1980, there were approximately 4,000 robots in the United States and 26,000 robots worldwide. By the mid-1980s, there were approximately 17,000 industrial robots in the United States. The average price was approximately $60,000, life expectancy increased to fifteen years, operating costs were in the range of $5.50 per hour, and—again using the automotive industry as a comparative example—labor rates were escalated to over $14.50 per hour.

By the end of 1997, RIA estimates, some 84,000 robots were in operation in U.S. factories, placing the United States second in the world to Japan. According to Dave Lavery, manager of the robotic program at NASA, there are some 650,000 robots at work today worldwide, and the average price is approximately $72,000, life expectancy over seventeen years, and operating costs in the range of $7.00 per hour as compared to average wages of over $24.00 in the automotive industry. Figure 1.2.1 shows the history of labor and robot cost in the automotive industry.

Over the past decade, highly selective applications for robots resulted in so-called "islands" of automation. With the development of more sophisticated automation concepts, such as Computer-Integrated Manufacturing (CIM) and Flexible Manufacturing Systems (FMS), users learned that industrial operations are usually best automated through the integration of robots with machines, which

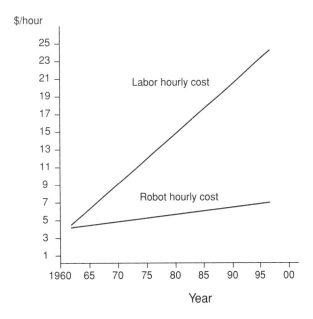

Figure 1.2.1 Hourly cost of a robot versus human labor in the automotive industry

are referred to as a **workcell**. This integration, which causes the need for knowledge about robots, has become very important in automated manufacturing today.

Robots can be used in any industry providing work and services, and can also be adapted easily to numerous job functions with uncanny skill and unmatched endurance.

These factors and many others, such as reducing production cost, improving quality, and increasing productivity, to name just a few, have contributed to the growth of robots and will continue to impact their evolution, both in pace and direction.

Robot application areas as a percentage of total robot population for the year 1996 are shown in Figure 1.2.2. Robot shipments from U.S.-based companies for the period 1993–1997 are shown in Figure 1.2.3.

Two reasons are given for selecting a robot to operate in a production line: (1) to reduce labor costs, and (2) to perform repetitive work that is boring, unpleasant, or hazardous for human beings.

Computer-controlled robots were commercialized in the early 1970s, with the first robot controlled by a minicomputer appearing in 1974. Table 1.1 presents a chronological list that summarizes the leading developments related to current robot technology.

The primary purpose of the robot as a machine is controlled motion: If it does not move, it is not a robot. All robotic design endeavor has controlled, sensitive, and intelligent motion as its collective goal. The variety of uses for robots is increasing, although their main use still seems to reside in automotive manufacturing.

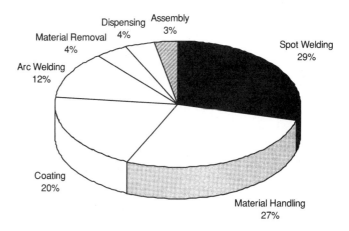

Figure 1.2.2 A comparison of leading robot applications in 1996. *(Source: RIA)*

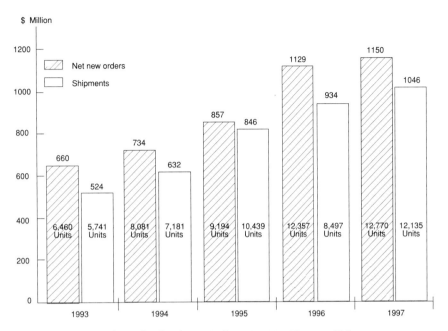

Figure 1.2.3 U.S. robot sales for the period 1993–1997 (*Source: RIA*)

A decade ago the projection was that robots would begin to grow in popularity about the year 1998 after the general public—as well as engineers and scientists—had learned to routinely accept them in their work environments, and that no robot would act or feel like a human being in the foreseeable future. These projections are still true. The primary driving force behind the renewed interest in robots will be lower unit cost, greater reliability, and simpler operation.

1.3 THE TECHNOLOGY OF ROBOTS

A robot is a machine constructed as an assemblage of joined links so that they can be articulated into desired positions by a programmable controller and precision actuators to perform a variety of tasks. Robots range from simple devices to very complex and "intelligent" systems by virtue of added sensors, computers, and special features. Figure 1.3.1 illustrates the possible components of a robot system.

There are several hundred types and models of robots. They are available in a wide range of shapes, sizes, speeds, load capacities, and other characteristics. Care must be taken to select a robot to match the requirements of the tasks to be done. One way to classify them is by their intended application. In general, there are industrial, laboratory, mobile, military, security, service, hobby, home, and personal robots. Figure 1.3.2 illustrates the general configuration and operating parameters of an industrial robot.

Table 1.1 Chronological Developments Related to Current Robot Technology

1801   Joseph Jacquard invented a textile machine that was operated by punch cards. The machine was called a programmable loom and went into mass production. It was later accepted as a forerunner of numerical control.

1805   H. Mailardet constructed a mechanical doll capable of drawing pictures.

1892   In the United States, Seward Babbitt designed a motorized crane with gripper to remove ingots from a furnace.

1921   The first reference to the word *robot* appeared in a play opening in London. The play, written by Czechoslovakian Karel Capek, introduced the word *robot* from the Czech *robota*, which means a serf or one in subservient labor. From this beginning, the concept of a robot took hold.

1938   Americans Willard Pollard and Harold Roselund designed a programmable paint-spraying mechanism for the DeVilbiss Company.

1946   George Devol patented a general-purpose playback device for controlling machines. The device used a magnetic process recorder. The U.S. patent was issued in 1952.

Also in 1946 the computer emerged for the first time. American scientists J. Presper Eckert and John Mauchly built the first large electronic computer called the ENIAC at the University of Pennsylvania. A second computer, the first general-purpose digital computer, called Whirlwind, solved its first problem at the Massachusetts Institute of Technology (MIT).

1948   Norbert Wiener, a professor at MIT, published *Cybernetics*, a book that describes the concept of communications and control in electronic, mechanical, and biological systems.

1949   John Parsons and MIT started the development of the numerical control machine, which was later accepted as a benchmark of automation.

1951   A teleoperator-equipped articulated arm was designed by Raymond Goertz for the Atomic Energy Commission.

1952   The prototype numerical control machine was demonstrated at the Massachusetts Institute of Technology after three years of development. Part-programming language called APT (Automatically Programmed Tooling) was subsequently developed and was released in 1961.

1954   The first programmable robot was designed by George C. Devol, Jr., who coined the term "Universal Automation." The U.S. patent was issued in 1961. Devol was joined by Joseph F. Engelberger in 1956; they shortened the name to "Unimation," which became the first successful robot manufacturing company.

British inventor C. W. Kenward applied for a robot design patent. The British patent was issued in 1957.

1959   Planet Corporation marketed the first commercially available robot. It was controlled by limited switches and cams.

1960   Unimation was purchased by the Condec Corporation, and the development of Unimate Robot Systems began.

American Machine and Foundry, later known as AMF Corporation, marketed a robot, called the Versatran, designed by Harry Johnson and Veljko Milenkovic.

1961   A Unimate robot was installed at Ford Motor Company for tending a die casting machine.

1962   General Motors installed the first industrial robot on a production line. The robot selected was a Unimate.

1968   Stanford Research Institute built and tested a mobile robot with vision capability, called Shakey.

1970   At Stanford University a robot arm was developed and became a standard for research projects. The arm was electrically powered and became known as the Stanford Arm.

1973   The first commercially available minicomputer-controlled industrial robot was developed by Richard Hohn for Cincinnati Milacron Corporation. The robot was called the T^3, The Tomorrow Tool.

The first computer-type robot programming language was developed at Stanford Artificial Intelligence Laboratory and was named "Stanford Arm," also called WAVE. It was followed by the language AL in 1974. The two languages were subsequently developed into the commercial VAL language for Unimation by Victor Scheinman and Bruce Simano.

1974   Professor Scheinman, the developer of the Stanford Arm, formed Vicarm Inc. to market a version of the arm for industrial applications. The new arm was controlled by a minicomputer. Asea introduced the all-electric drive IRb6 robot.

Kawasaki, under Unimation license, installed an arc-welding operation for motorcycle frames.

1975   The Olivetti "Sigma" robot was used in assembly operations—one of the very first assembly applications of robotics.

Table 1.1 Chronological Developments Related to Current Robot Technology (*continued*)

1976   The Remote Center Compliance (RCC) device for part insertion in assembly was developed at Charles Stark Draper Labs in the United States.

1977   Asea Brown Boveri Robotics Inc., a European robot company, offered two sizes of electric-powered industrial robots with a microcomputer controller.

1978   With support from General Motors, Unimation developed the Programmable Universal Machine for Assembly (PUMA) robot using technology from Vicarm Inc.

The Cincinnati Milacron T^3 robot was adapted and programmed to perform drilling and routing operations on aircraft components, under Air Force ICAM (Integrated Computer-Aided Manufacturing) sponsorship.

1979   The **SCARA**-type robot (Selective Compliance Assembly Robot Arm) was developed at Yamanashi University in Japan for assembly. Several commercial SCARA robots were introduced around 1981. Japan became the world's largest user of robots.

1980   The robot industry started a period of rapid growth, with a new robot or company entering the market every month. A bin-picking robotic system was demonstrated at the University of Rhode Island. Using machine vision, the system was capable of picking parts in random orientations and positions out of a bin.

1981   A "direct-drive robot" was developed at Carnegie Mellon University. It used electric motors located at the manipulator joints instead of the usual mechanical transmission linkages.

1982   IBM introduced the RS-1 robot for assembly, based on several years of in-house development. It was a box-frame robot, using an arm consisting of three orthogonal slides. The robot language AML, developed by IBM, was also introduced to program the RS-1. Bendix and G.E. entered the robotics business.

1983   The robot industry entered a maturing period as industry recognized that robots and the other automation hardware must be integrated into a unified system. The number of major robot manufacturers fell to twenty-five or fewer.

A report was issued on research at Westinghouse Corp. under a National Science Foundation sponsorship on the "adaptable-programmable assembly system" (APAS), a pilot project for a flexible automated assembly line using robots.

1984   Direct-drive robot arms were introduced by Adept Corporation with electric-drive motors connected directly to the arms eliminating the need for intermediate gear or chain drives. Several off-line programming systems were demonstrated at the Robots 8 show. Typical operation of these systems allowed the robot program to be developed using interactive graphics on a personal computer and then downloaded to the robot.

1986   Robot applications and installations continued to grow but with increased emphasis on the integration of the robot into workcell, FMS, and CIM systems.

1988   CNN News designed "SCAMP," the first robot pet with personality and feelings.

1990   Asea Brown Boveri Robotics Inc. purchased the robotics division of Cincinnati Milacron, and all future robots would be Asea machines.

1991   Robots changed dramatically. New technologies not only impact the way we live but affect the global economy and environment. The emphasis on producing high-quality competitive products at low cost and better service was the consumer's demand.

This new world order sparked new interests in the robotic and manufacturing industry by emerging more robots in the integrated and automated production systems.

1992   After General Electric made Fanuc Robotics Corporation the sales leader in the United States (General Motors Fanuc annual sales of $200 million), G.E. sold its interest to them.

1994   Robot sales in the United States set new records in the market at an annual total value of $765 million. The number of producers of robots decreased considerably, so that at that time there were only between five and ten manufacturers left, the biggest market share being occupied by ABB (Asea Brown Boveri) and by Fanuc Robotics Corporation.

1996   Dan Rather, of CBS news, announced on June 24 that Japan would abandon common robots (without **AI**) and would substitute human beings because common robots were too expensive to adjust for new jobs.

10 Introduction

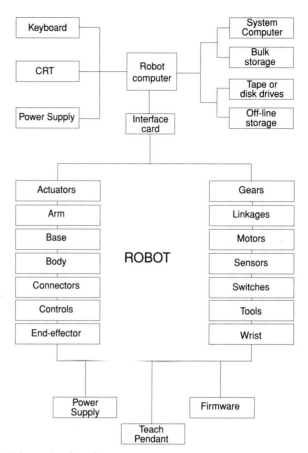

Figure 1.3.1 Schematic of a robot system

In the last three decades, the robotics field benefited considerably from the advancement of microelectronics, computer science, and improved design of electrical, electromechanical, and hydromechanical servo systems.

The industry continues to grow and expand. Currently, there are approximately thirty robot manufacturers in the United States and over five hundred worldwide. The annual growth rate of the industry is approximately 35 percent per year, and continued market expansion is expected. RIA estimates that annual sales volumes for 1998 will be in the $2 billion range; for the year 2000 it is predicted to be about $7 billion. Spot-welding still remains the largest application area for robots today.

Competitive forces are beginning to segment the market with many manufacturers focusing on specific industries or applications. This specialization approach will speed technological advancements and enhance robot capabilities in specific areas. More attention is being paid lately by manufacturers to sensor

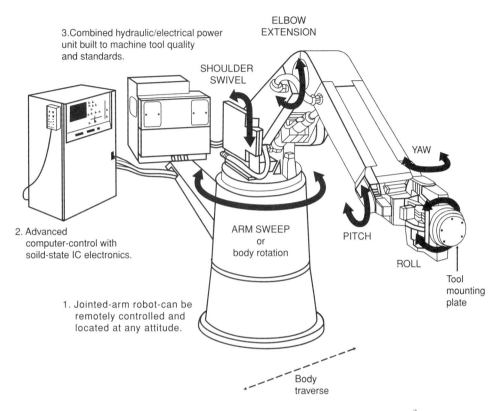

Figure 1.3.2 The general configuration and operating parameters of the Cincinnati Milacron T^3 series industrial robot

integration. More and more robots are sold with standard or optional capabilities, such as vision and tactile sensors and even fuzzy logic controls.

More-intelligent robots will be the result of such efforts. The robot manufacturers will continue to quickly implement the computer-industry improvements and expanded capabilities to raise the "IQ" of robots to handle more data and process it faster. The move is well under way to utilize more AC electric servo systems to power robot motions. It is expected that this trend will continue along with CAD/CAM integration, which is a logical direction for the users of robotic technology.

Computer graphics technology is also rapidly advancing to provide us with "simulation" capabilities to analyze manufacturing approaches and methods prior to implementation. "Off-line" programming, a necessary component to realize full CAD/CAM capability benefits, is now being offered. The "universal robot controller" finally emerged in 1998 into the market to eliminate those nuisance programming languages that are different for each robot model. This will replace all factory controllers with this new one and reduce the unnecessary islands of automation.

Mobility will also allow robots the necessary freedom of movement so that users can more fully appreciate and utilize their inherent flexibility throughout a manufacturing facility.

A dramatic decrease in cost of robots can be expected in the near future if the following conditions exist:

- Fewer manufacturers will be supplying robots to a larger market.
- Mass production methods with advanced technologies are applied to the manufacture of robots.

In reality, robot technology has an exciting future. We can expect to see the robot move out of the factory and foundry and enter the domestic and business worlds. The domestic robot will appear as an electronic pet and soon will develop the ability to perform useful tasks in the home. The sensory ability of all robots will greatly improve. In the long run, robots will acquire the capabilities they have been described as having in the movies and science fiction books. Self-reproducing factories may be placed on the moon or on other planets to help meet our growing need for energy and goods. Medical robots will be able to produce "six-million-dollar men" and "bionic women." Exciting changes are on the way as robots become the helpers that humans have always dreamed of.

Finally, the current and future applications of robots are moving in the direction to provide us with more and more capabilities like those of humans, which are going to increase our quality of life and the global economy.

1.4 ECONOMIC AND SOCIAL ISSUES

Although it is certainly true that robots can relieve humans of the need to perform what have been called "4 D jobs" (dull, dirty, dangerous, and difficult), the fact remains that manufacturing plant managers are extremely concerned with the "bottom line." A survey of robot users conducted in the 1980s by the Carnegie Mellon University Robotics Institute indicates that the primary reason for selecting a robot is to reduce labor costs. Table 1.2 shows how potentially beneficial robots may appear to be with respect to humans. If robots cannot be justified economically, they should not be purchased or used in production lines. The purpose of this section is to present briefly some simple techniques that have been used to demonstrate how robots can easily be shown to be an economically justifiable capital expenditure. It is not the intention, however, to develop sophisticated economic theories—that is beyond the scope of this book.

Today, the price of a single industrial robot ranges from about $5,000 to well over $100,000. To this must be added the cost of the associated tooling and fixturing that are to be used within the robot workcell and the total installation cost. It has been found that approximately 55 percent of the overall system cost is for the robot, 30 percent is for the additional tooling, and about 15 percent is for installation. Consider a system with a total cost of $100,000 broken down as follows:

Table 1.2 Reasons for Using Robots

Ranking	Potential Benefits
1	Reduce labor costs
2	Eliminate dangerous jobs
3	Increase output rate
4	Improve product quality
5	Increase product flexibility
6	Reduce material waste
7	Comply with OSHA
8	Reduce labor turnover
9	Reduce capital cost

- Material-handling robot $55,000
- Tooling and fixturing $30,000
- Installation $15,000

It should be noted that the figure used for this type of robot is about the current average in the United States. Also, the tooling and fixturing figure includes engineering development costs.

To determine the economics of such a robot, we need to know the cost of labor and of the operation of the robot itself.

It was estimated that in 1995, an automobile worker earned about $17 per hour, including fringe benefits. In addition, the Draper Laboratory at MIT has estimated that it costs about $6 per hour to run a robot based on operating sixteen hours per day (i.e., two shifts per day) and a useful life of about eight years. (Although other sources suggest a figure of $2 per hour, many robot manufacturers use the more conservative number.) Because a worker will normally put in about two thousand hours per year (forty hours per week for fifty weeks), it can be seen that the $11 per hour differential in labor costs ($17 − $6) produced by the robot results in a yearly "saving" of about $22,000. Thus, it will take about 2.8 years to pay back the original cost of the robot ($55,000/$22,000). After this time, the user will be "making" $22,000 per year or, more correctly, will be experiencing a positive cash flow. If we assume a two-shift-per-day activity, the payback period will be only 1.4 years, after which time a cash flow of $44,000 per year will occur.

Obviously, this analysis is an oversimplification because it does not look at all economic factors, such as the cost of money and the escalation of labor costs. Nevertheless, it does provide one with the idea that robots can be justified economically, and rather easily at that. Now refine the analysis somewhat by including such factors as corporate tax rates, depreciation, and the savings resulting from using less material in a particular process. It can be shown that the payback period Y can be calculated from the following equation:

$$Y = \frac{(P + A + I) - C}{(L + M - O)H(1 - TR) + D(TR)} \quad \text{(Equation 1.4.1)}$$

Where Y = number of years required to break even
P = price of the robot = $55,000
A = cost of the tooling and fixturing = $30,000
I = installation cost = $15,000
C = investment tax credit (assumed to be 10%) = $10,000
L = hourly cost of labor, including fringe benefits = $17
M = hourly savings in the cost of materials = $1
O = cost of running and maintaining the robot system = $6
H = number of hours per year per shift = 2,000
D = annual depreciation assuming an 8-year "tax life," the straight-line method, and a salvage value of $10,000 = ($100,000 − $10,000)/8 = $11,250
TR = corporate tax rate assumed to be 40% (= 0.4).

Substituting these values into the equation gives a payback period of 2.7 years for a double-shift operation. Another economic yardstick that is often used in determining whether a particular capital expenditure is warranted or not is the return on investment (ROI).

$$ROI = \frac{total\ annual\ savings}{total\ investment} 100\% \quad \text{(Equation 1.4.2)}$$

In terms of the quantities defined above, this can be expressed as

$$ROI = \frac{(L + M - O)H - D}{P + A + I - C} 100\% \quad \text{(Equation 1.4.3)}$$

Using our example values in this equation indicates that the ROI is only 14.2% for a single-shift operation. However, this figure increases to an impressive 40.8% when the robot is used two shifts per day. When compared with the 9% cost of borrowing money today, it appears that a robot used in a single or multiple-shift application is clearly a good investment. It is important to realize that a more or less favorable result will be obtained if different assumptions are made concerning labor and/or robot costs. For example, if we use $2 per hour for the running and maintenance of a system, the one- and two-shift ROIs become 23.1% and 58.6%, respectively.

One final point is worthy of mention. The quantitative measures just described do not take into consideration the economic benefits that can be derived from using a robot to produce a product that is of a consistently high quality.

The important results of the example given here demonstrate clearly that from an economic point of view, robots seem to make a great deal of sense. However, what about the human element? What will be the impact on the workers themselves of introducing these devices into the workplace?

The problems created by the introduction of automation into the workplace are not new. There is no question that the principal purpose of robots and manufacturing automation is to make processes more efficient.

Robots are usually justified on the basis of labor-cost savings, and this means that robots are doing jobs that were formerly performed by human workers. Even in new processes for which no human worker has been employed, the use of a robot means that one or more human worker(s) will not be employed in that job. The replacement rate is not one-to-one either. Robots can often perform an operation at speeds equal to or faster than the human operator and can work three shifts per day, seven days per week, which is the equivalent of four human workers. Even considering downtime for maintenance, a robot can often replace more than one worker.

There are, of course, many reasons for employing robots and manufacturing automation other than the drive for higher efficiency. Improving quality, enhancing safety, and removing tedious, fatiguing, or boring tasks are all worthwhile objectives, but are they worth the price of sacrificing workers' livelihoods? On the other hand, without automation, whole industries and all jobs in these industries can be lost to competition. The savings from these productivity improvements exceed total profits, and if the savings had not been there, the company would not have survived.

If we look at the larger picture and include a count of workers worldwide, we would undoubtedly find that automation is reducing the number of jobs required to sustain a given level of output. But because automation is here, failure to use it by any one company is almost surely to result in failure of the company and elimination of all jobs in that company. Therefore, without automation, industries in countries such as the United States would have little chance of survival against industries employing low-cost labor in foreign countries.

Early in the twentieth century, an automobile was as extravagant a luxury as the personal airplane is today. But Henry Ford's assembly lines and Detroit-style automation made automobiles available to the average person and resulted in huge demands for automobiles in the 1920s. Along with the huge volumes came millions of jobs for persons employed by the automobile and related industries. Today, automation breakthroughs, such as the computer industry, are increasing volumes of sales of other products and at the same time creating more jobs and ensuring better quality of life for our people.

1.5 PRESENT AND FUTURE APPLICATIONS

The state of the art of robotic applications is, in some ways, paralleling the development of digital computers. When they were first introduced, computers were used for tasks that had previously been performed by people. This was a natural application, for it was obvious that the new device would be able to perform such jobs much faster and even more reliably than people could perform them. However, as time progressed, it was recognized that tasks that had previously been rejected as being impossible to undertake because of excessive personnel and/or time requirements were now possible to attempt.

In the 1960s, it was feared that computers would reduce the number of white-collar jobs, such as accounting clerks. Instead, the computer has increased the number of these jobs by creating such new occupations as computer operator, computer programmer, and systems analyst. By making more information available, computers have generated and increased desire for information in our society. Now in the 1990s, there is fear that robots will reduce the number of blue-collar jobs, but it is more likely that robots will cause a total increase in jobs. Some of these may be white-collar jobs. New human jobs such as robot supervisor, robot programmer, robot setup person, robot trainer, and robot repair person are sure to emerge from the widespread use of robots in industry. Right now, new types of manipulator joints, actuators, and grippers are under development. Japanese researchers are experimenting with **shape-memory alloy,** a new actuator for humanlike hands. A wire made of shape-memory alloy (composed of nickel and titanium) is used.

At present, robots have little or no judgment and decision-making capacity, and their sensory capabilities are quite poor. Consequently, they are not yet ready to perform most complicated tasks. The advancement of the robot to the state shown in such films as *Star Wars* and *Star Trek* must await breakthroughs in the areas of artificial intelligence, voice interfacing, and vision and touch sensors. Considerable research is going on in these areas, and it is only a matter of time before the objects of today's laboratory curiosity become economical enough for use in robots.

Robots would advance much faster if the future products were needed now. Economics is a major driving force in research and development, as is the government, through its military and space programs. The government was responsible for much of the development of electronics and the electronic computer, for instance. Today, the Atomic Energy Commission is sponsoring research in robotics for advanced nuclear reactors, where robots will have to work in extremes of temperature, humidity, and radiation level and will have to be able to climb over obstacles. The military is hoping to make up for its shortage of personnel by using robots to make the human forces more efficient. Even in their present state of development, robots are finding more and more jobs. They have already been successfully used as brain surgeons, window washers, lab technicians, and in hospital applications. One of their newest uses occurred on February 22, 1988, when ABC began employing robots to run the television cameras for its national news broadcasts. This allows the cameras to be operated by remote control from the control booth by a single person.

The automated robot cook, maid, butler, or gardener is still far from becoming a reality. However, a robot chauffeur can be built with today's technology, if someone is willing to pay the high cost. The navigational computers and sensors necessary for getting around safely in city traffic already exists. As it merges with the electronic computer, the robot is evolving toward the point where it could be viewed as a new **life-form.**

Although artificial intelligence is still in its infancy, it has already shown great promise for robots. Work has begun on the development of self-reproducing machines. Robert A. Freitas has written an article on the prospects for a self-replicating factory to be practical by the early twenty-first century. Located on the moon and running on solar power, this factory would use the moon's raw materials to reproduce itself and to manufacture additional solar cells, which could then be hauled away for a nearby solar-powered generation station satellite to beam to earth as electrical power. NASA has proposed a four-part program along these lines, where part one is a robot in a warehouse assembling other robots and part four is a self-reproducing factory the size of a football field.

Robots are currently used in education as tools for teaching various topics. The show robot is very useful for working with abused children whose bad experiences often make it hard for them to trust and talk to adults. Show robots are also useful for working with shy children. Programmable mobile devices such as the Big Trak tank (robot) can be used to teach programming; because they deal with motion rather than with numbers, they are easier for children to relate to. Educational robots are used to teach applications programming for industrial robots.

Looking forward may show us future consequences of current policies and actions, which may help us make informed decisions now. The future is controlled by events from the past and present, and by our imaginations. Important products of imagination include art, fiction, science literature, cinema, and children's toys.

Computer memory devices have been quadrupling in capacity and speed and the price reduced by half every three years. The personal computers of today have more memory and run faster than did the mainframe computers five years ago.

Jules Verne predicted many of the mechanical inventions of the twentieth century in the middle of the nineteenth century. Hugo Gernbeck predicted many of the electronic devices—including radar and solar energy devices—of the late twentieth century during the early twentieth century.

As recently as ten years ago, it was estimated that 50 percent of the adult population of the United Sates suffered from **cyberphobia**—an unreasonable fear of computers. Children are not afraid of computers, however, because they are being brought up with them. The same will be true of children's attitudes toward robots when these become widely available as toys. However, there will be a short-term problem with reassigning persons who are displaced from simple manual-labor jobs by robots. Some will find more interesting and rewarding jobs, others may be given early retirement, and still others will require retraining at the company's or government's expense. But, in time, the problem will disappear because no one new will be joining the workforce in these areas.

1.6 SUMMARY

Automation and robots are two closely related technologies. Both are concerned with the use and control of production operations. Robotics (the study of robots) is a form of industrial automation. There are three types of industrial automation. Fixed automation is used for high production volume and utilizes expensive special equipment to process only one product. Flexible automation is used for medium production volume and utilizes a central computer to control the process of different products at the same time. Programmable automation is used for low production volume operated under control of a program. It processes one batch of similar products at a time. When one batch is completed, the equipment is reprogrammed to process the next batch.

Robots are examples of programmable automation; however, they are often used in flexible or even fixed automation systems.

Tracing the sequence of events that led to the development of industrial robots currently available is presented in Table 1-1.

The two reasons for selecting a robot to operate in a production line are first to reduce labor costs, and second to perform work that is boring, unpleasant, or hazardous for human beings.

Robots can perform repetitive tasks at a steady pace, be programmed to achieve and perform different unpleasant tasks, operate for long hours without rest or break periods, and respond in automated manufacturing operations on a continuous basis.

Without automation and robots our industries would be lost to competition. Automation reduces unit cost and increases volumes of sales, which creates jobs on other products, and ensures better quality of life.

Finally, the current and possible future applications of robots are moving in the direction to provide us with more and more capabilities like those of humans.

1.7 REVIEW QUESTIONS

1.1 Discuss the differences between fixed, flexible, and programmable automation.
1.2 What other terms are used to describe flexible automation?
1.3 What is the RIA definition of a robot?
1.4 Identify an early design that demonstrates the mechanical operation found in later industrial robots.
1.5 What is the ISO definition of a robot?
1.6 What company was the first to control a robot with a minicomputer?
1.7 Discuss at least three reasons for using robots instead of humans to perform a task.
1.8 Discuss several reasons why robots should be used in the workplace even though human beings may initially lose some jobs.
1.9 Explain briefly the CMU Robotics Survey in respect to human beings.
1.10 Explain why automation breakthroughs create more jobs and ensure better quality of life.
1.11 How might robots be advanced faster in the workplace?
1.12 Explain how the future of robots is controlled and what applications can be foreseen.

1.8 PROBLEMS

1.1 A new production machine costs $90,000 installed and is expected to generate revenues of $50,000 per year for 7 years. It will cost $20,000 per year to operate the machine. At the end of 10 years the machine will be scrapped at zero savage value. Determine the payback period for this investment.

1.2 Two production methods, one manual and the other automated, are to be compared. The data for the manual method is the same as Problem 1.1. For the automated method the initial cost is $135,000, the annual operating cost is $5,000, and the service life is expected to be 5 years. In addition, the equipment associated with the alternative will have a salvage value of $70,000 at the end of the 5 years. Revenues for either alternative will be $50,000 per year. Compare which method is more profitable.

1.3 A batch of 50 parts is to be processed through the factory for a particular customer. Raw materials and tooling are supplied by the customer. The total time for processing the parts is 100 hours. Direct labor cost is $12.00 per hour. The factory overhead rate is 125% and the corporate overhead rate is 160%. Compute the cost of the job.

1.4 A manually-operated production machine costs $66,063. It will have a service life of 7 years with an anticipated salvage value of $5,000 at the end of its life. The machine will be used to produce parts at a rate of 20 units per hour. The annual cost to maintain the machine is $2,000. A machine overhead rate of 15% is applicable to capital cost and maintenance. Labor to run the machine costs $10.00 per hour and the applicable overhead rate is 30%. Determine the profit break-even point if the sale price of parts is $1.00 per unit.

1.5 Two alternative production methods have been proposed; one with manual feeding, the other with a robot. Data are given in the following table. Select the more economical alternative method.

	Manual	Robot
Initial cost	$15,000	$96,000
Annual operating cost	$30,000	$10,000
Salvage value	0	$15,000
Service life (years)	10	8

1.6 The cost of a robot is $60,000, and the price of tooling and fixturing is $15,000. The installation cost is $20,000. Use 10% as the rate of investment. The cost of labor is $18 per hour, and the savings in the cost of materials is $2. The actual cost of running and maintaining the robot system is $5. The number of hours per year per shift is assumed to be 2,800.

Assume an annual depreciation on an 8-year tax life and the savage value to be $10,000. Use a corporate tax rate of 40% to determine the number of years that are required to break even.

1.7 Solve Problem 1.6 to determine if the return of investment is justified with 14% interest on the borrowing money.

1.8 A proposed robot is to be used exclusively to assemble one work part in a production line. The initial cost of the robot is $50,000, and its expected service life is 3 years with a salvage value of $20,000 at the end of the 3 years. The robot will be operated 4,000 hours per year (two shifts) at $8.00 per hour (labor, power, maintenance, and the like). Its production rate is 10 units per hour. Excluding raw material costs, compute the production cost per unit using a rate of return of 25%.

1.9 REFERENCES

Asfahl, C. R. *Robots and Manufacturing Automation.* 2d ed. New York: John Wiley and Sons, Inc., 1992.

Craig, J. J. *Introduction to Robotics: Mechanics and Control.* Reading, MA: Addison Wesley, 1989.

Engleberger, J. F. *Robotics in Service.* Cambridge, MA: MIT Press, 1989.

Holzbock, W. G. *Robotic Technology: Principles and Practice.* New York: Van Nostrand Reinhold, 1986.

Lee, M. H. *Intelligent Robotics.* New York: Halsted Press, 1989.

Masterson, J. W., R. L. Towers, and S. W. Fardo. "Robotics Technology," South Holland, IL: Goodheart-Willcox Co., Inc. 1996.

Miller, R. K. *Industrial Robot Handbook.* Auburn, GA: The Fairmont Press, Inc., 1987.

Pessen, D. W. *Industrial Automation.* New York: John Wiley and Sons, Inc., 1989.

Rehg, J. A. *Introduction to Robotics in CIM Systems.* 2d ed. Upper Saddle River, NJ: Prentice Hall, 1997.

Sandler, B. Z. *Robotics: Designing the Mechanisms for Automated Machinery.* Englewood Cliffs, NJ: Prentice Hall, 1991.

Sharon, D., J. Harstein, and G. Yantian. *Robotics and Automated Manufacturing.* Aulander, NC: Pittman, 1989.

Spiteri, C. J. *Robotics Technology.* Philadelphia, PA: Sanders College Publishing, 1990.

CHAPTER 2

Robot Technology

2.0 OBJECTIVES

After studying this chapter, the reader should:

1. Realize the fundamentals of robot technology
2. Know the general characteristics of robots
3. Understand the basic components of robots
4. Recognize robot anatomy
5. Be informed of robot generations
6. Be aware of robot selection

2.1 FUNDAMENTALS

Robot technology is an applied science that is referred to as a combination of machine tools and computer applications. It includes such diverse fields as machine design, control theory, microelectronics, computer programming, artificial intelligence, human factors, and production theory.

Research and development are proceeding in all these areas to improve the way robots work and behave. Advancements in technology will enlarge the scope and future applications of robots.

To describe the technology of a robot, it is necessary to define a variety of technical features about the way a robot is constructed and the way it works. To accomplish this, the following topics as applied to the industrial robot are discussed:

- General characteristics
- Basic components
- Robot anatomy
- Robot generations
- Robot selection

Although these topics are discussed in detail later in the book, a concise description is necessary here to help the reader become familiar with the terms and practical aspects of this subject.

2.2 GENERAL CHARACTERISTICS

The development of the **industrial robot** represents a logical evolution of automated equipment, combining certain features of fixed automation and human labor. Robots can be thought of as specialized machine tools with a degree of flexibility that distinguishes them from fixed-purpose automation. By the addition of sensory devices, robots are gaining the ability to adapt to their work environment and modify their actions based on work-condition variations. Industrial robots are becoming "smarter" mechanical workers and are now widely accepted as valuable productivity-improvement tools.

Industrial robots are properly thought of as machines or mechanical arms. It is inappropriate to think of them as mechanical people. A robot is essentially a mechanical arm that is bolted to the floor, a machine, the ceiling, or, in some cases, the wall, fitted with its mechanical hand, and taught to do repetitive tasks in a controlled, ordered environment. In most cases, it possesses neither the ability to move about the plant nor the ability to see or feel the part it is working on. Exceptions to these general rules exist in certain instances. However, even with these limitations, robots make outstanding contributions toward the improvement of manufacturing operations. Robots fill the gap between the specialized and limited capabilities normally associated with fixed automation and the extreme flexibility of human labor.

Robots offer many benefits simply because they are machines. As such, they are not as susceptible to fatigue, discomfort, boredom, or similar factors that negatively impact a human worker's job performance in harsh, noisy, hot, or hazardous environments. Robots can perform well and consistently where strenuous, dangerous, dirty, or repetitive work is required.

Robots have the ability to move their mechanical **arm** (or arms) to perform work. The set of points representing the maximum extent or reach of the robot hand or working tool in all directions is its **work envelope**. The motion characteristics of robots vary depending upon their mechanical design. There are five distinct design configurations for robots, which are discussed later in the robot anatomy section.

Robots interface with their work environment once a mechanical hand (end effector) has been attached to the robot's **tool-mounting plate**. **End effectors** (also commonly called end-of-arm-tooling, **EOAT**) are the **grippers**, tools, special devices, or **fixtures** attached to the robot arm that actually perform the work. The ability to carry, continuously and satisfactorily, a given maximum weight at a given speed defines a robot's **payload**, usually expressed in pounds or kilograms. Payload, in other words, is the weight that the robot is designed to lift, hold, and position repeatedly with the same accuracy.

The maximum **speed** at which the tip of a robot is capable of moving at full arm extension is its **velocity**, usually expressed in inches or millimeters per second. There are two components of its speed: its **acceleration** and **deceleration** rate, and its **slew rate**. The acceleration/deceleration rate is the time it takes to go from rest to full speed and the time it takes to go from full speed to a complete stop. The slew rate is the velocity once the robot is at full speed.

Cycle time is the time it takes for the robot to complete one cycle of picking up a given object at a given height, moving it to a given distance, lowering it, releasing it, and returning to the starting point.

Accuracy defines a robot's ability to position the end effector at a specified point in space upon receiving a control command without previously having attained that position. **Repeatability** refers to the ability of a robot to return consistently to a previously defined and achieved location. **Resolution** of a robot is the smallest incremental change in position that it can make or its control system can measure.

Size refers to the physical size of a robot, which influences its capacity and its capabilities. There are robots as large as gantry cranes and as small as grains of salt—the latter being made by micromachining, which is the same process used to make integrated circuits and computer chips. Some robots intended for light assembly work are designed to be approximately the size of a human so as to make the installation of the robot into the space vacated by the replaced human as easy and as undisruptive as possible. In general, industrial robots vary widely in size, configuration, and capabilities, yet they share a common family structure. All robots are made up of four basic components.

2.3 BASIC COMPONENTS

The most complex robot can be broken down into a few basic parts. This section provides an overview of the parts that make up an industrial robot and their function. It also introduces much of the terminology related to these parts and explains their origin.

The basic components of an industrial robot, labeled in Figure 2.3.1, are the **manipulator**, the end effector (which is part of the manipulator), the **power supply**, and the **controller**. Figure 2.3.2 illustrates clearly the relationship of these four components at a typical industrial robot installation.

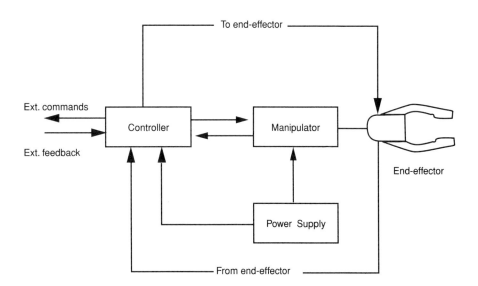

Figure 2.3.1 Basic components of an industrial robot

24 Robot Technology

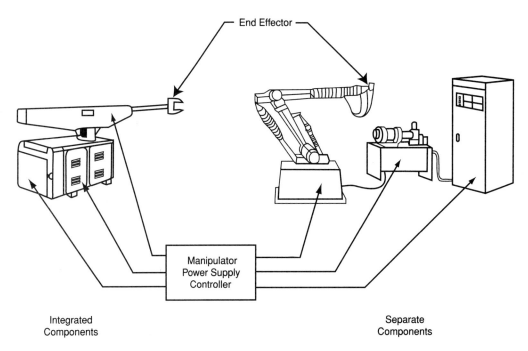

Figure 2.3.2 Industrial robot with integrated and nonintegrated components

The manipulator, which is the robot's arm, consists of segments jointed together with axes capable of motion in various directions allowing the robot to perform work. The end effector, which is the gripper tool, a special device, or fixture attached to the robot's arm, actually performs the work. The power supply provides and regulates the energy that is converted to motion by the robot actuators, and it may be either electric, pneumatic, or hydraulic. The controller initiates, terminates, and coordinates the motions and sequences of a robot. Also, it accepts the necessary inputs to the robot and provides the outputs to interface with the outside world.

Manipulator

The manipulator is a mechanical unit that provides motion similar to that of a human arm. Its primary function is to provide the specific motions that will enable the tooling at the end of the arm to do the required work.

A robot's movements can be divided into two general categories: arm and body (shoulder and elbow) motions, and wrist motions. The individual **joint** motions associated with these two categories are referred to as **degrees of freedom**. Each axis is equal to one degree of freedom. Typically, industrial robots are equipped with 4–6 degrees of freedom. The wrist can reach a point in space with

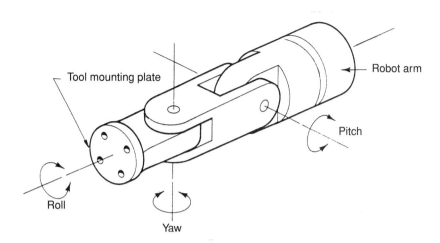

Figure 2.3.3 The three degrees of freedom associated with the robot wrist: roll, pitch, and yaw

specific orientation by any of three motions: a **pitch**, or up-and-down motion; a **yaw**, or side-to-side motion; and a **roll**, or rotating motion. The joints labeled pitch, yaw, and roll are called **orientation axes**. Figure 2.3.3 shows these three motions as associated with the robot wrist.

The manipulator, therefore, is the part of the robot that physically performs the work. The points that a manipulator bends, slides, or rotates are called joints or **position axes**. Manipulation is carried out using mechanical devices, such as linkages, gears, actuators, and feedback devices. Position axes are also called the world coordinates. The **world coordinate system** is identified as being a fixed location within the manipulator that serves as an absolute frame of reference. Figure 2.3.4 shows the location of the world coordinate system. The x axis of travel moves the manipulator in an in-and-out motion; the y axis motion causes the manipulator to move side-to-side; the z axis motion causes the manipulator to move in an up-and-down motion.

The mechanical design of a robot manipulator relates directly to its work envelope and motion characteristics. Figure 2.3.5 shows the parts of a manipulator.

End Effector

A robot can become a production machine only if a **tool** or device has been attached to its mechanical arm by means of the tool-mounting plate. Robot tooling is referred to by several names. The most frequently used is the end effector, but the term end-of-arm-tooling (EOAT) is commonly used both by industry and in publications. If the end effector is a device that is mechanically opened and closed, similar to the one shown in Figure 2.3.6, it is called a gripper. If the end effector

26 Robot Technology

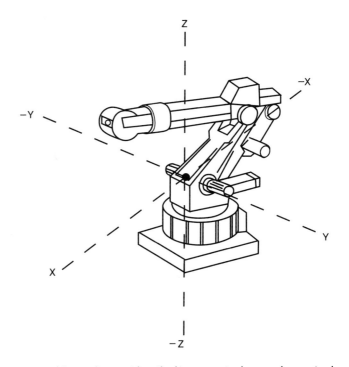

Figure 2.3.4 World coordinates identified on an articulator-style manipulator

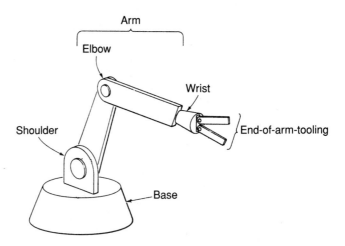

Figure 2.3.5 Parts of a manipulator: The industrial robot manipulator has a body, arm, and wrist. Names match those of the corresponding human parts.

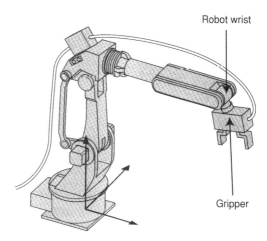

Figure 2.3.6 The Motoman-SK16 electrically controlled industrial robot illustrates here its six degrees of freedom and gripper. It is an articulated robot primarily designed for welding, assembly, and material-handling operations. *(Courtesy of Motoman, Inc.)*

is a tool or a special attachment similar to the ones shown in Figure 2.3.7, it is called **process tooling**.

Depending on the type of operation, conventional end effectors are equipped with various devices and tool attachments, as follows:

- Grippers, hooks, scoops, electromagnets, vacuum cups, and adhesive fingers for materials handling
- Spray gun for painting
- Attachments for spot and arc welding and arc cutting
- Power tools, such as drills, nut drivers, and burrs
- Special devices and fixtures for machining and assembly
- Measuring instruments, such as dial indicators, depth gauges, and the like.

Figure 2.3.8 illustrates various devices and tools attached to the end effector to perform a variety of operations.

The **tool center point (TCP)** is the origin of the coordinate system or the point of action of the tool attached to the robot arm. Notice in Figure 2.3.9 that the origin of the coordinate system is located on the tool-mounting plate of the manipulator. From this location, we can see the direction of the x, y, and z axes. All movements of the manipulator are referenced from this location in space. In Figure 2.3.10 a tool has been added to the mounted plate. Therefore the origin of the coordinate system moved to a new location, which is called the tool center point.

28 Robot Technology

Figure 2.3.7 The Cincinnati Milacron T³ 776 industrial robot illustrates here its wrist and process tool. It is primarily designed for heavy-payload process applications.

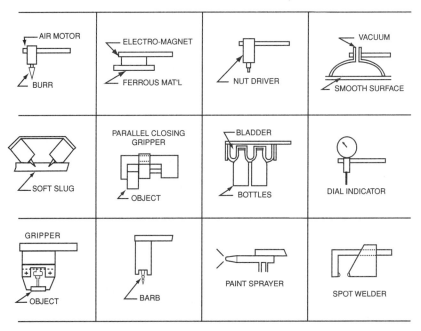

Figure 2.3.8 Various devices and tools attached to end effectors to perform a variety of operations *(Courtesy of Mack Corp.)*

Figure 2.3.9 The origin of the coordinate system is located at the center of the tool-mounting plate of the manipulator.

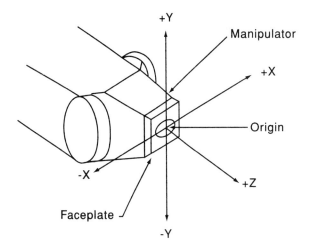

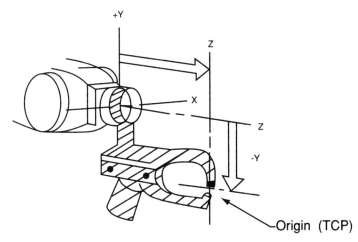

Figure 2.3.10 The location of origin is changed by the addition of a tool to the manipulator. This new origin is called the tool center point (TCP).

End effectors are generally custom-made to meet special handling requirements. Mechanical grippers are the most commonly used and are equipped with two or more fingers. The selection of an appropriate end effector for a specific application depends upon factors such as the payload, environment, reliability, and cost. (Chapter 5 is devoted to the various types of end effectors with their classification, selection, and applications.)

Power Supply

The function of the power supply is to provide and regulate the energy that is required for a robot to be operated. The three basic types of power supplies are electric, hydraulic, and pneumatic.

30 Robot Technology

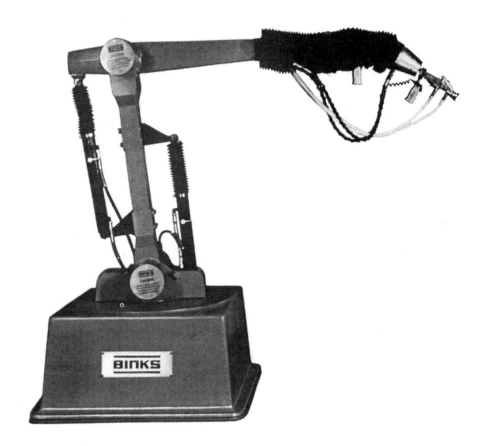

Figure 2.3.11 Binks Manufacturing Co. robot model 88-800 with power supply: Hydraulic on first three axes, electric on last three axes and pneumatic for the end effector. Maximum speeds 60°/s on θ_1, 45°/s on θ_2, 32°/s on θ_3, 90°/s on λ_4 and θ_5, and 150°/s on λ_6. Lead-through controller for spray painting applications. *(Source: Binks Manufacturing Co.)*

Electricity is the most common source of power and is used extensively with industrial robots. The second most common is **pneumatic**, and the least common is **hydraulic** power. Some robot systems require a combination of the three sources. For example, the robot shown in Figure 2.3.11 requires hydraulic power for the first three axes, electric power for the last three axes, and pneumatic power for the end effector. This robot is manufactured by Binks Manufacturing Co., and its application is for spray painting.

The power supply has a direct relation to the payload rating of a robot. Each source of energy and the **actuators** and controls involved have their own characteristics, advantages, and limitations. (Robot power supplies are discussed in detail in Chapter 3.)

Controller

The controller is a communication and information processing device that initiates, terminates, and coordinates the motions and sequences of a robot. It accepts the necessary inputs to the robot and provides the output drive signals to a controlling motor or actuator to correspond with the robot movements and outside world.

Controllers vary greatly in complexity and design. They have a great deal to do with the functional capabilities of a robot and, therefore, the complexity of tasks that robots must be able to fulfill.

The block diagram in Figure 2.3.12a illustrates the many different parts of a robot controller. Figure 2.3.12b illustrates the layout of an actual system installation of a Yamaha loading robot model structure.

The heart of the controller is the computer and its solid-state memory. In many robot controllers, such as the one shown in Figure 2.3.13, the computer includes a network of microprocessors.

The **input** and **output** section of a control system must provide a communication interface between the robot controller computer and the following parts:

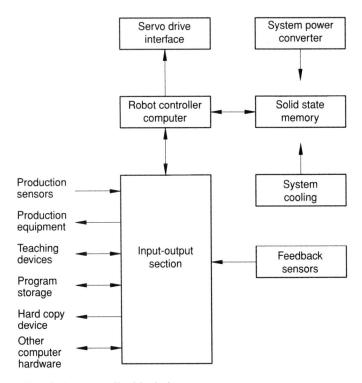

Figure 2.3.12a Robot controller block diagram.

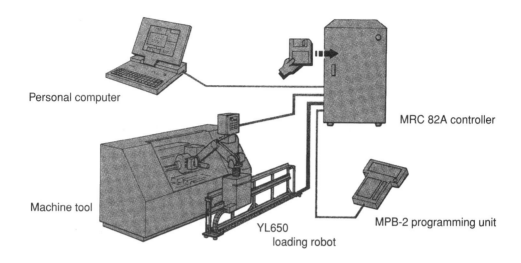

Figure 2.3.12b The Yamaha robot model YL650 is composed of the loading robot, controller, programming box (MPB-2), and I/F equipment for loading and unloading operations. *(Courtesy of Yamaha Robotics Corp.)*

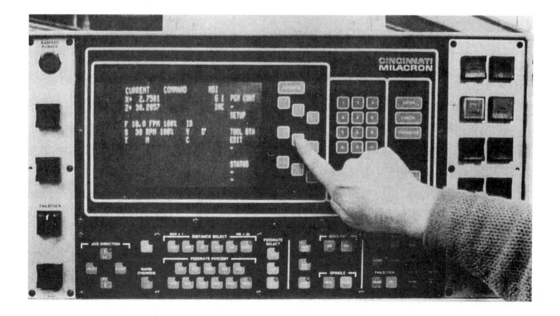

Figure 2.3.13 Microprocessor controller *(Source: Cincinnati Milacron)*

- Feedback sensors
- Production sensors
- Production machine tools
- Teaching devices
- Program storage devices
- Hard copy devices
- Other computer-device hardware

The computer controls the motion of the robot arm by means of drive signals that pass through the drive interface to the actuators on the arm.

In the United States, robots are often classified under three major categories, according to the type of control system used:

1. Nonservo
2. Servo
3. Servo-controlled

The **nonservo** is an **open loop** system, whereas the **servo** is a **closed loop** system. In an open loop system, the output signal is not dependent upon the output of the system, whereas in the closed loop system, the output of the control is constantly compared with the input through **feedback** devices so that the two quantities can be used simultaneously to achieve the desired performance. **Servo-controlled** robots are closed loop systems with continuously controlled path.

Nonservo robots are the simplest form in construction and operation. They are often referred to as limited-sequence robots, pick-and-place, fixed-stops, or bang-bang robots. Feedback devices, such as **transducers**, are also an important part of the controller. They transmit information to the controller on the position of various robot joints and linkages. A robot with either a closed-loop or an open-loop system controls the motion of its arm as it moves through a **programmable** path.

The primary difference between the three categories is that an open loop system has no sensors on the robot arm to provide feedback. Feedback signals indicate the arm position to the controller. Therefore, in an open loop system the controller continues adjusting the arm until it reaches a permanent stop.

Nonservo (open loop). A nonservo robot system is shown in Figure 2.3.14. The diagram is used to represent a four-axis pneumatic robot. At the beginning of the cycle, the controller starts to move the robot through its various sequential steps. At the first step, the controller sends a signal to the control value of the manipulator.

As the control valve opens, air is allowed to pass to the actuator or cylinder causing the rod of the cylinder to move. As long as the valve remains open, this segment of the manipulator continues to move until it is restrained by the end stops on the rod of the cylinder. After the rod of the cylinder reaches

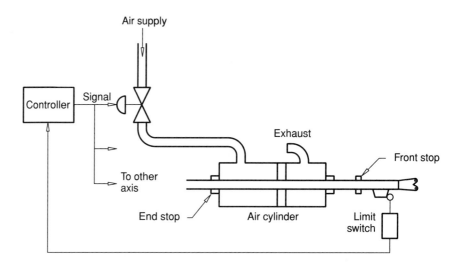

Figure 2.3.14 Nonservo robotic system (open loop)

its length of travel, a limit switch is activated. This tells the controller to close the control valve.

The controller then sends a signal to the control valve to close it, and from there moves to the next step in the program and initiates the necessary signals. The process is repeated until all the steps in the program have been completed.

This is a simple controller—an open loop device, commonly one that relies on sequences and mechanical stops to control the end-point positions along each axis. These robots have no provision for trajectory control between the end points.

For each motion in a nonservo robot, the manipulator members move to a full tilt until the limits of travel are reached. The robot's arm stops at that achieved position by one of the following methods:

- Actuator
- Fixed stop
- Variable stop
- Stepper motors

The actuator is a device in robots that converts energy into motion. Such devices are hydraulic and pneumatic cylinders and linear electric solenoids or motors in which their stroke from full retraction to full extension positions the robot's arm. A linear actuator is shown in Figure 2.3.15 and Figure 2.3.16.

Pneumatic and hydraulic actuators are both powered by moving fluids. In the first case, the fluid is compressed air; in the second case, the fluid is usually pressurized oil. The operation of these actuators is generally similar except in their ability to contain the pressure of the fluid. Pneumatic systems typically operate

2.3 Basic Components **35**

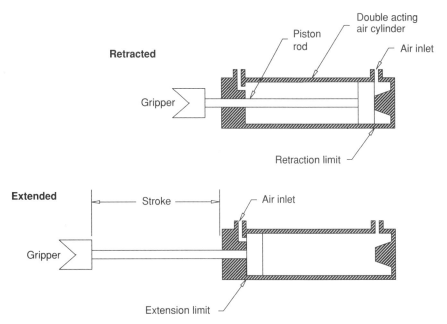

Figure 2.3.15 Linear actuator for motion position.

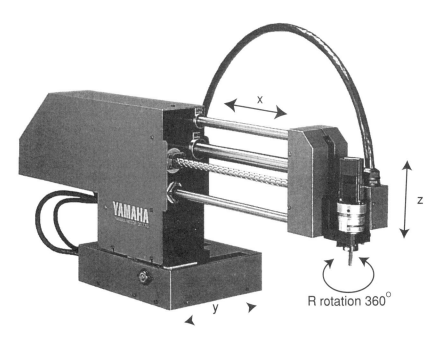

Figure 2.3.16 The Yamaha model YP304A Pick & Place-type robot is a four-axis unit that uses actuator stroke to position the end-of-arm tooling. *(Courtesy of Yamaha Robotics, Inc.)*

with pressure at about one hundred pounds per square inch and hydraulic systems at one thousand to three thousand pounds per square inch.

There are two relationships of particular interest when discussing actuators: the piston velocity of the actuator and the force output of the actuator with respect to the input power. The force output and piston velocity of double-acting cylinders are not the same for extension and retraction strokes. This phenomenon is due to the effect of the rod and is defined by Equations 2.3.1 through 2.3.4.

Extension stroke:

$$\text{force (lb)} = \text{pressure (psi)} \times \text{piston area (in.}^2) \quad \text{(Equation 2.3.1)}$$

$$\text{velocity (ft/sec)} = \frac{\text{input flow (ft}^3\text{/sec)}}{\text{piston area (ft}^2)} \quad \text{(Equation 2.3.2)}$$

Retraction stroke:

$$\text{force (lb)} = \text{pressure (psi)} \times [\text{piston area (in.}^2) - \text{rod area (in.}^2)] \quad \text{(Equation 2.3.3)}$$

$$\text{velocity (ft/sec)} = \frac{\text{input flow (ft}^3\text{/sec)}}{\text{piston area (ft}^2) - \text{rod area (ft}^2)} \quad \text{(Equation 2.3.4)}$$

The horsepower developed by a cylinder can be found using Equations 2.3.5 or 2.3.6.

$$\text{horsepower} = \frac{\text{piston velocity (ft/sec)} \times \text{force (lb)}}{550} \quad \text{(Equation 2.3.5)}$$

$$\text{horsepower} = \frac{\text{input flow (gpm)} \times \text{pressure (lb/in.}^2)}{1714} \quad \text{(Equation 2.3.6)}$$

Example: A hydraulic cylinder has a piston diameter of 2.0 in., a fluid pressure of 1200 lb/in.^{2}, a flow rate of 142 in.^{3}/min, and a stroke of 10 in. Find the force and the velocity generated by the piston.

Solution
Using Equation 2.3.1:

$$\text{force} = \text{pressure} \times \text{piston area}$$
$$= 1200 \times (0.785 \times 2^2)$$
$$= 3770 \text{ lb}$$

Using Equation 2.3.2:

$$\text{velocity} = \frac{\text{input flow}}{\text{piston area}} = \frac{142}{3.14} = 45.2 \text{ inch/min} = 0.0628 \text{ ft/sec}$$

Note: The length of the stroke has no bearing on the operating force and velocity.

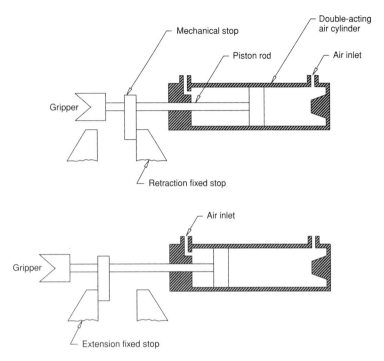

Figure 2.3.17 Fixed stops arrangement for motion position

Fixed stops are often blocks used to stop the extension or retraction of an actuator before it reaches its full stroke. Figure 2.3.17 shows such an installation.

Variable stops are often screws, collars, or sliding blocks that can be adjusted so that they can vary the stroke of the actuator. Figure 2.3.18 shows an example of this type of robot stop.

A **stepper motor** is a DC motor that rotates through a cycle or through part of a cycle in response to an electrical pulse. Stepper motors are designed to output incremental motion with readily available range from 90° down to 0.72°. Their positioning accuracy is 1%–5% of the step angle, with 3% being a common stated accuracy. Stepper motor speed is usually specified as steps per second (sps) as opposed to revolutions per minute (rpm). Equations 2.3.7 to 2.3.9 show the conversion from step angle to steps per revolution and from sps to rpm:

$$\text{Steps / revolution} = \frac{360}{\text{step angle (degrees)}} \qquad \text{(Equation 2.3.7)}$$

$$\text{rpm} = \frac{60 \text{ (sps)}}{\text{steps/revolution}} = 1/6 \text{ (sps) (step angle)} \qquad \text{(Equation 2.3.8)}$$

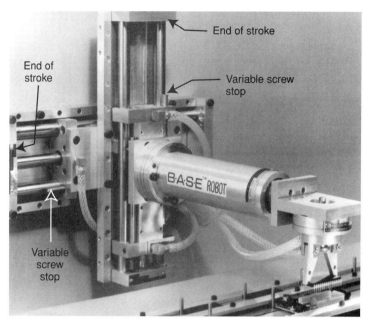

Figure 2.3.18 Robot with variable hard stops *(Courtesy of Mack Corp.)*

$$\text{sps} = \frac{6 \text{ (rpm)}}{\text{(step angle)}} \qquad \text{(Equation 2.3.9)}$$

Stepper motors are available to run as fast as 5,000 sps with an output torque of 6.25 inch-lbs. A stepper motor is often used to extend a robot arm for gripper positioning. In this case, the motor's rotary motion is converted to linear motion by means of a gear drive or a belt and pulley system. Because the stepper motor has a low torque output, considerable mechanical advantage must be gained from the gear or pulley system in order to move the arm and its tooling effectively. Figure 2.3.19 shows such a stepper motor with a translator control.

The advantages of nonservo control are:

- Low cost
- Ease of operation
- High repeatability within 0.010 in. for small units
- High speed
- Control simplicity

The disadvantages of nonservo control are:

- Lack of speed control and therefore less accuracy.
- Time lost through mechanical changes.
- Position stops of robot require accurate placement.

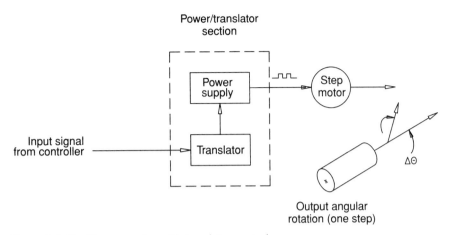

Figure 2.3.19 Stepper motor with translator control

Over half the robots currently in use fall into the nonservo (open loop) classification. Primary application areas served include material handling and machine tending.

Servo (closed loop). The servo robot is a more sophisticated system. The signal from the controller is dependent upon the output of the system. A **servomechanism** is a control system used to detect and correct errors. The system automatically measures the position of each joint and compares this to where it should be. It then utilizes feedback to drive the system to its proper position. Because the system has a self-correcting capability, the desired position of the end effector is stored in the controller memory. This difference allows robots under closed loop control to be programmed to stop at any point in their work envelope. Figure 2.3.20 illustrates a typical **servo system**.

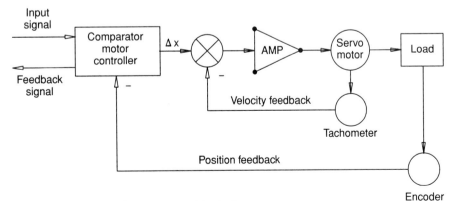

Figure 2.3.20 Typical servo system block diagram

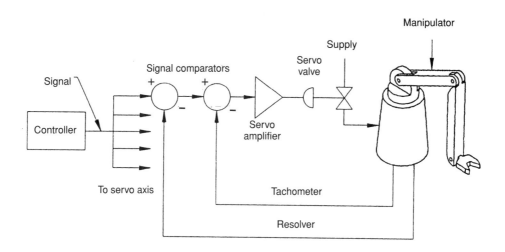

Figure 2.3.21 Actual servo robotic system (closed loop)

Figure 2.3.21 is used to explain the operating principle of the **servo robot**. The diagram is a simplified version of a six-axis robot with a hydraulic power supply.

When the cycle is initiated, the controller addresses the first desired location and interprets the actual locations of the various axes. The desired location signal generated by the controller is compared with the feedback signals from the resolver. The difference in the signals, known as the error signal, is amplified and applied to the servo valve. The valve opens proportionally to the level of the command signal generated by the amplifier. The open valve admits fluid to the actuator on the manipulator. The actuator then moves the manipulator. New signals are generated as the manipulator moves. When the error signal reaches zero, the servo control valve closes, shutting off the flow of fluid. The manipulator comes to rest at the desired position. The controller then addresses the next point in memory. The process is repeated until all steps of the program are completed. A tachometer is used in conjunction with the controller to control acceleration and deceleration of movements.

Servo-controlled (closed loop with controlled path). Servo-controlled robots are directed by a controller that memorizes a sequence of arm and end-effector positions to follow a programmed trajectory or contoured surface. Hundreds or thousands of points can be stored in the computer memory, and the velocity and acceleration along the path can be controlled.

Transducers, called **encoders**, are constructed into the robot arm to convert position data into electrical signals that provide position feedback for each joint.

These feedback signals are compared with the desired position data to direct the robot motion. Velocity data may be computed from the encoder signals and used as an additional feedback signal to assure servo stability and smooth motion.

The advantages of closed loop control are:

- Higher positional accuracy
- Higher speeds
- Higher torque
- Flexible program control
- Ease of changing programmed points
- Capability for complex manufacturing tasks
- Multiple program storage and executions

The disadvantages of closed loop control are:

- Large capital investment
- Sophisticated programming
- User training
- High-skill maintenance

The nonservo robot is further classified as a nonintelligent robot and the servo robot as either intelligent or highly intelligent. The difference between an intelligent and a highly intelligent robot is its level of awareness of its enrichment. (Control systems are discussed in detail in Chapter 8.)

2.4 ROBOT ANATOMY

Robot anatomy is concerned with the physical construction and characteristics of the body, arm, and wrist, which are components of the robot manipulator. Most robots today are mounted on a base. The body is attached to the base and the arm assembly to the body. At the end of the arm is the wrist, which consists of a number of components that allow it to be oriented in a variety of positions. Movements between the various components of the body, arm, and wrist are provided by a series of joints. These joint movements usually involve either rotation or sliding motions (which are described in detail in Chapter 3).

Attached to the robot's wrist is the end effector (or end-of-arm-tooling) that performs the work. The end effector is not considered a part of the robot's anatomy. (The end effector will be discussed in much greater detail in Chapter 5.) The body and arm joints of the manipulator are used to position the end effector, and the wrist joints of the manipulator are all used to orient the end effector.

Robot Configurations

Industrial robots are available in a wide range of shapes, sizes, speeds, load capacities, and other capabilities. The motion characteristics of robots vary, depending upon their mechanical design. The vast majority of today's commercially available robots possess five distinct design configurations:

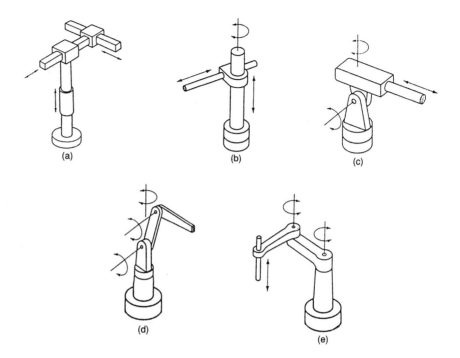

Figure 2.4.1 Five common anatomies of commercial industrial robots: (a) Rectangular, (b) Cylindrical, (c) Spherical, (d) Jointed-arm, and (e) SCARA, or Selectively Compliant Assembly Robot Arm

- Rectangular (or Cartesian)
- Cylindrical (or Post-type)
- Spherical (or Polar)
- Jointed-arm (Articulated or **revolute**)
- SCARA (Selective Compliance Assembly Robot Arm)

The five configurations are illustrated in Figure 2.4.1.

The rectangular configuration, illustrated in (a), uses three perpendicular slides to construct the x, y, and z axes. By moving the three slides relative to one another, the robot is capable of operating within a rectangular work envelope.

The cylindrical configuration, illustrated in (b), uses a vertical column and a slide that can be moved up and down along the column. The robot arm is attached to the slide so that it can be moved radially with respect to the column. By rotating the column, the robot is capable of retrieving a cylindrical work envelope.

The spherical configuration, illustrated in (c), uses a telescoping arm that can be raised or lowered about a horizontal pivot point. The pivot point is mounted on a rotating base and gives the robot its vertical movement. These various joints provide the robot with the ability to move its arm within a spherical envelope.

The jointed-arm configuration, illustrated in (d), consists of two straight components whose shoulder and elbow joints rotate about horizontal axes corresponding to the human forearm and upper arm. A wrist is attached to the end of the forearm to provide additional joints. Its work envelope is of irregular shape. When it is viewed from the top, it is circular; when viewed from the side, it is circular with an inner scalloped surface due to the limits of the joints.

The SCARA configuration, illustrated in (e), is a special version of the jointed-arm robot whose shoulder and elbow joints rotate about vertical axes instead of horizontal. Its work envelope is cylindrical and much larger than all other configurations, which provides a substantial rigidity in the vertical direction for many essential tasks.

Robots may be attached permanently to the floor of a manufacturing plant, may move along overhead rails (a **gantry** robot), or may be equipped with wheels to move along the factory floor (a mobile robot).

2.5 ROBOT GENERATIONS

For years, people have predicted that robots are about to experience a market growth rate similar to that of the electronic computer. The new generation of computers currently under development should greatly increase the power of artificial intelligence programs. These programs, when applied to robotics, should greatly increase the new robots' abilities.

The trends in sensors are also moving toward miniaturization for compatibility with logic systems and less emphasis on electromechanical systems. New developments include sensor fusion and **smart sensors**, which are capable of microcomputer-based calibration, computation, and decision making.

Expert systems, which are computer programs, utilize artificial intelligence and a knowledge base acquired from expert data to solve problems and make decisions. They are also being used to simplify maintenance and diagnostics for robots. These systems are being used to program **robotic** workcells and to route products dynamically between work cells and other manufacturing systems.

In 1984, the growth rates of the work force and the population throughout the world began to slow down. We are now approaching the time when the size of the work force will actually decline. When this happens, it will become much more difficult to find workers to fill various less-attractive jobs—jobs that robots are ready and waiting to fill. It is now estimated by the RIA that the U.S. robotics industry should see a 25–35 percent growth rate each year.

Voice-activated robots are now being used experimentally to help the physically disabled. They will soon be generally available for helping disabled people in a multitude of ways.

New types of robot actuators will also appear. These might rely on shape-memory alloy wire or other types of artificial muscles.

The fully automated factory has been under development in Japan for some time and should be completed within the near future. Such factories will use robots in areas that require flexibility.

The Japanese government is also working on the fifth generation of electronic computers, in which thousands of microcomputers will work in parallel. It is hoped that this design will greatly increase the intelligence capacity of artificial intelligence programs. Companies in the United States are working on fifth-generation computers as well. Such computer designs will also lead to more-intelligent robots.

The five generations of robot controllers after the high-tech inception in 1960 are described as follows:

1. *First Generation:* Repeating robots. These were generally pneumatically powered "pick and place" robots, with mechanical sequences defining stop points. Revolving drums with cam-and-follower control provided the programming. To reprogram the robot, a new precision cam was installed. It is estimated that more than 90 percent of the early robots belong to this category.
2. *Second Generation:* Hardwired (patch board) controllers provided the first programmable units. In "pick and place" robots, signals were derived from limit switches, proximity switches, and similar devices. These controllers were also applicable to servo control. The electrical system consisted of a bank of relays and the reprogramming for any new job that required rewiring. These controls are still used in simpler pick-and-place robots, and will probably always play a role in robotics as the most economical solution for situations involving only simple motion requirements.
3. *Third Generation:* **Programmable logic controllers (PLC)**, introduced into industry over thirteen years ago, provided a microprocessor-based robotic controller that is easy to reprogram. The controller primarily serves to direct the sequence of robot motions, stop points, gripper actions, and velocity.
4. *Fourth Generation:* When control beyond a PLC is required, a microcomputer may control the entire system, including other programmable machinery in a robot workcell. Whereas PLCs are limited in their programming, minicomputers may utilize a special robot programming language or standard language (such as BASIC, C, or C++) for more-advanced **off-line programming** or **CAD/CAM** and **CIM** interface. Minicomputer-type robots based on artificial intelligence became commercially available at the end of 1980. These controllers now allow integration with vision or tactile sensors.
5. *Fifth Generation:* Robot controllers will involve complete artificial intelligence (AI), miniaturized sensors, and decision-making capabilities. Some rudimentary efforts have already been made in this direction, as previously explained, with some AI algorithms (such as for bin picking and search routines for gripper positioning) now available. An artificial biological robot might provide the impetus for sixth and higher generation robots. A visualization of these overlapping generations of robots is shown in Figure 2.5.1.

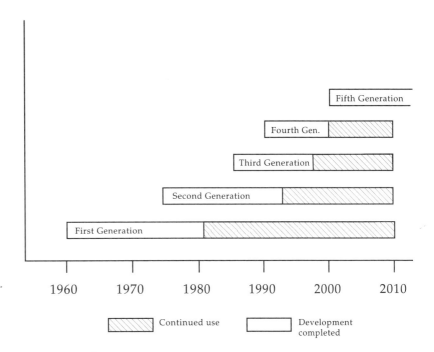

Figure 2.5.1 Robot generations

2.6 ROBOT SELECTION

As with human workers, robots have individual characteristics, features, and capabilities. Therefore, it is imperative that robots be matched properly by capabilities to task requirements. An objective approach to robot selection provides fewer restrictions in system design by allowing for the optimum system design to be achieved, regardless of the specific robot needed. The system design, when accomplished, establishes the requirements for the robot within the workplace. The ideal approach then is not to choose a robot first, but to wait until all task requirements have been defined before the final choice of robot is made.

Process Knowledge

In designing systems, we can observe the manual operations for guidance in the concepts for an automated approach. The point here is *not* necessarily to attempt to duplicate all that the human worker does, but rather to analyze the work elements in light of human capabilities and how they will need to be restructured for a robot. In task restructuring, we can exploit the capabilities of the robot.

To arrive at the proper conclusion concerning application development, system design, and task restructuring, it is necessary to start with process knowledge. With a thorough knowledge of the part and the manner in which it is processed, attention can focus on improved efficiency within the process that may be achieved

Table 2.1 Criteria for Robot Selection

Technical Issues	Nontechnical issues
1. Type	1. Cost & benefit consideration
a. Nonservo	2. Commonality of equipment
b. Servo	3. Training & maintenance requirements
c. Servo-controlled	4. Reliability
2. Work envelope	5. Service
a. Rectangular	6. "Systems" Help
b. Cylindrical	7. Safety
c. Spherical	
d. Jointed-arm	
e. SCARA	
3. Payload	
4. Cycle time	
5. Repeatability	
6. Drive	
a. Electric	
b. Pneumatic	
c. Hydraulic	
d. Any combination	
7. Unique capabilities	

through part design modifications. Alternatively, the efficiency improvements may come from reversing or combining processes, implementation of better production planning and scheduling, better material flow by grouping successive operations together, grouping families of parts to be manufactured in flexible manufacturing cells, and utilizing other group technology approaches.

Special attention should be given to end-effector concepts as the system elements are designed during stages of application development. Keeping in mind the end-effector workplace, it is appropriate to rank tooling design high on the list of critical system elements.

Determining the proper robot type and options necessary for the task can best be made by following the consideration of all the aforementioned issues. Thinking in generic terms about the robot best suited for a task based on such technical issues as work envelope, control system, power supply, payload, velocity, accuracy, and repeatability, and such nontechnical issues as cost, reliability, vendor support, and experience, robot selection is simplified, as shown in Table 2.1.

(To evaluate the potential of a robot selection in detail, see Chapter 12.)

Safety

Safety is of such importance that it is not mentioned as a phase of process knowledge because human workers and robots interface issues need to be addressed in all design, manufacture, and implementation phases of a robot project. Properly designed systems do not allow a human worker and a robot to share the same work space unless the greatest amount of care has been taken to ensure protection of the human worker through proper presence/position-detection sensors. Physical barriers are the safest of all possible approaches. (Robot safety is discussed extensively in Chapter 11.)

2.7 SUMMARY

Robot technology is an applied science that is referred to as a combination of machine tool fundamentals and computer applications. To describe this technology, we must define the variety of technical features about the way a robot is constructed and works and the factors that influence its selection.

The basic components of an industrial robot are:

1. The manipulator, which provides motions similar to the human arm and hand and physically performs the work
2. The end effector, which is equipped with various types of attachments to meet special handling requirements to actually perform the work
3. The power supply, which is the prime mover to drive the robot by independent actuators and controls using electric, pneumatic, or hydraulic sources
4. The control system, which is the brain of the robot and stores data to initiate and terminate movements of the manipulator

The control systems used in the United States to position the tooling are classified into three major categories: nonservo, servo, and servo-controlled.

Robot anatomy is concerned with the physical construction and operation of the manipulator and has five basic configurations: rectangular, cylindrical, spherical, jointed-arm, and SCARA.

So far, there have been five generations of robot controllers, and we are merging now to sixth, seventh, and even higher generations. Robots with increasing intelligence, sensory capability, dexterity, and sophisticated control systems have become the dominant factor in modern manufacturing.

The three factors that influence the selection of robots in manufacturing are:

1. Dynamic properties and performance
2. Economics
3. Safety

2.8 REVIEW QUESTIONS

2.1 What are the general characteristics of an industrial robot?
2.2 What are the basic components of an industrial robot?
2.3 What is the primary function of a manipulator?
2.4 What is mounted on the end of a manipulator?
2.5 What kind of power sources are used for industrial robots?
2.6 What controls the motion of the robot's manipulator and how is it accomplished?

2.7 What are the number of joints or movable axes present in a manipulator called?
2.8 What are the three orientation axes in the wrist called?
2.9 What is the area that represents the maximum extent of the robot's manipulator called?
2.10 What is the measure of how well the controller can drive the robot's arm back to a taught point called?
2.11 What is the maximum rate at which the robot can move the TCP to a specified location called?
2.12 List six end-effector attachments.
2.13 What methods of path motion are used in industrial robots?
2.14 What are feedback devices and how are they operated?
2.15 What are the five design configurations of industrial robots?
2.16 List the three factors that influence the selection of robots in manufacturing.

2.9 PROBLEMS

2.1 An air cylinder is to be used to actuate the linear arm joint for a polar configuration robot. The piston diameter is 2.0 in., the air pressure is 100 psi, and the airflow rate is 0.273 feet^3/min. Determine the force generated by the piston and the velocity during the forward stroke.

2.2 A double-acting hydraulic cylinder is operated at 1000 psi pressure. The cylinder has a 6-in. bore and a 12-in. stroke. The piston rod is 1.5 in. in diameter. Find the "push" and "pull" force exerted by the piston rod.

2.3 A pump supplies oil at 20 gpm to a 2-in. diameter double-acting hydraulic cylinder. If the load is 1000 lbs (extending and retracting) and the rod diameter is 1 in., find:
   a. The hydraulic pressure during the extending stroke
   b. The piston velocity during the extending stroke
   c. The cylinder horsepower during the extending stroke
   d. The hydraulic pressure during the retraction stroke
   e. The piston velocity during the retraction stroke
   f. The cylinder horsepower during the retraction stroke

2.4 A stepper motor runs at 1500 rpm with a step angle of 1.8°. Determine the motor's speed in steps/sec.

2.5 A stepper motor runs at 2500 sps with a step angle of 1.8°. Determine the motor's speed in rpm.

2.10 REFERENCES

Asfahl, C. R. *Robots and Manufacturing Automation.* 2d ed. New York: John Wiley and Sons, Inc., 1992.

Caldwell, D. G., A. Wardl, O. Kocak, and M. Goodwin. "A Twin-Armed Mobile Robot." *IEEE Robotics and Automation Magazine* (September 1996): 29–38.

Dario, P., E. Guglielmelli, B. Allotta, and M. C. Carozza. "Robotics for Medical Applications." *IEEE Robotics and Automation Magazine* (September 1996): 44–56.

Groover, M. P., M. Weiss, R. N. Nagel, and N. G. Odrey. *Industrial Robotics.* New York: McGraw-Hill Book Co., 1986.

Hoekstra, R. L. *Robotics and Automated Systems.* Cincinnati, OH: South-Western Publishing Co., 1986.

Lewis, F. L., C. T. Abdallah, and D. M. Dawson. *Control of Robot Manipulators.* New York: Macmillan Publishing Co., 1993.

Mandow, A., J. M. Gomez-de-Gabriel, J. L. Martinez, V. F. Munoz, A. Ollero, and A. Garcia-Cerezo. "The Autonomous Mobile Robot." *IEEE Robotics and Automation Magazine* (December 1996): 18–28.

Parkin, R. E. *Applied Robotic Analysis.* Englewood Cliffs, NJ: Prentice Hall, 1991.

Rehg, J. A. *Introduction to Robotics in CIM Systems.* 2d ed. Upper Saddle River, NJ: Prentice Hall, 1997.

Rosheim, M. E. *Robot Evolution.* New York: John Wiley and Sons, Inc., 1994.

Sandler, B. Z. *Robotics: Designing the Mechanisms for Automated Machinery.* Englewood Cliffs, NJ: Prentice Hall, 1991.

Schilling, R. J. *Fundamentals of Robotics: Analysis and Control.* Englewood Cliffs, NJ: Prentice Hall, 1990.

Spiteri, C. J. *Robotics Technology.* Philadelphia, PA: Saunders College Publishing, 1990.

CHAPTER 3

Robot Classification

3.0 OBJECTIVES

After studying this chapter, the reader should:

1. Be aware of robot classification
2. Be acquainted with the manipulator arm geometry
3. Understand the degrees of freedom of a robotic system
4. Recognize the types of power sources used in current robots
5. Be familiar with types of motion
6. Know a robot's path control
7. Understand the intelligence level of robots

3.1 CLASSIFICATION

Industrial robots can be classified into six categories according to their characteristics:

1. Arm geometry: rectangular; cyclindrical; spherical; jointed-arm (vertical); joined-arm (horizontal)
2. Degrees of freedom: robot arm; robot wrist
3. Power source: electrical; pneumatic; hydraulic; any combination
4. Types of motion: slew motion; joint-interpolation; straight-line interpolation; circular interpolation
5. Path control: limited-sequence; point-to-point; continuous path; controlled-path
6. Intelligence level: low-technology (nonservo); high-technology (servo)

3.2 ARM GEOMETRY

The robot manipulator may be classified according to the type of axis movement needed to complete a task. Because we live in a three-dimensional world, the general robot must be able to reach a point in space within three axes by moving forward and backward, to the left and right, and up and down. This can be accomplished in several ways. The simplest way is by identifying those movements as robot arm geometry and describing them in the coordinate system as follows.

Rectangular-Coordinated

A rectangular- or **cartesian-**coordinated robot manipulator has three linear axes of motion or coordinates. The first coordinate, x, might represent left and right motion; the second, y, may describe forward and backward motion; the third, z, generally is used to depict up-and-down motion. The disadvantage of this design

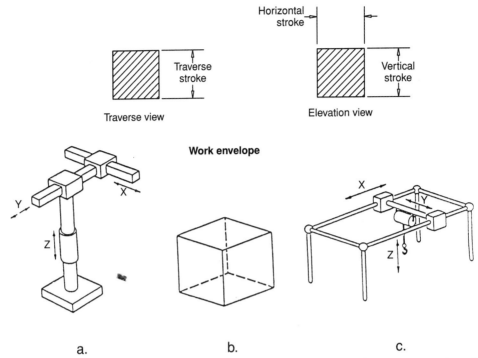

Figure 3.2.1 Rectangular or Cartesian-coordinated robot: (a) A rectangular coordinated arm moves in three linear axes. (b) The box-shaped work envelope within which a rectangular manipulator operates. (c) Overhead crane movements are similar to those of a rectangular-coordinated arm.

is that the motion of each axis is limited to one direction and makes it independent of the other two. However, equal increments of motion may be achieved in all axes by using identical actuators.

The work envelope of a rectangular robot is a cube or rectangle, so that any work performed by the robot must only involve motions inside this space. The work envelope of a robot is the outline of the work volume region. When a robot is mounted from above in a bridgelike frame, it is referred to as a gantry robot, otherwise called traverse-type. Figure 3.2.1 shows (a) a typical rectangular-coordinated robot, (b) its work envelope, and (c) its work area similar to an overhead crane, referred to as a gantry robot. An actual rectangular-coordinated robot is shown in Figure 3.2.2.

The power for movement in the x, y, and z directions is provided by linear actuators in small robots or by ball-screw-drives in large systems. Rectangular-coordinated robots have the following advantages:

- They can obtain large work envelopes because traveling along the x axis, the volume region can be increased easily.

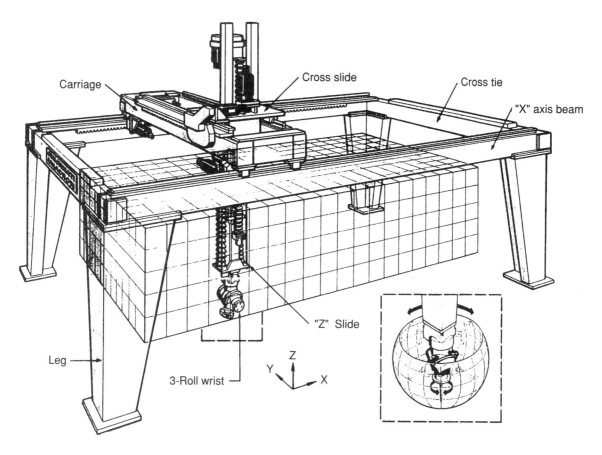

Figure 3.2.2 An actual rectangular gantry robot model T³ 886, manufactured by Cincinnati Milacron. This unit can lift loads up to 198 pounds and has six electrically actuated servo-controlled axes. *(Source: Cincinnati Milacron)*

- Their linear movement allows for simpler controls.
- They have a high degree of mechanical rigidity, accuracy, and repeatability due to their structure.
- They can carry heavy loads because the weight-lifting capacity does not vary at different locations within the work envelope.

Rectangular-coordinated robots have the following disadvantages:

- They makes maintenance more difficult for some models with overhead drive mechanisms and control equipment.
- Access to the volume region by overhead crane or other material-handling equipment may be impaired by the robot-supporting structure.
- Their movement is limited to one direction at a time.

Typical applications for rectangular robots include:

- Pick-and-place operations
- Adhesive applications (mostly long and straight)
- Advanced munitions handling
- Assembly and subassembly (mostly straight)
- Automated loading CNC lathe and milling operations
- Inspection
- General machining operations
- Nuclear material handling
- Remotely operated decontamination
- Robotic X-ray and neutron radiography
- Surface finishing
- Welding
- Waterjet cutting

Cylindrical-Coordinated

A cylindrical- or post-type-coordinated robot has two linear motions and one rotary motion. Robots that have one rotational capability or degree of freedom and two translational (linear) degrees of freedom can achieve variable motion. The first coordinate describes the angle θ of base rotation, perhaps about the up-down axis. The second coordinate may correspond to a radical or y in-out motion at whatever angle the robot is positioned. The final coordinate again corresponds to the up-down or z position.

The **cylindrical-coordinated** robot shown in Figure 3.2.3 can reach any point in a cylindrical volume of space, although a central portion of the space must be devoted to the robot and limits to the full rotation may also be imposed.

Its rotational ability gives the advantage of moving rapidly to the point in the z plane of rotation. An actual cylindrical-coordinated robot is shown in Figure 3.2.4, where its smallest possible increment of change, which is called resolution, is not usually equal in its three axes of motion. The resolution of the base rotation is expressed in terms of an angular measurement, and the resolution of the linear axis is expressed in terms of linear increments.

A cylindrical-coordinated robot generally results in a larger work envelope than a rectangular robot manipulator. These robots are ideally suited for pick-and-place operations.

Some advantages of the cylindrical-coordinated robots are:

- Their vertical structure conserves floor space.
- Their deep horizontal reach is useful for far-reaching operations.
- Their capacity is capable of carrying large payloads.

Some disadvantages of cylindrical-coordinated robots are:

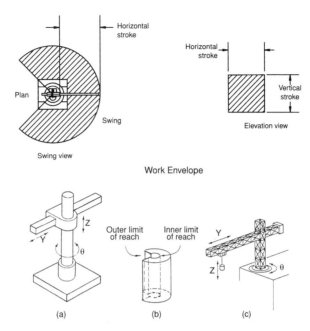

Figure 3.2.3 Cylindrical-coordinated robot: (a) A cylindrical-coordinated arm rotates about its base, moves in and out, and up and down. (b) The space between the two cylinders shown is the work envelope occupied by a cylindrical-coordinated manipulator. (c) The movements of a construction crane on top of a tall building are similar to those of a cylindrical-coordinated manipulator.

- Their overall mechanical rigidity is lower than that of the rectilinear robots because their rotary axis must overcome the inertia.
- Their repeatability and accuracy are also lower in the direction of rotary movement.
- Their configuration requires a more sophisticated control system than the rectangular robots.

Typical applications for cylindrical robots include:

- Assembly
- Coating applications
- Conveyor pallet transfer
- Die casting
- General material handling
- Foundry and forging applications
- Inspection molding
- Investment casting
- Machine loading and unloading
- Meat packing
- Pick-and-place operations

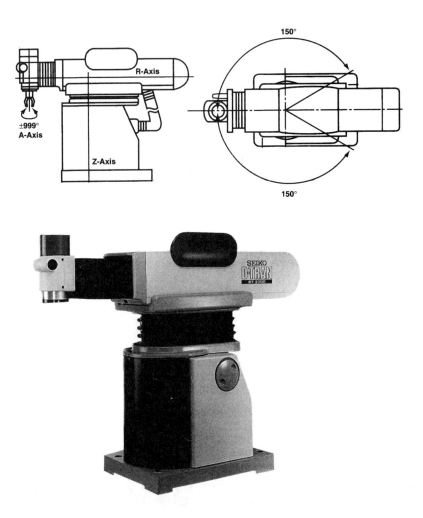

Figure 3.2.4 A high-speed cylindrical robot model RT3300, manufactured by Seiko Instruments. This robot is a four-axis multitasking unit with AC servo-controlled motors and can handle loads up to twenty-seven pounds. *(Courtesy of Seiko Instruments USA, Inc.)*

Spherical-Coordinated

A spherical- or polar-coordinated robot has one linear motion and two rotary motions. The work volume or envelope is shaped like a section of a sphere with upper and lower limits imposed by the angular rotations of the arm. A central core of the work volume is omitted to accommodate the robot base. A pie-shaped section may also be omitted to accommodate the rearward motion of the arm or to provide a safe operating position for the operator.

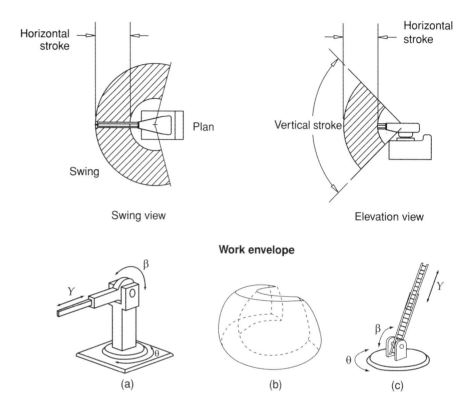

Figure 3.2.5 Spherical- or polar-coordinated robot: (a) A polar- or spherical-coordinated manipulator rotates about its base and shoulder and moves linearly in and out. (b) The work envelope of a polar-coordinated manipulator is the space between the two hemispheres. (c) A ladder on a hook-and-ladder truck has movements similar to those of a polar-coordinated manipulator.

The **spherical-coordinated** robot, shown in Figure 3.2.5, reaches any point in space through one linear and two angular motions. The first motion corresponds to a base rotation about a vertical axis. The second motion corresponds to an elbow rotation. The third motion corresponds to a radial, or in-out, translation. The two rotations can point the robot in any direction and permit the third motion to go directly to a specified point. The points that can be reached by the spherical-coordinated robot include the volume of a globe or sphere. An actual spherical-coordinated robot is shown in Figure 3.2.6. A spherical-coordinated robot generally provides a larger work envelope than the rectilinear or cylindrical robot. The design is simple and gives good weight-lifting capabilities. This configuration is suited for applications where a small amount of vertical movement is adequate, such as loading and unloading a punch press.

58 Robot Classification

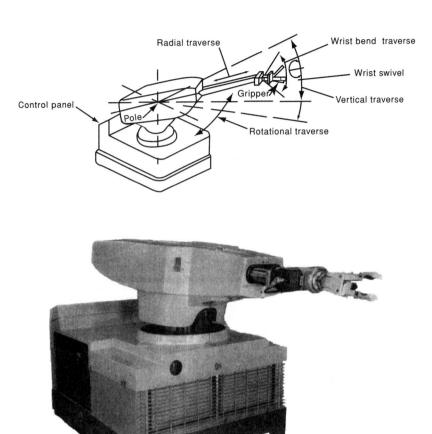

Figure 3.2.6 An actual spherical-coordinated robot, series 2000, manufactured by Unimation Inc., with six servo axes and a payload capacity of three hundred pounds. It is used for die casting, injection molding, forging, machine tool loading, heat treating, glass handling, dip coating, press loading, and material transfer.

The advantages and disadvantages listed for cylindrical-coordinated robots can also be applied to spherical, with the following exceptions: Cylindrical is more vertical in structure, whereas spherical yields a low and long machine size to provide the horizontal reach; also, their vertical movement is limited.

Typical applications for spherical robots include:

- Die casting
- Dip coating
- Forging

- Glass handling
- Heat treating
- Injection molding
- Machine tool loading
- Material transfer
- Parts cleaning
- Press loading
- Stacking and unsticking

Jointed-Arm-Coordinated

A jointed-arm or revolute (often referred to as **anthropomorphic** or human-like) coordinated robot performs in an irregularly shaped work envelope and comes in two basic configurations: vertical and horizontal.

Vertical. The vertical jointed-arm robot has three rotary motions to reach any point in space. This design is similar to the human arm, which has two links, the shoulder and the elbow, and positions the wrist by rotating the base about the z axis. The first rotation, therefore, is about the base; the second rotation is about the shoulder in a horizontal axis; and the final motion is a rotation of the elbow, which may be about a horizontal axis, but the axis may be at any position in space determined by the base and shoulder rotations.

The work envelope is circular when viewed from the top of the robot. When looked at from the side, the envelope has a circular outer surface with an inner scalloped surface due to the limits of the joints. The vertical jointed-arm robot, typically called tear drop, is shown in Figure 3.2.7.

This type of robot can move at high speeds in various directions and has a greater variety of angles of approach to a given point; therefore, it is very useful for painting and welding applications. An actual anthropomorphic robot is shown in Figure 3.2.8.

Horizontal. The horizontal jointed-arm robot generally reaches any point in space through one linear (vertical) motion and two rotary motions. Also called the SCARA (Selective Compliance Assembly Robot Arm), this robot has two horizontally jointed-arm segments fixed to a rigid vertical member (base) and one vertical linear motion axis. The first rotary motion corresponds to the shoulder about its vertical axis. The second rotary motion corresponds to the elbow also about its vertical axis, and the third (linear) motion corresponds about the vertical up-down z axis.

The horizontal jointed-arm robot (SCARA) is shown in Figure 3.2.9. One particularly attractive feature of this robot is that it is extremely useful in assembly operations where insertions of objects into holes are required. The SCARA robot was developed at Yamanashi University in Japan in 1978 for parts assembly work.

The jointed-arm robot has several advantages. It is by far the most versatile configuration and provides a larger work envelope than the rectangular-, cylin-

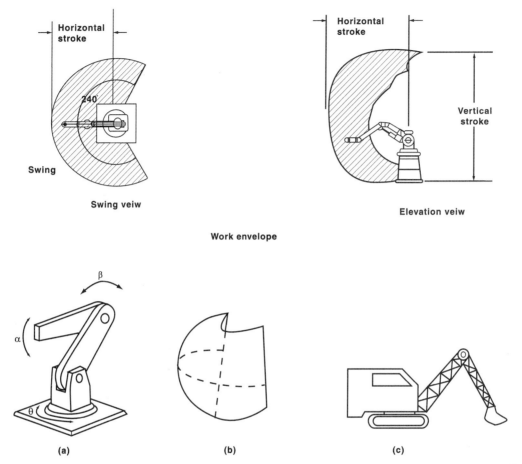

Figure 3.2.7 A jointed-arm or articulated-coordinated robot: (a) A jointed-arm-coordinated manipulator has all three axes rotational. (b) The area between the sphere and the column (representing the base support) is the work envelope for the jointed-arm manipulator. The jointed-arm manipulator can reach above and below an obstacle. (c) A power shovel has movements similar to those of a jointed-arm manipulator.

drical- or spherical-coordinated robots. It occupies minimum floor space and achieves deep horizontal reach. The high positioning mobility at the end-of-arm-tooling allows the arm to reach into enclosures and around obstructions. An actual SCARA robot is shown in Figure 3.2.10.

The drawback of this type of robot is that it requires a very sophisticated controller because the programming is more complex. Also, different locations in the work envelope can affect accuracy, load-carrying capacity, dynamics, and repeatability of movement. This configuration can also become less stable when the arm approaches its maximum reach, but this can be overcome by the addition of feedback controls.

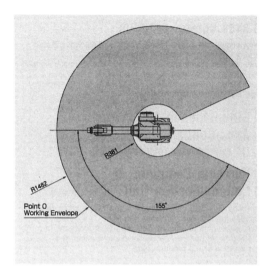
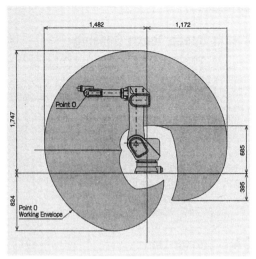

Figure 3.2.8 An actual vertical jointed-arm or articulated-coordinated robot model VR-008A manufactured by Matsushita Industrial Equipment Co., Ltd., with six rotational axes and exceptional wide working envelope for flexible, quick, smooth, and vast variety of welding operations *(Courtesy of Panasonic Factory Automation, Division of Matsushita Corp.)*

Figure 3.2.9 A horizontal or SCARA robot: (a) A SCARA robot rotates in two axes in the horizontal plane and moves linearly up and down. (b) The work envelope for the SCARA manipulator is the space between two cylinders, which can reach around obstacles. (c) A folding lamp has movements similar to those of a SCARA manipulator.

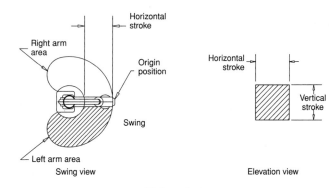

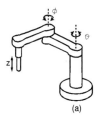
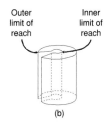

(a)   (b)   (c)

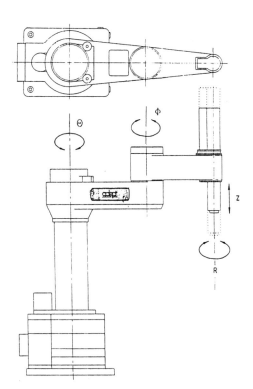

Figure 3.2.10 An actual SCARA model AdeptOne robot manufactured by Adept Technology, Inc., with a playload of 20 lb, resolution of ±0.0005 in. and repeatability of ±0.001 in. This is an electrically powered manipulator robot with maximum velocity of an end effector of 30 ft/s. It is used in accurate assembly (insertion, sealing, fitting), loading/unloading, molding inserts, palletizing, inspection, gauging, and fine-part soldering. *(Source: Adept Technology Inc.)*

Typical applications for articulated robots include:

- Automatic assembly
- Die casting
- In-process inspection
- Machine loading and unloading
- Machine vision
- Material cutting
- Material removal
- Multiple-point light machining operations
- Paint and adhesive applications
- Palletizing
- Thermal coating
- Welding

Table 3.1 summarizes the five robot manipulator configurations with their respective advantages and disadvantages.

Table 3.1 Summary of the Five Basic Robot Manipulator Configurations

Configuration	Advantages	Disadvantages
Rectangular coordinates (x, y, z-base travel, reach, and height)	Three linear axes Easy to visualize Rigid structure Easy to program off-line Linear axes make for easy mechanical stops	Can only reach in front of itself Requires large floor space for size of work envelope Axes hard to seal
Cylindrical coordinates (θ, y, z-base rotation, reach, and height)	Two linear axes, one rotating axis Can reach all around itself Reach and height axes rigid Rotation axis easy to seal	Cannot reach above itself Base rotation axis is less rigid than a linear axis Linear axis is hard to seal Won't reach around obstacles Horizontal motion is circular
Spherical coordinates (vertical) (θ, y, β-base rotation, elevation angle, reach angle)	One linear axis, two rotating axes Long horizontal reach	Can't reach around obstacles Generally has short vertical reach
Revolute (or jointed-arm) coordinates (vertical) (θ, β, α-base rotation, elevation angle, reach angle)	Three rotating axes Can reach above or below obstacles Largest work area for least floor space	Two or four ways to reach a point Most complex manipulator
SCARA coordinates (horizontal) (θ, Φ, z-base rotation, reach angle, height)	One linear axis, two rotating axes Height axis is rigid Large work area for floor space Can reach around obstacles	Two ways to reach a point Difficult to program off-line Highly complex arm

3.3 DEGREES OF FREEDOM

The degrees of freedom or grip of a robotic system can be compared to the way in which the human body moves. For each degree of freedom, a joint is required. The degrees of freedom located in the arm define the configuration. Each of the five basic motion configurations discussed previously utilizes three degrees of freedom in the arm. For applications that require more flexibility, additional degrees of freedom are used in the wrist of the robot. Three degrees of freedom located in the wrist give the end effector all the flexibility. A total of six degrees of freedom is needed to locate a robot's hand at any point in its work space. Although six degrees of freedom are needed for maximum flexibility, most robots employ only three to five degrees of freedom. The more degrees of freedom, the greater the complexity of motions encountered.

In comparison, the movement of the human hand is controlled by thirty-five muscles. Fifteen of these muscles are located in the forearm. The arrangement of the muscles in the hand provides great strength to the fingers and thumb for grasping objects. Each finger can act alone or together with the thumb. This enables the hand to do many intricate and delicate tasks. Some of the grips of the human hand for moving objects are difficult for robotic systems to duplicate.

The human hand has twenty-seven bones: the eight bones of the carpus, or wrist, arranged in two rows of four; the five bones of the metacarpus, or palm, one to each digit; and the fourteen digital bones, or phalanges, two in the thumb and three in each finger. The carpal bones fit into a shallow socket formed by the bones of the forearm. There are twenty-two degrees of freedom (joints) in the hand, with seven in the wrist. From Figure 3.3.1, it can be seen that the hand is a very complex multipurpose tool. Hands can be used to perform various repetitive tasks.

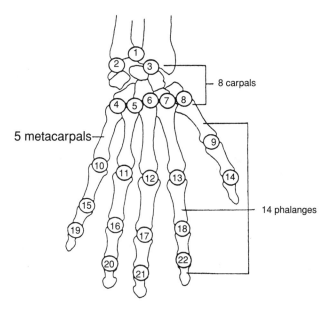

Figure 3.3.1 Degrees of freedom of the human hand

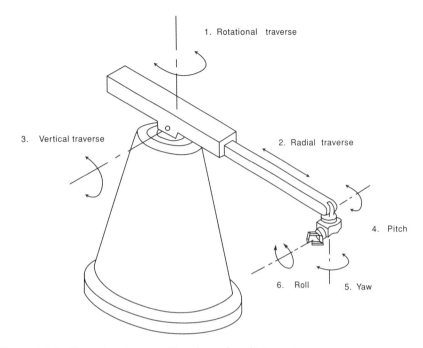

Figure 3.3.2 Six major degrees of freedom of a robotic system

The bones and joint arrangement give the human hand dexterity not found in robotic systems. The movements of a robotic system seem awkward and clumsy, because the robot is usually accomplishing these movements with only six degrees of freedom.

The three degrees of freedom located in the arm of a robotic system are (1) the rotational traverse, (2) the radial traverse, and (3) the vertical traverse. The rotational traverse is the movement of the arm assembly about a rotary axis, such as the left-and-right swivel of the robot's arm about a base. The radial traverse is the extension and retraction of the arm or the in-and-out motion relative to the base. The vertical traverse provides the up-and-down motion of the arm of the robotic system.

The three degrees of freedom located in the wrist, which bear the names of aeronautical terms, are (4) pitch, (5) yaw, and (6) roll, as previously discussed.

The pitch or bend is the up-and-down movement of the wrist. The yaw is the right-and-left movement of the wrist. The roll or swivel is the rotation of the hand. Figure 3.3.2 illustrates the six basic degrees of freedom of a robotic system.

3.4 POWER SOURCES

A 1990 survey of manufacturers of robots in the United States, Europe, and Japan, based upon nearly two hundred suppliers and not including a number of small and very specialized robot manufacturers, indicated that models available were de-

Table 3.2 Load-Carrying Capacity of Robots

Load Capacity		Percentage of Robots
Pounds	Kilograms	Models offered
300–2,300	136–1,043	11
100–299	45–136	16
50–99	23–45	15
20–49	9–22	21
10–19	5–9	17
less than 10	less than 5	20

signed to handle loads ranging from about 1 pound (0.5 kilogram) upward to about 2,300 pounds (1,043 kilograms).

Applications listed in the survey for robots included die casting, forging, plastic molding, machine tools, investment castings, a general and miscellaneous category, spray painting, welding, and machining. It is well established that the use of robots in light manufacturing and inspecting operations, such as are found in the electronics industry, has increased markedly during the past few years. In terms of models available, the breakdown was approximately that shown in Table 3.2.

A recent survey indicated that, in terms of total robots made, electric drives account for about one half of the robot drives used; pneumatic drives, about one third of the total; and hydraulic drives, about one sixth of the total. Some authorities believe that these ratios will hold rather steady; others profess a solid trend toward electric drives. Electric servo units lately have been advanced in power and durability.

Electric Power Source

All robot systems use electricity as the primary source of energy. Electricity turns the pumps that provide hydraulic and pneumatic pressure. It also powers the robot controller and all the electronic components and peripheral devices.

In all electric robots, the drive actuators, as well as the controller, are electrically powered. Most electric robots use **servomotors** for axes motion, but a few open loop robot systems utilize stepper motors. The majority of robots presently are equipped with DC servomotors, but eventually will be changed to AC servomotors because of their higher reliability, compactness, and high performance. Most new model robots appear to be with an AC servomotor and an encoder, which simplifies wiring, reduces maintenance, and increases performance. Therefore, AC servomotors are gaining confidence and importance in the robot industry. Electric motors provide the greatest variety of choices for powering manipulators, especially in the low- and moderate-load ranges, and for low-speed high-load operations.

Motors generally operate at speeds that far exceed those desirable for manipulator joints; therefore, speed reducers are required. The ability to accelerate and decelerate the working load quickly is a very desirable attribute. Also required is the ability to operate at variable speeds. An example of an electrically powered robot is shown in Figure 3.4.1.

Because electric robots do not require a hydraulic power unit, they conserve floor space and decrease factory noise. Direct drive models provide very quick response. No energy conversion is required because the electric power is applied directly to the drive actuators on the axes. In an electric manipulator, the motors generally provide rotational motion and, therefore, must use rack-and-pinion gears or ball-screw drives to change to linear movements, for direct drives are connected to the joints through some kind of mechanical coupling, such as a lead screw, pulley block, spur gears, or harmonic drive.

Permanent magnet DC motors have proved a good choice for medium- and small-size manipulators in the past. The brushless, electronically commutated versions have very long lives.

Printed-circuit motors have high torque relative to their rotor inertia and, therefore, fast response times. These motors are capable of driving at low speeds without the need of speed reducers.

The disadvantages of electric drives are that the payload capability is limited to three hundred pounds or less, and the operation in explosive environments poses problems.

Pneumatic Power Source

Pneumatic drives are generally found in relatively low-cost manipulators with low load-carrying capacity. When used with non-servo controllers, they usually require mechanical stops to ensure accurate positioning. Pneumatic drives have been used for many years for powering simple stop-to-stop motions. Most often used configurations are a linear single or a double-acting piston actuator. Rotary actuators also are used. In converting linear actuation to rotary motion, a drive pulley connected to the actuator by a cable may be used, thus avoiding the non-linearities of joint motion inherent in linkwork conversion of linear to rotary motion.

An advantage of the pneumatic actuator is its inherently light weight, particularly when operating pressures are moderate. This advantage, coupled with readily available compressed air supplies, makes pneumatics a good choice for moderate to low load applications that do not require great precision. Because of the light weight, pneumatics are often used to power end effectors even when other power sources are used for the manipulator's joints.

The principal disadvantages of pneumatic actuators include their inherent low efficiencies, especially at reduced loads; their low stiffness (even at the high end of practical operating pressure); and problems of controlling them with high accuracy. An example of a pneumatically powered robot can be seen in Figure 2.3.16. Also, a simple pneumatic circuit is shown in Figure 3.4.2.

68 Robot Classification

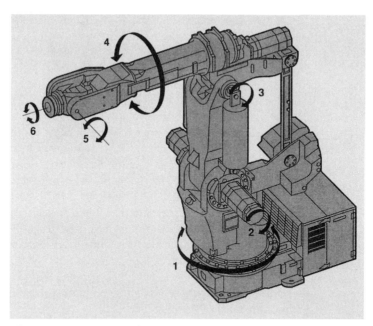

Items		S-420iF	S-420iF	S-420iL	S-420iS	S-420iW
Maximum reach		2.40m	2.85m	3.00m	2.25m	2.40m
Load capacity		120kg (265 lbs)	120kg (265 lbs)	75kg (165 lbs)	80kg (176 lbs)	155kg (342 lbs)
Axis 1	Range	360°	360°	360°	360°	360°
	Speed	100°/sec	90°/sec	100°/sec	70°/sec	90°/sec
Axis 2	Range	142°	142°	142°	142°	142°
	Speed	110°/sec	100°/sec	110°/sec	110°/sec	100°/sec
Axis 3	Range	135°	135°	135°	135°	135°
	Speed	100°/sec	90°/sec	100°/sec	100°/sec	75°/sec
Axis 4	Range	600°	600°	600°	480°	600°
	Speed	210°/sec	210°/sec	210°/sec	210°/sec	140°/sec
Axis 5	Range	260°	260°	260°	260°	260°
	Speed	150°/sec	150°/sec	150°/sec	150°/sec	110°/sec
Axis 6	Range	720°	720°	720°	720°	720°
	Speed	210°/sec	210°/sec	210°/sec	210°/sec	140°/sec
Moments	Axis 4	60kgf·m	60kgf·m	53kgf·m	54kgf·m	86kgf·m
	Axis 5	60kgf·m	60kgf·m	53kgf·m	54kgf·m	86kgf·m
	Axis 6	30kgf·m	30kgf·m	27kgf·m	28kgf·m	50kgf·m
Load Inertia	Axis 4	306kgf·cm·sec^2	306kgf·cm·sec^2	387kgf·cm·sec^2	376kgf·cm·sec^2	486kgf·cm·sec^2
	Axis 5	306kgf·cm·sec^2	306kgf·cm·sec^2	387kgf·cm·sec^2	376kgf·cm·sec^2	486kgf·cm·sec^2
	Axis 6	77kgf·cm·sec^2	77kgf·cm·sec^2	104kgf·cm·sec^2	100kgf·cm·sec^2	164kgf·cm·sec^2
Repeatability		±0.4mm (±0.016")	±0.4mm (±0.016")	±0.4mm (±0.016")	±0.4mm (±0.016")	±0.4mm (±0.016")
Mounting method		Upright	Upright	Upright	Upright	Upright
Mechanical brakes		All axes	All axes	All axes	All axes	All axes
Weight		1500 kg (3300 lbs)	1600 kg (3520 lbs)	1600 kg (3520 lbs)	1500 kg (3300 lbs)	1500 kg (3300 lbs)

Figure 3.4.1 The FANUC S-420i line of articulated robots used in the automotive industry for spot welding, body assembly, and general material-handling applications. The six-axis AC servo drive and single board control maximizes integration with other equipment. Various models are listed in the table. *(Courtesy of FANUC Robotics North America, Inc.)*

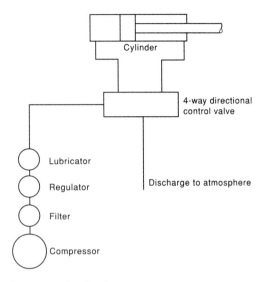

Figure 3.4.2 Simple pneumatic circuit

Hydraulic Power Source

Hydraulic drives are either linear piston actuators or a rotary vane configuration. If the vane type is used as a direct drive, the range of joint rotation is limited to less than 360 degrees because of the internal stops on double-acting vane actuators. Hydraulic actuators provide a large amount of power for a given actuator.

The high power-to-weight ratio makes the hydraulic actuator an attractive choice for moving moderate to high loads at reasonable speeds and moderate noise levels.

Hydraulic motors usually provide a more efficient way of using energy to achieve a better performance, but they are more expensive and generally less accurate.

A major disadvantage of hydraulic systems is their requirement for an energy storage system, including pumps and accumulators. Hydraulic systems also are susceptible to leakage, which may reduce efficiency or require frequent cleaning and maintenance. The working fluid must always be kept clean and filter-free of particles. Fluid must be kept at a constant warm temperature (100°F–110°F). Also, air entrapment and cavitation effects can sometimes cause difficulties. One of the chief concerns with hydraulic power is the environmental issue. Oil that is contaminated is costly to remove, and any leakage is considered an environmental contamination problem. During the 1996 National Robotics Safety Convention in Detroit, some facts about this issue were presented by the three big auto companies. They feel that electric servomotor power is the solution to avoid the hazards of oil contamination.

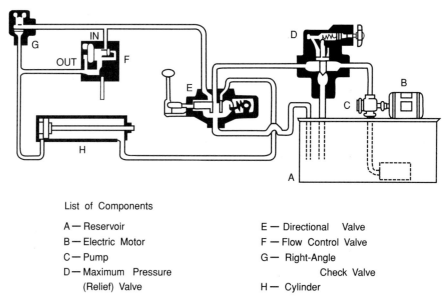

Figure 3.4.3 A hydraulic power supply system with linear actuator (cylinder). *(Courtesy of Vickers, Inc.)*

List of Components

A — Reservoir
B — Electric Motor
C — Pump
D — Maximum Pressure (Relief) Valve
E — Directional Valve
F — Flow Control Valve
G — Right-Angle Check Valve
H — Cylinder

In paint spraying and other applications, where the environment may present an explosion hazard, the robot must be either explosion-proof or intrinsically safe. In such cases, the hydraulically driven robot has obvious advantages over its electric counterpart.

Hydraulic power lends itself to some robot applications because energy can be easily stored in an accumulator and released when a burst of robot activity is called for. A typical hydraulic power system with linear hydraulic actuator (cylinder) is shown in Figure 3.4.3, and two actual different-sized power units are shown in Figure 3.4.4. Also, a simple (single-line) hydraulic circuit is shown in Figure 3.4.5. Virtually all hydraulic circuits are essentially the same regardless of the application. There are six basic components required in a hydraulic circuit (refer to Figure 3.4.3):

1. A tank (reservoir) to hold the liquid, which is usually hydraulic oil.
2. A pump to force the liquid through the system.
3. An electric motor or other power source to drive the pump.
4. Valves to control liquid direction, pressure, and flow rate.
5. An actuator to convert the energy of the liquid into mechanical force or torque to do useful work. Actuators can be either cylinders to provide linear motion, such as shown in Figure 3.4.3, or motor (hydraulic) to provide rotary motion.
6. Piping, which carries the liquid from one location to another.

3.4 Power Sources 71

Figure 3.4.4 (above) Two actual different-sized, complete, hydraulic power units *(Courtesy of Continental Hydraulics)*

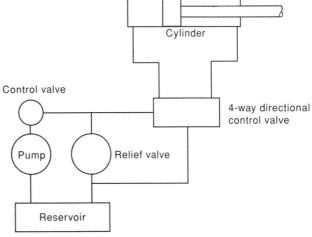

Figure 3.4.5 (right) Simple hydraulic circuit

Electromechanical Power Source

Electromechanical power sources are used in about 20 percent of the robots available today. Typical forms are servomotors, stepping motors, pulse motors, linear solenoids and rotational solenoids, and a variety of synchronous and timing belt drives.

The primary use of AC servomotors in robot joint movements is for fast, accurate positioning, high stall torque, small frame size, and light weight. Pneumatically driven robots, because of the compressibility of air, normally are found in light-service, limited-sequence, and pick-and-place applications. Hydraulic robots usually employ hydraulic servo valves and analog resolvers for control and feedback. Digital encoders and well-designed feedback control systems can provide hydraulically actuated robots with an accuracy and repeatability generally associated with electrically driven robots.

3.5 TYPES OF MOTION

A robot manipulator can make four types of motion in traveling from one point to another in the workplace:

1. Slew motion
2. Joint-interpolated motion
3. Straight-line interpolation motion
4. Circular interpolation motion

Slew motions represent the simplest type of motion. The robot is commanded to travel from one point to another where each axis of the manipulator travels at a default speed from its initial position to the required final destination.

Joint-interpolated motion requires the robot controller to calculate the time it will take each joint to reach its destination at the commanded speed. Then it selects the maximum time among these values and uses it as the time for the other axes. The advantage of joint-interpolated motion compared to slew motion is that the joints are driven at lower velocities, and therefore the maintenance problems are much less for the robot.

Straight-line interpolation motion requires the end of the end effector to travel along a straight path determined in rectangular coordinates. This type of motion is the most demanding for a controller to execute, except for a rectangular-coordinated robot. Straight-line interpolation is very useful in applications such as arc welding, inserting pins into holes, or laying material along a straight path.

Circular interpolation motion requires the robot controller to define the points of a circle in the workplace based on a minimum of three specified positions. The movements that are made by the robot actually consist of short straight-line segments. Circular interpolation, therefore, produces a linear approximation of the circle and is more readily available using a programming language rather than manual or **teach pendant** techniques.

3.6 PATH CONTROL

Commercially available industrial robots can be classified into four categories, according to their path **control** system:

1. Limited-sequence
2. Point-to-point
3. Controlled-path
4. Continuous-path

Limited-sequence robots do not use servo-control to indicate relative positions of the joints. Instead, they are controlled by setting limit switches and/or mechanical stops together with a sequencer to coordinate and time the actuation of the joints. With this method of control, the individual joints can only be moved to their extreme limits of travel. This has the effect of severely limiting the number of distinct points that can be specified in a program for these robots. Therefore, their control system is intended for simple motion cycles, such as pick-and-place applications where each axis is normally limited to two end points. However, some pick-and-place robots also include one or two intermediate stops; therefore, they can be called stop-to-stop or sometimes bang-bang.

Pick-and-place robots were named by the job they normally perform in industry. They pick up parts or materials from one location and place them in another location. A pick-and-place robot can be used to unload a conveyor or a transfer line. It also can be used for simple press loading and unloading applications. Limited-sequence robots use pick-and-place motion. A pick-and-place robot is shown in Figure 3.6.1.

Pick-and-place robots are the simplest of all robots. This is not to say that because they are simple they lack value. Pick-and-place robots are an excellent

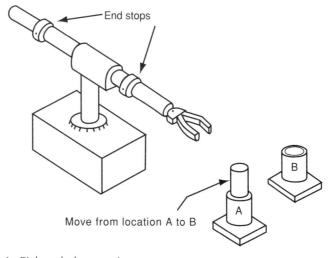

Figure 3.6.1 Pick-and-place motion

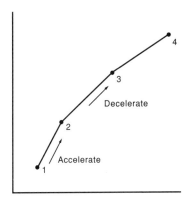

Figure 3.6.2 Point-to-point motion

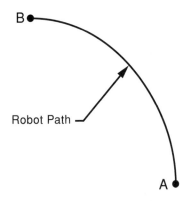

Figure 3.6.3 Illustration of the path of the manipulator of a point-to-point robot as it moves from one point to another (combined horizontal and vertical movement)

choice for simple jobs. They have the lowest level of control but are the least expensive of the four types. There is no reason to buy more capacity than is necessary to do a job.

Pick-and-place robots also have the advantage of being the easiest to maintain. They can normally be served by the plant electrical or machine repair personnel. In general, a limited-sequence robot will operate longer between failures, and when it does fail, the repairs will normally be simple and fast. These types of robots are normally pneumatically actuated.

Point-to-point robots are the most common of the four classifications and can move from one specified point to another but cannot stop at arbitrary points not previously designated. Point-to-point robots driven by servos are often controlled by potentiometers set to stop the robot arm at a specified point. Point-to-point robots can be programmed (taught) to move from any point within the work envelope to any other point within the work envelope. This versatility greatly expands their potential applications. Therefore, these robots can be used in simple machine loading and unloading applications as well as more-complex applications, such as spot welding (resistance welding), assembly, grinding, inspection, palletizing, and depalletizing.

Point-to-point motion involves the movement of the robotic system through a number of discrete points, as shown in Figure 3.6.2. The programmer uses a combination of the robot axes to position the end effector at a desired point. These positions or points are recorded and stored in memory. During the playback mode, the robot steps through the points recorded in memory.

The point-to-point robot can move more than one of its axes at a time. For example, the spherical-coordinated robot can rotate about its base when at the same time it is reaching other axes. In a more complex application, it is possible to have the robot moving all of its major axes and all of its minor axes at the same time.

Although the point-to-point robot can move to any point within its work envelope, it does not necessarily move in a straight line between two points. Figure 3.6.3 shows the path that a point-to-point robot might take between points A and B. The vertical movement is much shorter than the horizontal movement required to move from point A to point B. When directed by the control, the robot's manipulator will begin to move from point A toward point B. The manipulator will begin to rise on the vertical axis and then reach the horizontal axis. Because the distance the manipulator must travel along the vertical axis is much shorter than the distance it must travel along the horizontal axis, the manipulator will complete the required vertical movement long before it completes the horizontal movement. The resulting path will be some form of an arc, but the exact shape of the arc is not predictable during the programming of the robot.

To program a point-to-point robot, the programmer must push buttons on a teach pendant, as shown in Figure 3.6.4. The teach pendant is much like the control used to move an overhead crane in a factory. To rotate the robot about its base, a button is pushed and the robot turns. To extend the manipulator, another button is pushed. When the robot has been led to the desired point, the program-

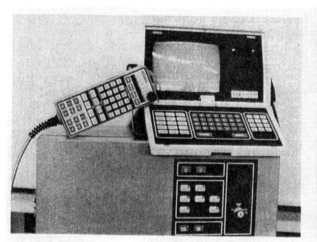 

Figure 3.6.4 Teach pendant used for programming a point-to-point robot. *(Source: Cincinnati Milacron)*

mer pushes a button to record that point in the robot's memory. The programmer then moves the manipulator to the next desired point by pushing the appropriate buttons. The path the programmer takes to get the manipulator to the next point is not remembered by the robot. When the manipulator is finally brought to the desired point, the button to record the point is again pushed, and the second point is recorded into the robot's memory. When the program is played back, the robot will move from the first point in its memory to the second point, then to the third point, and so forth, until it has moved to all the points it has been taught. After the last point has been reached by the robot, the controller moves the manipulator back to the first point in the memory and the entire program is repeated. With some robots, the arm may be pulled roughly to a desired location manually, but the final location is fine-tuned by using the teach pendant. The simple control method of the point-to-point robot makes it difficult, if not impossible, to predict the exact path of the manipulator between two taught points.

Controlled path is a specialized control method that is part of the general category of a point-to-point robot but with more-precise control. The controlled-path robot ensures that the robot will describe the right segment between two taught points. Figure 3.6.5 shows a multiple-exposure photograph of a robot that is under the influence of a controlled-path controller. As can be seen in this illustration, the robot moves in a straight line between the two taught points.

Controlled-path is a calculated method and is desired when the manipulator must move in a perfect path motion. Controlled-path robots can generate straight lines, circles, interpolated curves, and other paths with high accuracy. Paths can be specified in geometric or algebraic terms in some of these robots. Good accuracy can be obtained at any point along the path. Only the start and finish coor-

76 Robot Classification

Figure 3.6.5 A Cincinnati Milacron T^3 566 robot in a vivid demonstration of straight-line, controlled-path motion. The dotted line shows the path of the robot manipulator. *(Source: Cincinnati Milacron)*

dinates and the path definition are required for control. Although assembly operations can be accomplished simply by point-to-point control, programming with a controlled-path robot can perform such operations more easily. For example, if the robot is to put a shaft into a bearing, any deviation from a straight line could cause the shaft to score the bearing or could bend the shaft. Using a controlled-path robot will ensure that the robot will slip the shaft into the bearing in a straight line. On the other hand, if you desire to use the point-to-point robot in a straight-line operation, it is necessary to program the robot with many points along the path. The more points programmed, the straighter the path on the point-to-point robot will be.

Other applications that are simplified through the use of a controlled-path robot are arc welding, drilling, polishing, and assembly.

The method for programming the controlled-path robot is identical to that for programming the point-to-point robot except that the points must be calculated.

The difference between the execution of a point-to-point controlled-path and a point-to-point noncontrolled-path program is illustrated in Figure 3.6.6. The figure shows two programmed points for a two-axis (x and y) robot.

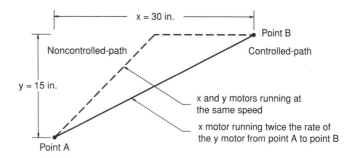

Figure 3.6.6 Comparison of controlled-path and noncontrolled-path operation

Notice that all the axis drivers on a noncontrolled-path robot move at the same time, whereas the axis drivers on a controlled-path robot move at a speed proportional to the length of the axis that is produced by a controlled-path controller.

For the controlled-path the x-axis motor runs at twice the speed of the y-axis motor, and therefore both finish their moves simultaneously at point B, which is a complete straight-line motion.

Continuous-path motion is an extension of the point-to-point method. The difference is that continuous path involves the utilization of more points and its path can be an arc, a circle, or a straight line. A continuous-path program can have several thousand points. Because more points are used, the distances between points are extremely close, as shown in Figure 3.6.7.

Because of the large number of points, the robot is capable of producing smooth movements that give the appearance of continuous or contour movements. Continuous-path motion is more concerned with control of the path movement than with end-point positioning. Programming of the path of motion is accomplished by an operator physically moving the end effector of the robot through its path of motion. While the operator is moving the robot through its motion, the positions of the various axes are recorded on some constant time frame. Programs are generally recorded on magnetic tape or a magnetic disk.

The continuous-path robot is programmed differently than the point-to-point robot and the controlled-path robot. Rather than leading the robot to the point desired by pushing buttons on a teach pendant, the manipulator of a continuous-path robot is programmed by grabbing hold of the robot's arm and actually leading the arm through the path that we wish the robot to remember. The robot remembers not only the exact path through which the programmer moves the manipulator but also the speed at which the programmer moves the manipulator.

If the programmer should move the arm too slowly, the speed can be adjusted at the control console. Changing the speed does not affect the path of the robot's arm.

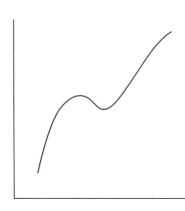

Figure 3.6.7 Continuous path motion

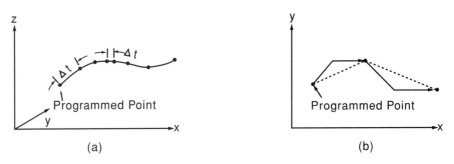

Figure 3.6.8 (a) In continuous path, real-time programming points are automatically programmed. (b) In point-to-point, the path generated is not easily predicted.

The continuous-path robot is really a form of the point-to-point robot. As the robot is taught the desired program, the control examines the location of the manipulator hundreds of times per second and stores each point in memory for playback at a later time. It is like programming thousands of individual points into the memory of a standard point-to-point robot. The continuous-path robot is often used for spray painting, arc welding, or any other operation that requires constant control of the robot's path.

The major difference between the continuous path control and the standard point-to-point control is the control's ability to remember thousands of programmed points in the continuous path, whereas the point-to-point control is limited to several hundred points of memory. Figure 3.6.8 illustrates the difference between the two methods.

Off-line programming is another method that has become very popular today. It's a simple software-driven program that is done either on a computer or by the use of a programmable controller. (Details about off-line programming will be discussed in Chapter 9.)

3.7 INTELLIGENCE LEVEL

Robot systems are usually classified as low-technology and high-technology groups. Although the distinction between these two groups is often unclear, the classification serves two purposes. First, it emphasizes the broad range of systems and features available. Second, it indicates to the potential purchaser the degree of training that will be required for new operators and maintenance technicians.

Low-technology robots do not use servo control to indicate relative positions of the joints. Instead, they are controlled by setting limit switches and/or mechanical stops, together with a sequencer, to coordinate and time the actuation of the joints. Their control system is intended for simple motion cycles, such as pick-and-place applications where each axis is normally limited to two end points. Figure 3.6.1 illustrates such an application; however, some pick-and-place robots

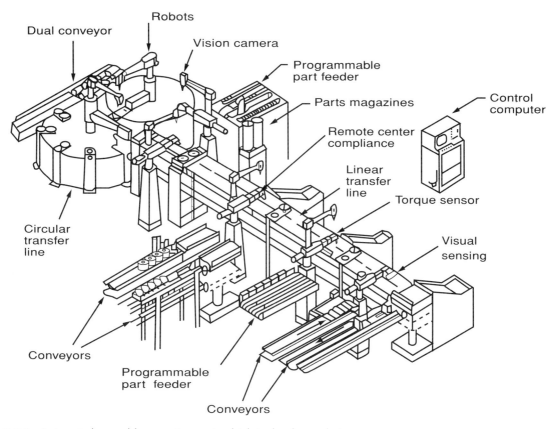

Figure 3.7.1 Automated assembly operations using high-technology robots

also include one or two intermediate stops; therefore, they can be called stop-to-stop or sometimes bang-bang.

High-technology robots are servo-controlled systems; they accept more-sophisticated sensors and complex programming languages. Figure 3.7.1 illustrates such a system. There are two types available in this category: numerically controlled and intelligent control.

The numerically controlled robot is programmed and operates much like a **numerical control (NC)** machine (see Chapter 8) in which the mechanical sections are controlled and programmed by coded alphanumerical data. The robot is servo controlled by digital data, and its sequence of movements can be changed with relative ease. The two basic types of controls are point-to-point and continuous-path.

Point-to-point robots are easy to program and have a higher load-carrying capacity and a much larger work envelope. Continuous-path robots have greater repeatability than point-to-point, but lower load-carrying capacity. Some advanced robots have a complex system of path control, which enables them to have high-speed movements with great accuracy.

The intelligent control robot is capable of performing some of the functions and tasks carried out by human beings. It can detect changes in the work environment by means of sensory perception. Also, an **intelligent robot** is equipped with a variety of sensors and sensor apparatus providing visual (computer vision) and tactile (touching) capabilities to respond instantly to variable situations.

Much like humans, the robot observes and evaluates the immediate environment by perception and pattern recognition. It then makes appropriate decisions for the next movement and proceeds. Because its operation is so complex, powerful computers are required to control its movements and more-sophisticated sensing devices to respond to its actions.

Extensive research has been and still is concerned with how to equip robots with seeing "eyes" and tactile "fingers." **Artificial intelligence (AI)** that will enable robots to respond, adapt, reason, and make decisions to react to change is also an inherent capability of the intelligent robot.

Significant developments still continue with the assumption that robots will behave more and more like humans, performing tasks such as moving among a variety of machines and equipment on the shop floor and avoiding collisions; recognizing, picking, and properly gripping the correct raw material or workpiece; transporting a workpiece to a machine for processing or inspection; and assembling the components into a final product. It can be appreciated that in such tasks, the accuracy and repeatability of the robot's movements are important considerations, as are the economic benefits to be gained. The potential application of intelligent robots seems limited only by human imagination and creativity. (Chapter 10 discusses advanced robots in detail.)

3.8 SUMMARY

This chapter introduced the general concepts of robot classification. It provided an overview of all types of robot arm geometry and styles, considering degrees of freedom, power sources, control systems, and path control.

The arm geometry, which is also described in the previous chapter, is currently available in five basic configurations: rectangular, cylindrical, spherical, jointed-arm, and SCARA. For each configuration, the chapter described the geometry of its work envelope, the advantages and disadvantages of its axes, and its typical applications. Table 3.1 summarizes the five basic configurations.

The three degrees of freedom located in the arm of a robot system are the rotational traverse, the radial traverse, and the vertical traverse. The three degrees of freedom located in the wrist are pitch, yaw, and roll.

The four power sources used in current robots are electric, hydraulic, pneumatic, and electromechanical.

There are four types of motion that a robot manipulator can make in traveling from one point to another in the workplace: slew motion, joint-interpolated motion, straight-line interpolation motion, and circular interpolation motion.

There are four types of path controls for robots: limited-sequence, point-to-point, controlled-path, and continuous path.

Robot systems are usually classified as low-technology and high-technology groups.

3.9 REVIEW QUESTIONS

3.1 Name the five styles of manipulators.
3.2 Describe the axes of the five styles of manipulators.
3.3 Compare the five styles of manipulators according to the advantages and disadvantages of each configuration, work envelope, and typical applications.
3.4 Describe the four types of power sources used in current robots and compare their advantages and disadvantages according to economics, reliability, and load-carrying ability.
3.5 Discuss the roles that the major and minor axes of a robot play in positioning a part in space.
3.6 Discuss the major differences between servo-controlled and nonservo-controlled robots.
3.7 Describe the six degrees of freedom that are needed to locate a robot's arm at any point in its work space.
3.8 What is the main difference between an intelligent and highly intelligent robot and what are its advantages compared to a nonservo type?
3.9 Name the four types of path control and compare them with this application.
3.10 How does the SCARA arm geometry differ from the vertical articulated arm?
3.11 Give three typical applications for each of the four-path controls.
3.12 Why is the SCARA arm more ideal for assembly applications?

3.10 PROBLEMS

3.1 Sketch the work envelopes of the robot configurations for each of the five types of manipulators.
3.2 In Problem 3.1, compare the work envelopes and explain which configuration gives the largest work envelope and why.
3.3 Which style of manipulator has the largest reach for the amount of floor space it occupies? Sketch and discuss your answer.
3.4 Compare the five basic robot configurations according to the work envelope, typical applications, and power sources.
3.5 What is meant by the degrees of freedom of a robot system and how do they compare to the human hand?
3.6 Sketch the degrees of freedom located in a SCARA robot's arm and a cylindrical robot's arm.
3.7 What are the four patterns of motion of industrial robotics systems? Discuss and sketch each pattern.
3.8 What are some factors to consider in the selection of a control unit for an industrial robot? Give an example.

3.11 REFERENCES

Asfahl, C. R. *Robot and Manufacturing Automation.* New York: John Wiley, 1992.

Fu, K. S., R. C. Gonzalez, and C. S. G. Lee. *Robotics: Control, Sensing, Vision and Intelligence.* New York: McGraw-Hill Book Co., 1987.

Fuller, J. L. *Robotics: Introduction, Programming and Projects.* Englewood Cliffs, NJ: Prentice Hall, 1991.

Groover, M. P. *Automation, Production Systems, and Computer Integrated Manufacturing.* Englewood Cliffs, NJ: Prentice Hall, 1987.

Horn, R. E. "Computed Path Control for an Industrial Robot." *Proceedings of the International Symposium on Industrial Robots,* Tokyo, October 1987, pp. 327–336.

Klafter, R. D., T. A. Chmielewski, and M. Negin. *Robotic Engineering: An Integrated Approach.* Englewood Cliffs, NJ: Prentice Hall, 1989.

Lewis, F. L., C. T. Abdullah, and D. N. Dawson. *Control of Robot Manipulators.* New York: Macmillan Publishing Co., 1993.

Masterson, J. W., R. L. Towers, and S. W. Fardo. *Robotics Technology.* South Holland, IL: Goodheart-Willcox Co. 1996.

Pessen, D. *Industrial Automation.* New York: John Wiley, 1989.

Rehg, J. A. *Introduction to Robotics in CIM Systems.* 2d ed. Upper Saddle River, NJ: Prentice Hall, 1997.

Rembold, V. (ed.) *Robot Technology and Applications.* New York: Marcel Dekker, 1990.

Sandler, B. Z. *Robotics: Designing the Mechanisms for Automated Machinery.* Englewood Cliffs, NJ: Prentice Hall, 1991.

Schilling, R. J. *Fundamentals of Robotics: Analysis and Control.* Englewood Cliffs, NJ: Prentice Hall, 1990.

Spiteri, C. J. *Robotics Technology.* Philadelphia, PA: Saunders College Publishing, 1990.

CHAPTER 4

Control System Analysis

4.0 OBJECTIVES

After studying this chapter, the reader should:

1. Be acquainted with robot operation
2. Recognize hierarchical control structure
3. Understand line tracking capabilities
4. Know the dynamic properties of a robot
5. Be familiar with modular robot components

4.1 ROBOT OPERATION

In the preceding chapters, we have briefly described the basic components and movements of the industrial robot, including the manipulator, power supply, controller, types of motion, and path control. We now consider the operation of the robot and its control systems.

All robots need an intelligence control section, which can range from a very simple and relatively limited type up to a very complex and sophisticated system that has the capability of continuously interacting with varying conditions and changes in its environment. Such a control system would include computers of various types, a multitude of microprocessors, memory media, many input/output ports, absolute and/or incremental encoders, and whatever type and number of sensors may be required to make it able to accomplish its mission. The robot may be required to sense the positions of its several links, how well it is doing its mission, the state of its environment, and other events and conditions. In some cases, the sensors are built into the robots; in others they are handled as an adjunct capability—such as machine vision, voice recognition, and sonic ranging—attached and integrated into the overall robotic system.

In general, control systems may be divided into two types, commonly called open-loop or closed-loop systems. Both are used in industrial robots. An example of an open-loop system is a stepper motor in which the control signals directly position the motor without feedback. Currently, two types of closed-loop systems are used. These two types are called nonservo and servo. Each one uses a feedback signal. These are the two types that we are going to describe in detail for the robot operation.

The word *servo* refers to a continuous position-controlling device. A nonservo robot may use a limit switch to indicate that the robot has reached the desired or end position. A servo system provides continuous positioning information along the path of the robot's movement.

Open and closed loop may also be used in reference to the overall robot manipulator. Five different classes of robot controllers are shown in the block diagrams in Figure 4.1.1. The first, shown in Figure 4.1.1a, is an open-loop axis controller that uses no feedback signals. The second, shown in Figure 4.1.1b, is called a nonservo controller and receives only on-off feedback signals. This type is in common use today and consists of an overall open-loop control, but with local closed-loop joint controls. The third type, shown in Figure 4.1.1c, uses servomotors and feedback for each axis and is called a servo controller. This popular type uses local environment sensors that sense position and velocity to provide feedback information. The fourth type, shown in Figure 4.1.1d, is a closed-loop, more intelligent control with sensors in both the local and global environments to provide an overall feedback control. This type of control has not yet been implemented on an industrial robot. The fifth type, shown in Figure 4.1.1e, is an intelligent closed-loop control in which both local and global sensors are used to create and modify the strategy of the robot. This type is also currently in the research stage. An intelligent robot can detect changes in the work environment by means of sensory perception (visual and/or tactile). Then using its decision-making capability, it can proceed with the appropriate operations.

Nonservo Robot Operation

The operation of a nonservo robot for controlling a single axis might be as follows: A controller is used to initiate signals to the control valve for the axis motion. The control valve opens, admitting air or oil to the actuator, which would be either pneumatic or hydraulic. The actuator starts the robot axis moving. The valve remains open, and the member continues to move until it is physically restrained by contact with an end stop. A limit switch placed at the end stop is used to signal the end of travel back to the controller, which then commands the control valve to close. If the controller is a sequencer or device capable of sending a sequence of control signals, it then indexes to the next step, and the controller again provides an output signal. These signals may go to the actuators on the robot manipulator or to external devices, such as the gripper. The process is repeated until the entire sequence of steps is completed.

Some features of this design are that the manipulator's members move until the limits of travel or end stops are reached. The number of stopping positions for each axis is at least two, providing the starting and stopping positions. It is possible to have an intermediate stopping position; however, there is a practical limit to the number of these that can be installed. This type of robot is thus limited in the number of positions in space that can be reached. If a six-axis robot has only start and end stops for each axis, then 2^6, or 64, positions can be reached. To soften the shock upon reaching a stop, a shock absorber or additional valving may be used to provide deceleration. The sequencer may be programmed and conditionally modified through the use of external sensors. However, the sequence for this class of robots is usually restricted to the performance of a single program, such

4.1 Robot Operation **85**

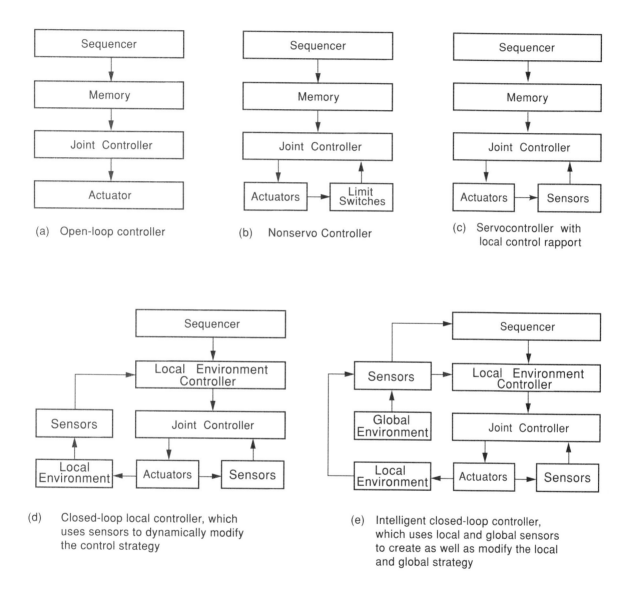

Figure 4.1.1 Five classes of robot controllers. (a) Open-loop controller in which drive signals are sent to actuators but no feedback is used. An example of this type of control is the stepper or servomotor actuators. (b) Bang-bang, on-off, or nonservo controller in which drive signals are sent to the actuators but a return signal is sent back to the drive motor when the desired position is reached. (c) Servo controller in which drive signals sent to the actuator are compared with measured signals from the axes to control the motion. (d) Closed-loop local controller that controls all axes in a coordinated manner. (e) Intelligent closed-loop controller in which the local and global sensors are used to create and modify the robot strategy and motion. *(Source: Digital Equipment Corp.)*

as that required to pick up an object at a fixed location and place it at a given location.

Several characteristics make the nonservo robot ideal for certain tasks. One such characteristic is the relatively high speeds achievable, because a control valve can provide the full flow of air or oil to the actuator. Manual speed adjustment may be provided by regulating this flow. These robots are also relatively low in cost; simple to operate, set up, and maintain; offer excellent repeatability; and have high reliability. They have been mainly used in materials-handling tasks for investment casting, die casting, conveyor unloading, palletizing, multiple parts handling, machine loading, and injection molding. A typical nonservo robot is shown in Figure 4.1.2.

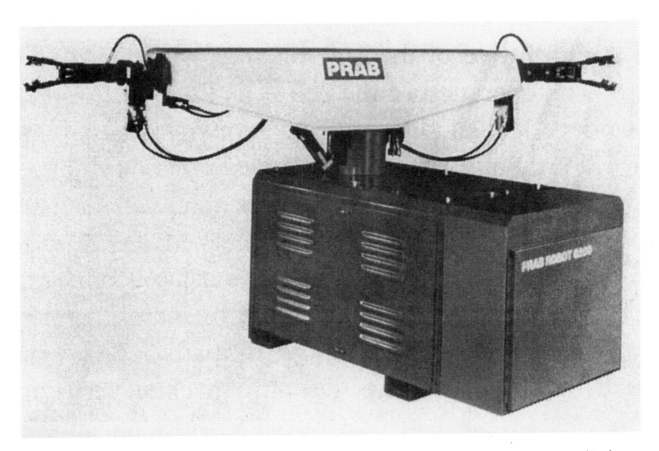

Figure 4.1.2 A nonservo controlled Prab robot, which is effectively used for many pick-and-place operations and is characterized by excellent repeatability and modest cost. Two arms, capable of working independently, qualify the new Prab Model 6200 robot for high-speed parts-transfer jobs, especially in metal stamping manufacturing. With five to nine axes of motion and ±0.008 in repeatability, the Prab Model 6200 handles payloads weighing from a few ounces to seventy pounds. *(Source: Prab Robots Inc.)*

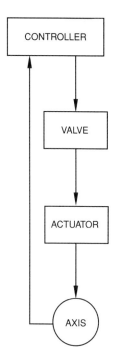

Figure 4.1.3 Control loop for a nonservo robot. The controller sends a drive signal to a valve that energizes the actuator. When the axis reaches the desired position, a signal is sent back to the controller to close the valve. This type of control can stop the robot only at preset positions.

The dual-gripper design would permit the robot to load and unload a machine tool without having to rotate completely. The control loop of a nonservo robot is shown in Figure 4.1.3. Note that the nonservo robot is limited to situations requiring little adaptability.

Servo Robot Operation

The servo robot has a basic controlled system that receives its reference position signal from the sequence controller. The axis position measurement device also provides a feedback signal proportional to its current location. The difference between the desired and current position is called the error signal. This signal is converted to the proper form and applied to the actuator. If there is a large difference, a large signal is applied to the actuator and it moves quickly. If the error signal is zero, no signal is applied to the actuator, because it is at the desired location. With proper design, the action of this feedback is very smooth and reliable.

88 Control System Analysis

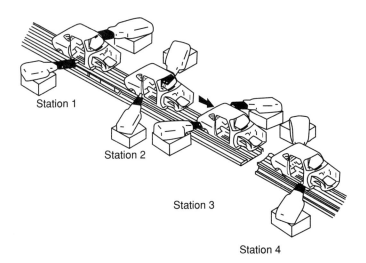

Figure 4.1.4 Examples of servo-controlled robots. The characteristic of this type of control is that the robot can be commanded to stop at any point in its work envelope.

This operation, with a more modern controller, can be described as follows: Upon start of execution, the controller addresses the memory location of the first command position and also reads the actual position of the axis from the position-measuring device. The desired and actual position signals are subtracted to form an error signal. The error signal is then amplified and converted to a velocity signal. The actual velocity signal is read from a velocity-measuring device, such as a tachometer. The difference between the desired and actual velocities is used as another error signal. This velocity error is fed to a compensation network, which serves to keep the controlled motion stable. The output of this network is amplified and used to control the actuating device that moves the robot arm. The position and velocity feedback signals are linked directly to the robot axis. Industrial examples of the use of servo-controlled robots are shown in Figure 4.1.4.

As the actuators move the manipulator's axis, the feedback signals are compared with the desired position data, generating new error signals that are used to command the robot. This process continues until the error signals are effectively reduced to zero, and the axis comes to rest at the desired position. The controller then addresses the next memory location and responds appropriately until the entire sequence or program has been executed A simple control loop for a servo-controlled robot is shown in Figure 4.1.5.

One of the main features of the servo-controlled robot is its versatility—it can move to any point within its work envelope. It is also possible to control the velocity, acceleration, and deceleration between program points. With many systems, one can specify the travel velocity between points, which permits dexterous movements. The repeatability can be varied by changing the magnitude of the

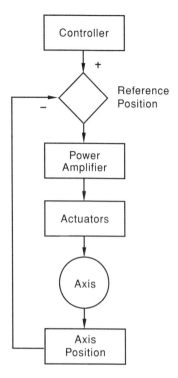

Figure 4.1.5 A control loop of a servo robot. The controller sends a reference-desired position. This desired position is compared with the actual position as measured by sensors, such as a shaft encoder and tachometer, where the error signal is used to drive the actuator through the power amplifier. When the error signal is zero, the actuator stops.

error signal, which is considered zero. This feature may be used to permit the robot to round off a path between control points. Servo-controlled systems have been used for hydraulic-, electric-, and, more recently, pneumatic-powered systems.

Programming may be accomplished either by teach pendant, which permits the manual insertion of program points, or by external control or off-line programming. Either the output of the feedback position devices or the location of the end effector may be stored in the memory of the controlling computer. The memory capacity of the computer is usually sufficient to store thousands of program points.

The characteristics of a servo-controlled robot may be observed in the smooth motions with control of speed and acceleration. This permits the controlled movement of heavy loads or delicate operations for sophisticated tasks. Because the servo robot can be positioned at any point within its working envelope, it has maximum resolution. Furthermore, most computer controllers permit the storage of main programs, as well as macros or subroutines, and permit program transfer based on tool conditions or external signals. Because of their complexity, servo-controlled robots may be more expensive than nonservo-controlled robots; however, the greater flexibility permits them to accomplish a greater variety of tasks.

4.2 HIERARCHICAL CONTROL STRUCTURE

The overall control of a robot is a complex problem consisting of strategy development, path planning, sensory information integration, position commands, and drive signals. One method of simplifying this problem and reducing programming time is to break it into levels of a **hierarchy**.

The concept of **hierarchical control** was proposed by the National Bureau of Standards in December 1977, adapted from Albus in 1981, and is represented here schematically in Figure 4.2.1.

The system has five levels, with each upper level showing increased sophistication in computer software but reduced effort in task programming.

Level 1 represents the basic servo control of a conventional robot. The inputs to this level of control are the desired joint positions, the current joint positions, and velocity feedback. The outputs are the joint errors that form the drive signals to the servo actuators.

The output from level 2, the primitive function control level, is the input to level 1. This control level generates trajectory information, handles sensory feedback, and allows the programmer to work with a real-world coordinator system, which is more applicable to the problem and easier to handle than the joint coordinate system of the robot. Typical programming commands to this level would be MOVE TO (X,Y,Z); level 1 interprets this global command into individual joint position commands like APPROACH, DETECT, or GRASP OBJECT. These are then sent to the first-level control. If the command GRASP OBJECT is sent, then tactile or force feedback from the end effector may be sent to the second-level controller to verify that a grasp has been accomplished. Proximity sensors may also be used to alter the approach speed of the gripper. The interpretative nature of the second- and higher-level controls often dictates a digital controller for implementation. The abilities to interface with external sensors and interpret commands are characteristics of the second-level control.

Level 3 in the hierarchy allows several primitive commands to be combined into a single command. A command such as GET PART A would be equivalent to the primitive commands APPROACH A, TEST B, and so on. The top level, level 5, is designed to work with system commands such as MAKE K1.

As an example, suppose that this last command, MAKE K1, is issued to a single assembly robot that is surrounded by a supply chamber of parts and is capable of assembling several models like model K1. The system control (level 5) sends a series of commands to the workstation control (level 4) to assemble and test for the individual components. One of these statements, ASSEMBLE A, is sent to the workstation control, which initiates a series of commands such as GET PART A, which is sent to the elemental move control (level 3), which, in turn,

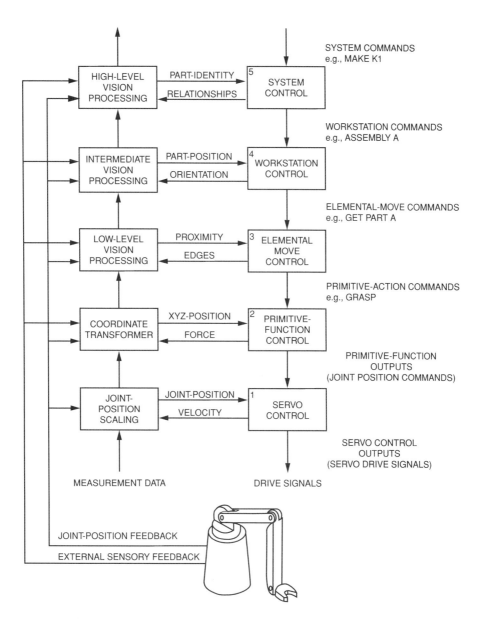

Figure 4.2.1 Hierarchical control structure of servo robots: The top of the hierarchy is divided into simple tasks, the simple tasks into elemental moves required to complete the task, and the elemental moves into primitive action commands in a world-coordinated system. These coordinated commands are transformed into robot joint commands; the joint commands then become the drive signals for the robot actuators. The sensors required to accomplish each level of a task accurately are shown on the left. Actuator joint positions and velocity are required by the servos. Overall measurement of position and force may be desired for the coordinate transformer. Low-level visual processing to determine position or proximity may be needed by the action generator. Higher-level visual or tactile sensing may be required to determine the position and orientation of the workpiece. Very high level computer vision may be required to determine the position of obstacles, to permit path planning, and to react to unusual events in the environment.

issues the commands APPROACH A, DETECT A, GRASP, and so on, to the primitive function control (level 2). The location of part A has been stored in the computer at the time of the initial setup. Using this information, the computer can generate the joint-position commands to the servo control (level 1), which can activate the robot motions.

Although this five-level robot control hierarchy is not available at present, the concept promises to reduce programming time significantly. Experiments conducted at the National Bureau of Standards show that programming a robot with three control levels of the hierarchy results in a reduction in programming time as great as two orders of magnitude.

In practice, the hierarchical control strategy for robots is described as follows: It starts with a level 1 control that receives joint position commands and sends drive signals to the robot's joint actuators. Level 1 control loops often have feedback from position and velocity indicators to ensure that the joint motion is controlled and stable. In implementation, the level 1 controller may be an analog, digital, or hybrid circuit. Stability, speed, and repeatability are key design specifications. The level 2 control receives as input such commands as MOVE TO (X,Y,Z) and interprets this global command into individual joint position commands. The level 3 control divides or separates higher-level commands into individually required actions for the level 2 control. For example, the command MOVE TO (X, Y, Z) AT VELOCITY 30 INCHES/SECOND, DECELERATE AT 12 INCHES FROM OBJECT, GRASP UNTIL FORCE EQUALS 1 POUND must be divided into motion commands, positioning commands, and grasping commands. Feedback from joint positions and velocity sensors, proximity sensors, and tactile force sensors is needed to execute this command.

Above level 3, each level takes more-complex task commands and higher-level sensory information to develop commands for the lower-level controllers. For example, in executing the preceding command, the position and orientation of the object being grasped must be known. This information may be determined by a vision sensor. If more than one type of object is available, then the part identity must be established. Also, the relation of the part to the other objects and its surroundings must be determined. If the object is covered by other objects, then these obstructing objects must be removed before the part can be grasped. Higher levels of control are needed for positioning commands. If there are obstacles in the environment, a strategy for avoiding them must be developed. If a path has been specified, it must be checked to see if it can actually be followed. If a path has not been specified, one must be determined. Even higher levels of control, which could take a single command like MAKE A WELD from a human and develop all the required lower-level commands, can be envisioned; however, to accomplish that, a great deal more research is needed before such a system can be implemented.

Today, a microprocessor-based controller is commonly used in robots as a control system hardware. The controller is organized in a hierarchical structure, as indicated in Figure 4.2.2, so that each joint has its own feedback control system,

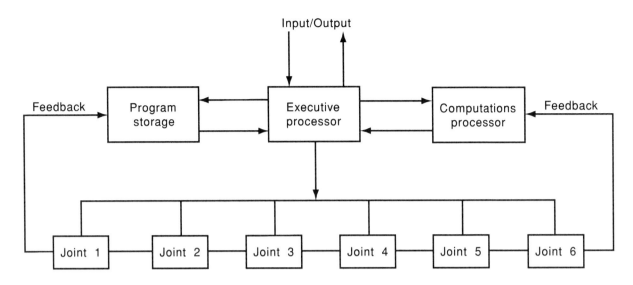

Figure 4.2.2 Hierarchical control structure of a robot with microprocessor-based controller

and a supervisory controller coordinates the combined actuations of the joints and sequences the motions according to the sequence of the robot program.

4.3 LINE TRACKING

The assembly line is a control concept in existing and planned future factories. It characterizes the division of a task into subtasks, such as the material processing required to transform a given raw material into a finished product. The stationary industrial robot is integrated into the assembly by line tracking. Line tracking is the ability of a machine to carry out operations on parts mounted on a continually moving conveyor. Without line tracking, expensive, high-speed, and often heavy-duty shuttle systems must be installed and maintained. These shuttle systems take parts from the main conveyor to a location for processing, then return them to the main line. If the processing time is too long or the process robot too limited, a shuttle system is required. However, a robot with line-tracking capability can eliminate this nonproductive shuttle operation and reduce the part cycle time.

Industrial robots may be used for line tracking in two ways. In the moving-base line-tracking system, the robot is mounted on a traverse base. Generally, this method gives the robot a longer time to work on a given part, but requires an expensive traverse system. The second method is stationary-base line tracking. This method gives the robot access to a larger portion of the object and eliminates the need for the traverse system. Special considerations are required with line tracking. In the simplest case, a position resolver is attached to the conveyor and calibrated, and a part in the range switch is installed. Teach programming is then performed on a stationary part. This taught program can then be executed on the moving part, even at variable speeds.

4.4 DYNAMIC PROPERTIES OF ROBOTS

The capacity of a robot to position and orient the end of its wrist with accuracy and repeatability is an important control attribute in nearly all industrial applications.

Among the important properties of a robot that properly regulate its motions are:

1. Stability
2. Control resolution
3. Spatial resolution
4. Accuracy
5. Repeatability
6. Compliance

To consider some of these factors in the design of a robot is innately complex because of the manner in which these properties are interrelated. Complex interrelationships also make optimization difficult. Therefore, in this section, we briefly describe these six dynamic properties of a robot and their associated performance characteristics.

Stability

Stability is associated with the oscillations in the motion of the robot tool (end effector). The fewer the oscillations present, obviously the more stable the operation of the robot. Negative aspects of oscillations include the following:

1. Additional wear is imposed on the mechanical, hydraulic, and other parts of the robot arm.
2. The tool will follow different paths in space during successive repetitions of the same movement, thus requiring more distance between the intended trajectory and surrounding objects.
3. The time required for the tool to stop at a precise position will be increased.
4. The tool may overshoot the intended stopping position, possibly causing a collision with some object in the system.

Oscillations may be damped or undamped. Damped (transient) oscillations will degrade and cease with time. Undamped oscillations may persist or may grow in magnitude (runaway oscillation) and are the most serious because of the potential damage they may cause to the surroundings or equipment.

Variations of inertial and gravitational loads on the individual joint servos (as the arm's posture changes) make the operation of the robot prone to oscillation. Furthermore, the servos must operate over a wide dynamic range of position (and

4.4 Dynamic Properties of Robots

in some cases, velocity) error. The servos must operate reliably in all situations despite the limits on velocity and acceleration imposed by the actuators used. Certain exceptional conditions can be extremely unstabilizing to a joint servo system, as, for example, when the load accidentally slips out of the end effector. This causes a step change in the gravity loading on one or more joints and can cause a poorly designed arm to go into oscillation. Motion of a joint can also exert various combinations of inertial, centrifugal, and Coriolis forces on the other joints. The reactions of the other joints to these forces can exert forces on the original joint, and this is another potential source of oscillation. Finally, the gain of a controller also determines a very important characteristic of a control system's response: the type of damping or instability that the system displays in response to a disturbance. The four general conditions are illustrated in Figure 4.4.1. As the gain of the controller is increased, the response changes in the following order:

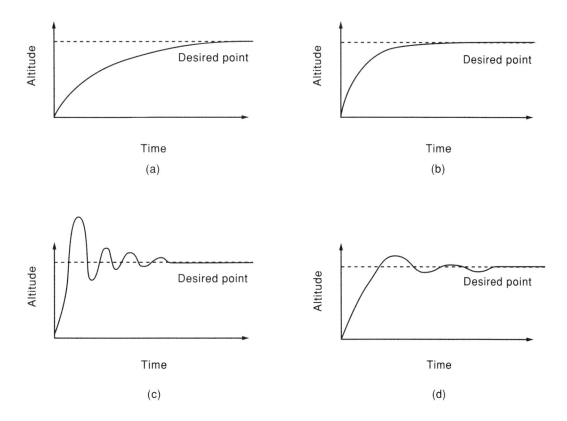

Figure 4.4.1 Four types of response changes and stability of robots: (a) Overdamped response; (b) Critically damped response; (c) Underdamped; (d) Slightly underdamped.

overdamped, critically damped, underdamped, and slightly underdamped. Obviously, neither the underdamped response nor the overdamped response satisfies the objective of minimizing the error. Therefore, the optimum response is either critically damped or slightly underdamped. Exactly how much damping is optimum depends on the requirements of the process.

Two solutions are suggested to eliminate oscillation problems: (1) In method one, by having the joint servos operate continuously. Some sophisticated servo designs, like in numerical control machines, prevent oscillations by eliminating start-and-stop motions regardless of the load carried. (2) In method two, by having the robot controller lock each joint independently the first time it reaches its set point. Special circuitry also decelerates the joint after it comes within a prescribed distance of that position. The joint in this type of robot may lock in any order. When the joints are all locked (total coincidence), the arm is stationary and it can then begin to move to the next position. If the position is held for more than a few seconds, the tool slowly creeps away from its programmed position (as oil leaks out of actuator cylinders). When the position error accumulates sufficiently, the joint servos are allowed to operate again to return the tool to the original position. Technically, this is a form of instability, inasmuch as the tool position can vary periodically (although the period may be on the order of thirty or sixty seconds). However, it is part of the machine's normal operation and causes no problems.

Control Resolution

Resolution is a function of the design of a robot control system and specifies the smallest increment of motion by which the system can divide its working space. This may be a function of the smallest increment in position the control can command, or it may be the smallest incremental change in position that its control system can measure.

Two manipulator positions that differ by only one increment of a single joint are called *adjacent*. A unit change in the position of a sliding joint will move the tool tip the same distance, regardless of where it is in the work space. Thus, a manipulator with an XYZ geometry thus has essentially a constant spatial resolution throughout its work volume. This consideration could be important if the arm is to be trained to perform a precise manipulation in one location of its work space and then repeat it elsewhere in the workpiece.

In contrast, a unit change in the position of a rotary joint will move the tool tip through a distance that is proportional to the perpendicular distance from the joint axis to the tool tip. There is an angular-position error on the final tool tip position that depends on how far the boom is extended. The farther the boom is extended, the larger the distance that the tool tip will move when the rotary joint moves to an adjacent position.

Spatial Resolution

Spatial resolution is the control resolution combined with mechanical inaccuracy. In order to determine spatial resolution, the range of each joint on the manipulator is divided by the number of control increments.

The spatial resolution of a robot is the smallest increment of movement into which the robot can divide its work volume. Spatial resolution depends on two factors: the system's control resolution and the robot's mechanical inaccuracies. It is easier to conceptualize these factors in terms of a robot with one degree of freedom.

The control resolution is determined by the robot's position control system and its feedback measurement system. It is the controller's ability to divide the total range of movement for the particular joint into individual increments that can be addressed in the controller. The increments are sometimes referred to as **addressable points**. The ability to divide the joint into increments depends on the bit storage capacity in the control memory. The number of separate, identifiable increments (addressable points) for a particular axis is given by the formula: 2^n, where n is the number of bits in the control memory.

For example, a robot with 8 bits of storage can divide the range into 256 discrete positions. The control resolution would be defined as the total motion range divided by the number of increments. We assume that the system designer will make all the increments equal.

Example 4.1 Using our robot with one degree of freedom as an illustration, we will assume it has one sliding joint with a full range of 1.0 m (39.37 in.). The robot's control memory has a 12-bit storage capacity. The problem is to determine the control resolution for this axis of motion.

Solution
The number of control increments can be determined as follows:

$$\text{Number of increments} = 2^{12} = 4{,}096$$

The total range of 1 m is divided into 4,096 increments. Each position will be separated by

$$1\text{m}/4{,}096 = 0.000244\text{m or } 0.244 \text{ mm}$$

The control resolution is 0.244 mm (0.0096 in.).

This example deals with only one joint. A robot with several degrees of freedom would have a control resolution for each joint of motion. To obtain the control resolution for the entire robot, component resolutions for each joint would have to be summed vectorially. The total control resolution would depend on the wrist motions as well as the arm and body motions. Because some joints are likely to be rotary while others are sliding, the robot's control resolution can be a complicated quantity to determine.

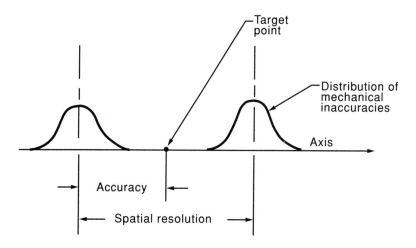

Figure 4.4.2 An illustration of accuracy and spatial resolution where mechanical inaccuracies are represented by a statistical distribution

Mechanical inaccuracies in the robot's links and joint components and its feedback measurement system (if it is a servo-controlled robot) constitute the other factor that contributes to spatial resolution. Mechanical inaccuracies come from elastic deflection in the structural members, gear backlash, stretching of pulley cords, leakage of hydraulic fluids, and other imperfections in the mechanical system. These inaccuracies tend to be worse for larger robots simply because the errors are magnified by the larger components. The inaccuracies would also be influenced by such factors as the load being handled, the speed with which the arm is moving, the condition of maintenance of the robot, and other similar factors.

The spatial resolution of the robot is the control resolution degraded by these mechanical inaccuracies. Spatial resolution can be improved by increasing the bit capacity of the control memory. However, a point is reached where it provides little additional benefit to increase the bit capacity further because the mechanical inaccuracies of the system become the dominant component in the spatial resolution. Figure 4.4.2 illustrates the robot's accuracy to be one half of its spatial resolution.

Accuracy

The accuracy of a robot is the difference between where its control point goes and where it is instructed or programmed to go. In other words, accuracy refers to a robot's ability to position and orient the end of its wrist at a desired target point within the work volume. The accuracy of a robot can be defined in terms

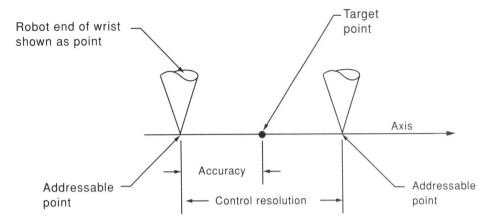

Figure 4.4.3 An illustration of accuracy and control resolution when mechanical inaccuracies are assumed to be zero

of spatial resolution because the ability to achieve a given target point depends on how closely the robot can define the control increments of reach of its joint motions. In the worst case, the desired point would lie in the middle between two adjacent control increments. By ignoring for the moment the mechanical inaccuracies, which would reduce the robot's accuracy, we could define accuracy to be under the worst-case assumption as one half of the control resolution. This relationship is illustrated in Figure 4.4.3.

The mechanical errors arise from such factors as gear backlash, deflection of the links, hydraulic fluid leaks, and a variety of other sources.

The definition of accuracy applies to the worst case, where the target point is directly between two control places. It also implies that the accuracy is the same anywhere in the robot's work volume. In fact, the accuracy of a robot is affected by several factors. First, the accuracy varies within the work volume, tending to be worse when the arm is in the outer range of its work volume and better when the arm is closer to its base. The reason for this is that the mechanical inaccuracies are magnified with the robot's arm fully extended. The term *error map* is used to characterize the level of accuracy possessed by the robot as a function of location in the work volume. Second, the accuracy is improved if the motion cycle is restricted to a limited work range. The mechanical errors tend to be reduced when the robot is exercised through a restricted range of motions. The robot's ability to reach a particular reference point within the limited work space is sometimes called its local accuracy. When the accuracy is assessed within the robot's full work volume, it is called global accuracy. A third factor influencing accuracy is the load being carried by the robot. Heavier workloads cause greater deflection of the mechanical links of the robot, resulting in lower accuracy.

Repeatability

Repeatability is the ability of the robot to reposition itself to a position to which it was previously commanded or taught. Repeatability is affected by resolution and component inaccuracy. Both short- and long-term repeatability exist. Long-term repeatability is of concern for robot applications requiring the same identical task to be performed over several months. Over a long time period, the effect of component wear and aging on repeatability must be considered. For some applications where the robot is frequently reprogrammed for new tasks, only short-term repeatability is important. Short-term repeatability is influenced most by temperature changes with the control and the environment, as well as by transient conditions between shutdown and start-up of the system. Factors that influence both short-term and long-term repeatability are referred to as *drift*.

Obtaining good repeatability is more difficult in a computer-controlled manipulator that records tool positions rather than joint positions because three additional data processing steps are required: (1) converting the several joint positions to a tool position and storing it (sometimes called back solution); (2) transforming a tool position in some useful way, such as by translating, rotating, or scaling; and (3) converting the transformed tool position back to a set of joint positions (sometimes called the arm solution). The kinematic equations used in the arm solution and back solution must accurately reflect the design of the manipulator. The accuracy of these computations depends on the accuracy with which the following values (joint parameters) are known: (1) joint extensions and rotations, (2) link lengths, (3) offset distances between successive joint axes, and (4) angles between successive joint axes.

The repeatability of a manipulator system is measured in the following way. The manipulator is moved by its control system from a reference position to a specified second position with a known nominal location of the center of the end effector. The distance of the actual position of the center of the end effector from its nominal position is measured and recorded. This value is denoted as δ, as shown in Figure 4.4.4.

The experiment is repeated a number of times and the largest value of (δ_{max}) is the manufacturer's value of repeatability, which is usually given as $\pm\delta_{max}$. Obviously, the results of such a test can be used only as a rough guide as to the accuracy one can expect to obtain. One frequently will be delivered a different machine, that is, not the same manipulator that was tested. Also, manufacturers rely on a nominal-size model and do not have the dimensions of the actual manipulator in its own controller. Thus, the actual travel, from reference to final position, may be different from that of the manufacturer's test. Furthermore, other important factors include ambient conditions, rigidity of the structure, and the tendency for the precision of the drive trains and transmissions to differ somewhat from one production model to the next. A manufacturer can improve the repeatability simply by decreasing the load rating below its true value.

Also, accuracy due to repeatability can range from several hundredths of an inch for a simple robot to several thousandths of an inch for a robot doing precision

4.4 Dynamic Properties of Robots

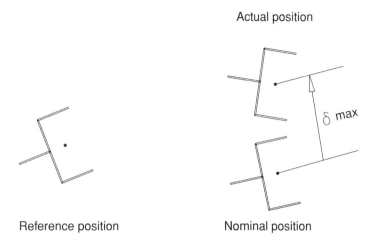

Figure 4.4.4 Determination of repeatability

assembly or handling small parts. In the case of a robot used in testing printed-circuit boards, the end effector must get close enough to the board to pick it up and close enough to the fixture on the tester to place the board on it with sufficient alignment. This could call for an accuracy of a few thousandths of an inch. A less accurate robot can perform adequately but requires guides to direct the board onto the end effector. Adding these extra features to a less accurate robot is usually more costly than purchasing a more expensive, more precise robot. Figures 4.4.5 and 4.4.6 both illustrate repeatability and accuracy of a robot.

A random scan of manufacturers' literature giving their claims of repeatability pertaining to widely procured contemporary robots yields the data given in Table 4.1.

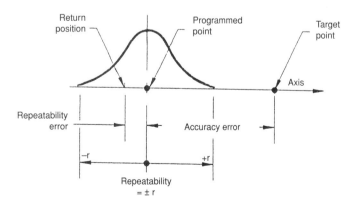

Figure 4.4.5 An illustration of repeatability and accuracy.

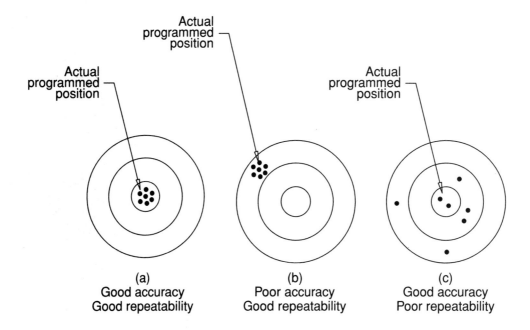

Figure 4.4.6 The difference between accuracy and repeatability.

Table 4.1 Load capacity and repeatability of robots

LOAD CAPACITY		REPEATABILITY (±)	
Pounds	Kilograms	Inches	Millimeters
5	2.2	0.004	0.10
14	6	0.004	0.10
22	10	0.008	0.20
35	16	0.001	0.03
66	30	0.002	0.05
110	50	0.020	0.50
132	60	0.020	0.50
150	68	0.010	0.25
176	80	0.020	0.50
200	90	0.010	0.25
264	120	0.040	1.00

Compliance

Compliance is a quality that gives a manipulator of a robot the ability to tolerate misalignment of mating parts. It prevents jamming, wedging, and galling of the parts and, therefore, it is essential for assembly of close-fitting parts. The force that applies may be a reaction force (torque) that arises when the manipulator pushes (twists) the tool against an object, or it may be the result of the object pushing (twisting) the tool. High compliance means that the tool moves a lot in response to a small force, and the manipulator is then said to be spongy or springy. If it moves very little, the compliance is low and the manipulator is said to be stiff.

Compliance is a complicated quantity to measure properly. Ideally, one would find the relationship between disturbances and displacements to be linear (displacement or rotation proportional to force or torque); isotropic (independent of the direction of the applied force); and diagonalized (displacement or rotation occurring only in the same direction as the force or torque); constant with time; and independent of tool position, orientation, and velocity.

In practice, a manipulator's compliance turns out to be none of these. It is a nonlinear, anisotropic tensor quantity that varies with time and with the manipulator's posture and motion. It is a tensor because a force in one direction can result in displacements in other directions and even rotations. A torque can result in rotation about any axis and displacement in any direction. A six-by-six matrix is a convenient representation for a compliance tensor. Time can affect compliance through changes in temperature and subsequently viscosity in a hydraulic fluid, for example. Furthermore, the compliance will often be found to be a function of the frequency of the applied force or torque. A manipulator may, for example, be very compliant at frequencies around 2 Hz but very stiff in response to slower disturbances.

Compliance may exhibit **hysteresis**. For example, the servos in one design of hydraulic manipulator turn off when the arm stops moving. In this condition, all the servo valves are closed, and the compliance has a value that is determined by the volume of incompressible hydraulic fluid trapped in the hydraulic hoses and the elasticity of the hoses. However, if an outside force on the tool should move any of the joints more than a given distance from the position at which they are supposed to remain, then the servos on all joints will turn on again. The compliance then changes to a completely different value (presumably stiffer in some sense).

Both electric and hydraulic manipulators have complicated compliance properties. In an electric manipulator the motors generally connect to the joints through a mechanical coupling. The sticking and sliding friction in such a coupling and in the motor itself can cause strange effects on the compliance measured at the tool tip. In particular, some of these couplings are not very back-drivable. For example, if one pushes on the nut of a leadscrew (back drive), the leadscrew will not turn (unless the screw's pitch is very coarse and ball bearings are used between the threads to reduce friction). But one can turn the screw easily, and the nut will move.

Most manipulators are operated as an open loop in the sense that they go blindly to a given point in space without regard to the actual position of the object in the environment or to any reaction forces (feedback) that those objects exert on the arm (or tool). In this case, less compliance than that of surrounding objects is advantageous because it means that contact with objects would cause high-frequency oscillations that can be filtered out without degrading overall response. Such filtering actually requires no special effort because the servo valves and actuators commonly used have relatively low bandwidths (1–2 Hz).

Tactile sensors that measure forces and moments exerted on the tool can allow the manipulator to track or locate objects. Even in such cases, however, oscillations may arise in the force-feedback control loop if the compliance at the point of sensing is too low (stiff). Examination of a particular servo design is required to reliably predict whether it will provide the kind of compliance needed for a specific task. There is no substitute for an actual test with the real tool on the manipulator.

4.5 MODULAR ROBOT COMPONENTS

Most industrial robots are custom-built to the specifications of the buyer. However, one other method is to use modular components and assemble the type of robot desired. Mack Corporation supplies nine basic components that can be combined to build a six-degree-of-freedom Cartesian robot. The nine basic components are the grippers; adapters; x-axis, y-axis, and z-axis transporters; roll, pitch, and yaw rotators; and position incrementers, as shown in Figure 4.5.1.

The grippers are provided in four sizes in two- and three-finger configurations for external or internal gripping, as well as soft blank fingers that can be easily modified for special shapes. The smallest size is about twice as large as a thimble and the largest about the size of a human hand. All units operate on the principle of a double-acting cylinder controlled through a simple four-way valve circuit. Fluid pressure opens or closes the fingers. Maximum operating pressure is 150 pounds per square inch (psi) in either hydraulic or pneumatic service. However, most applications use plant air at 80 psi for a reliable source of fluid power. At 80 psi, a pinch force between 5 and 50 pounds may be developed, depending on the model size.

The gripper may be mounted in several orientations. Perhaps the simplest way may be to mount the gripper in line with the first desired motion. This mounting is facilitated by the various adapters. The transporters are the next elements to consider. The x-axis transporter is an air cylinder that provides straight-line motion. Special features include adjustable travel stops and pistons keyed against rotation. The x-axis transporter may also be combined with y- and z-axis transporters. These transporters may also be powered with plant air and may be controlled by simple air logic, or by programmable controllers or computers when more sophistication is required. Rotators are also key components in the modular system. The rotators are nonservo, air-operated, vane-actuated units with a choice of 90- or 180-degree rotations in roll, pitch, and yaw. Adjustable stops provide close control over the angular positions.

4.5 Modular Robot Components

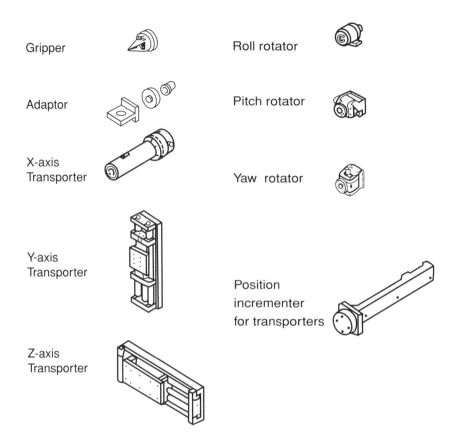

Figure 4.5.1 Basic components of the Mack Corporation modular robotic system. The components may be connected together to form a variety of different Cartesian robot designs. These robot components are powered by pneumatic or electric sources. The maximum load capacity is five pounds. The resulting manipulator, when combined with a power system and controller, can make a complete robot system custom-designed to an application. *(Courtesy of Mack Corporation)*

Many different designs can be constructed from the basic components, as shown in Figure 4.5.2. Combinations of the various components can provide a six-degree-of-freedom, rectangular robot with a payload of five pounds and nonservo operation. Controllers are also provided. However, it should be noted that the **modular robot** components described here are for a fixed-sequence, nonservo positioner.

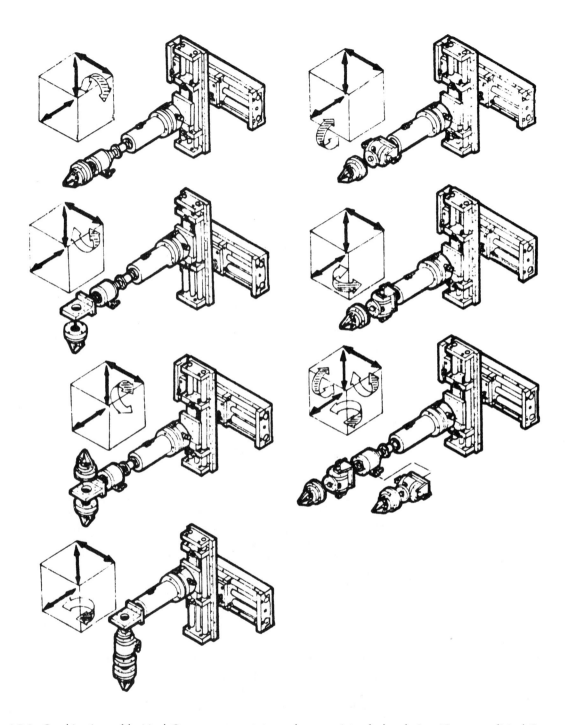

Figure 4.5.2 Combinations of the Mack Corp. components to produce a variety of robot designs (Courtesy of Mack Corporation)

4.6 SUMMARY

Control systems are divided into two types: open loop and closed loop. The open-loop control system (nonservo) does not use the results of its output as part of its input information. The closed-loop servo-controlled system uses the output of the system as part of the input to the control system in order to correct any errors in carrying out the commands.

A hierarchical control permits simultaneous computer programming to control on-line industrial robots.

In hierarchical systems the interpretation for succeeding tasks depends upon the selection of higher-ordered levels.

Line tracking is the ability of a robot to carry out operations on parts mounted on a continually moving system without failure.

The capacity of a robot to position and orient the end of its wrist with accuracy and repeatability is an important control attribute. Among the important properties of a robot to properly regulate its motions are:

1. Stability
2. Control resolution
3. Spatial resolution
4. Accuracy
5. Repeatability
6. Compliance

Modular robot components are available commercially to build a six-degree-of-freedom robot for a fixed-sequence, nonservo positioner.

4.7 REVIEW QUESTIONS

4.1 What is the primary difference between a servo and a nonservo robot system?
4.2 What are the advantages associated with nonservo robots?
4.3 Describe the advantages associated with servo-controlled robots.
4.4 What types of manufacturing applications would be best served by a nonservo system?
4.5 What characteristics of manufacturing applications require that the robot used have a closed-loop system?
4.6 What is the number of degrees of freedom required to position a robot manipulator at any point in a three-dimensional space?
4.7 In the hierarchical control strategy for a servo industrial robot, what is the function of the first-level control?
4.8 Discuss the two methods of assembly-line tracking used with industrial robots.
4.9 Define and explain the meaning of the following terms:
   (a) Stability
   (b) Control resolution
   (c) Spatial resolution
   (d) Accuracy
   (e) Repeatability

(f) Compliance
4.10 Describe the three compliance conditions normally found in robot applications.
4.11 What are the advantages and disadvantages of modular robot components?
4.12 What are the nine basic components that can be supplied commercially for modular robots?

4.8 PROBLEMS

4.1 One of the axes of a robot is a telescoping arm with a total range of 0.50 m (slightly less than 20 in.). The robot's control memory has an 8-bit storage capacity for this axis. Determine the control resolution for the axis.

4.2 Solve Problem 4.1 except that the robot has the following bit storage capacity in its control memory: (a) a 10-bit storage memory; (b) a 12-bit storage memory.

4.3 A large Cartesian coordinate robot has one orthogonal slide with a total range of 30 in. One of the specifications on the robot is that it have a maximum control resolution of 0.010 in. on this particular axis. Determine the number of bits of storage capacity that the robot's control memory must possess to provide this level of precision.

4.4 The mechanism connecting the wrist assembly is a twisting joint that can be rotated through eight full revolutions from one extreme position to the other. It is desired to have a control resolution of plus or minus 0.2 degrees of rotation. What is the required bit storage capacity in order to achieve this resolution?

4.5 The linear joint of a certain industrial robot is actuated by a piston mechanism. The length of the joint when fully retracted is 25 in., and when fully extended, it is 39 in. If the robot's controller has an 8-bit storage capacity, determine the control resolution for this robot.

4.6 In Problem 4.5, the mechanical errors associated with the linear joint form a normal distribution in the direction of the joint actuation with standard deviation of 0.003 in. Determine the spatial resolution, the accuracy, and the repeatability for the robot.

4.7 The revolving joint of an industrial robot has a range of 240° of rotation. The mechanical errors in the joint and the input/output links can be described by a normal distribution with its mean at any given addressable point, and a standard deviation of 0.25°. Determine the number of storage bits required in the controller memory so that the accuracy of the joint is as close as possible to, but less than, its repeatability. Use 6 standard deviations as the measure of repeatability.

4.8 A cylindrical robot has a twisting wrist axis that can be rotated a total of five rotations (each rotation is a full 360°). It is desired to be able to position the wrist with a control resolution of 0.5° between adjacent addressable points. Determine the number of bits required in the binary register for that axis in the robot's control memory.

4.9 One axis of a robot is a linear slide with a total range of 36 in. The robot's control memory has a 10-bit capacity. It is assumed that the mechanical errors associated with the arm are normally distributed with a mean at the given taught point and an isotropic standard deviation of 0.10 mm. Determine:
(a) The control resolution for the axis under consideration
(b) The spatial resolution for the axis
(c) The defined accuracy
(d) The repeatability

4.10 One of the joints of a certain industrial robot is a linear type with a range of 0.5 m. The bit storage capacity of the robot controller is 10 bits for this joint. The mechanical errors form a normally distributed random variable about a given taught point. The mean of the distribution is zero, and the standard deviation is 0.06 mm. The errors will be assumed to be isotropic (the same in all directions). Determine the control resolution, the spatial resolution, the accuracy, and the repeatability for this robot.

4.9 REFERENCES

Albertson, P. "Automated Evaluation Eases Robot-Performance Specification." *EDN* October 1984, 159–169.

Asada, H., and J. Slotine. *Robot Analysis and Control.* New York: John Wiley and Sons, 1986

Chen, C. T. *Control System Design.* New York: Saunders College Publishing, 1993.

Craig, J. J. *Introduction to Robotics: Mechanics and Control.* Reading, MA: Addison Wesley, 1989

Dawson, D. M., Z. Qu, F. L. Lewis, and J. F. Dorsey. "Robust Control for the Tracking of Robot Motion." *International Journal of Control* 52, no. 3 (1990): 581–95.

Dransfield, P. "Hydraulic Control Systems: Design and Analysis of Their Dynamics." No. 33. *Lecture Notes in Control and Information Sciences*, A. V. Belakrishnan and M. Thomas (Eds.). Berlin: Springer, 1981.

Hunt, V. D. *Smart Robots.* New York: Chapman and Hall, 1985.

Klafter, R. D., T. A. Chmielewski, and M. Negin. *Robot Engineering: An Integrated Approach.* Englewood Cliffs, NJ: Prentice Hall, 1989.

Kno, B. C. *Automatic Control Systems.* 5th ed. Englewood Cliffs, NJ: Prentice Hall, 1987.

Mathews, S. H., et al. Repeatability Test System for Industrial Robots. SME Technical Paper MS84-1045, 1984.

Nakano, E. *An Introduction to Robot Engineering* (in Japanese). Tokyo: Omnsha, 1983.

Ogata, K. *System Dynamics.* Englewood Cliffs, NJ: Prentice Hall, 1992.

Parkin, R. E. *Applied Robotic Analysis.* Englewood Cliffs, NJ: Prentice Hall, 1991.

Roth, B. "Introduction to Robots," in *Design and Application of Small Standardized Components.* Data Book 757, published by Stock Drive Products (New Hyde

Park, New York) and distributed by Education Products, Mineola, New York, 1983.

Snyder, W. E. *Industrial Robots: Computer Interfacing and Control.* Englewood Cliffs, NJ: Prentice Hall, 1985.

Stadler, W. *Analytical Robotics and Mechatronics.* New York: McGraw-Hill, 1995.

CHAPTER 5

Robot End Effectors

5.0 OBJECTIVES

After studying this chapter, the reader should:

1. Be acquainted with the types of end effectors
2. Understand standard, special, and other types of grippers
3. Be familiar with gripper selection, design, and force analysis
4. Recognize process tools and their classifications
5. Be able to define compliance systems and their applications
6. Be able to analyze multiple and complex end-effector systems

5.1 TYPES OF END EFFECTORS

Industrial robots today come in a variety of sizes, shapes, and capabilities. However, all have four basic components: a manipulator, an end effector (which is part of the manipulator), a computer controller, and a power supply, as shown in Figure 5.1.1.

The manipulator is a mechanism consisting of several segments or arms and composed of three sections, as shown in Figure 5.1.2:

- The major linkages
- The minor linkages (wrist components)
- The end effector (gripper or tool)

The major linkages are the set of joint-link pairs that out-position the manipulator in space. Usually, they consist of the first three sets (counting from the base of the robot). The minor linkages are those joints and links associated with the fine positioning of the end effector. They provide the ability to orient the tool-mounting plate and subsequently the end effector once the major linkages get it close to the desired position.

The end effector, which is mounted on the tool plate, is a device used to make intentional contact with an object or to produce the robot's final effect on its surroundings by performing a particular task.

The end effector may be a tool that does a function such as welding or drilling, or it may be some type of gripper if the robot's task is to pick up parts and transfer them to another location. A gripper may be a simple pneumatically controlled device that opens and closes or a more complex servo-controlled unit capable of exerting specified forces or of measuring the part within its grasp.

112   Robot End Effectors

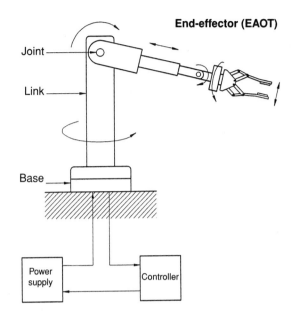

Figure 5.1.1 Components of a robot system

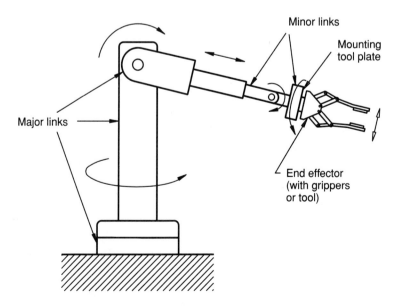

Figure 5.1.2 Components of a robot manipulator

There is a wide assortment of end effectors required to perform the variety of different work functions. The various types can be divided into two major categories: grippers and process tooling.

Grippers are end effectors used to grasp and hold objects. There may be two- or more-fingered devices designed to grasp an object or tool in a manner similar to the human hand and fingers.

Process tooling is an end effector designed to perform work on the part rather than to merely grasp it. By definition, the tool-type end effector is attached to the robot's wrist or tool-mounting plate. Process tooling may be any useful device, such as a spot-welding torch, a spray-painting gun, a vacuum cup, or a set of interchangeable tools.

Grippers may be designed as physical constraints or as friction devices. A physical constraint device might work like a spatula that slides under an object to enable one to lift it. A frictional device depends upon the frictional force between two materials to provide the gripping force.

Also, grippers can be classified as single grippers or double grippers. This classification applies best to mechanical grippers. The single gripper is distinguished by the fact that only one grasping device is mounted on the robot's wrist. A double gripper has two gripping devices attached to the wrist and is used to handle two separate objects. The two gripping devices can be activated independently. The double gripper is very useful in machine loading and unloading applications.

If two or more grasping mechanisms are fastened to the wrist, we call it a multiple gripper. Double grippers are a subset of multiple grippers. The occasions are somewhat rare when more than two grippers would be required in an application. There is also a cost-and-reliability penalty that accompanies an increasing number of gripper devices on one robot arm.

Another way of classifying grippers depends on whether the part is grasped on its exterior surface or its internal surface—for example, a ring-shaped part. The first type is called an external gripper if the part is grasped from the outside surface, and the second type is referred to as an internal gripper if the part is grasped from the inside surface.

Grippers are sometimes used to hold tools rather than work parts. The reason for using a gripper instead of attaching the process tooling directly to the robot's wrist is typically because the job requires several tools to be manipulated by the robot during the work cycle. An example of this kind of application would be a deburring operation in which several different tools must be used in order to reach all surfaces of the work part. In this case, the gripper serves as a quick-change device to provide the capability for a rapid changeover from one tool to the next.

A gripper, regardless of its type and capacity, must fulfill the following characteristics:

1. It must be capable of gripping, lifting, and releasing the part or family of parts required by the process.
2. Some grippers sense the presence of the part with their gripping action.

3. Tooling weight must be kept to a minimum.
4. Containment of the tooling part must be assured under conditions of maximum velocity and loss of holding power.
5. The gripper should be simple in design, accurate in operation, economical, and maintenance free.
6. It must be equipped with a collision sensor to accommodate overload conditions and safeguarding.

In addition to grippers and common process tooling, other types of fixturing and tooling are required in many industrial robot applications. These include holding fixtures, welding fixtures, alignment devices, and other forms of tooling to position the work part during the work cycle.

Therefore, the end effectors used on current robots can be classified further in the following three categories:

1. According to the method of gripping mechanism: standard grippers (mechanical), vacuum, magnetic, adhesive, and other miscellaneous devices.
2. According to process tooling and devices, including drills, welding guns and torches, paint sprayers, and grinders.
3. According to multiple-function capabilities, including special-purpose grippers and compliance devices currently in use.

5.2 MECHANICAL GRIPPERS

Mechanical grippers or standard grippers are end effectors that use **mechanical fingers** actuated by a mechanism to grasp an object. The fingers, sometimes called jaws, are the appendages of the grippers that actually make contact with the object. The fingers are either attached to the mechanism or are an integral part of the mechanism.

If the fingers are of the attachable type, they can be detached and replaced. The use of replaceable fingers allows for wear and interchangeability. Replaceable fingers also can be designed to accommodate different-part models. See Figure 5.2.1.

The function of the gripper mechanism is to translate some form of power input into the grasping action of the fingers against the part. The mechanism must be able to open and close the fingers and to exert sufficient force against the part to hold it securely.

The input power to the mechanism is supplied from the robot and can be pneumatic, hydraulic, electric, or mechanical (spring activated).

There are two ways of constraining the part in the gripper. The first is by physical constriction of the part within the fingers. In this approach, the gripper fingers enclose the part to some extent, thereby constraining the motion of the part. This is usually accomplished by designing the contacting surfaces of the fingers to be in the approximate shape of the part geometry. This method of constraining the part is illustrated in Figure 5.2.2. The second way of holding the part is by friction between the fingers and the work part. With this approach, the fingers must apply a force that is sufficient for friction to retain the part against

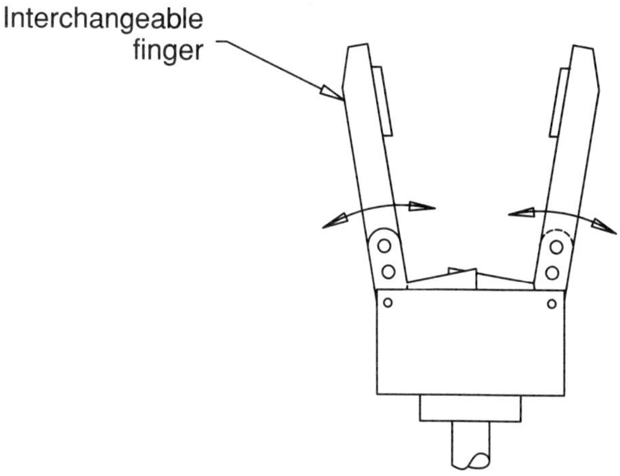

Figure 5.2.1 Interchangeable fingers used with the same gripper mechanism

gravity, acceleration, and any other force that might arise during the holding portion of the work cycle. The fingers, or the pads attached to the fingers that make contact with the part, are generally fabricated out of a material that is relatively soft. This tends to increase the coefficient of friction between the part and the contacting finger surface. It also serves to protect the part surface from scratching or other damage.

The friction method of holding the part results in a less complicated and, therefore, less expensive gripper design. However, there is a problem with the

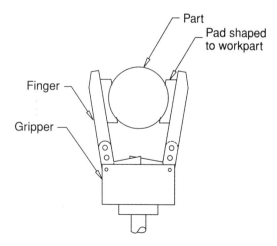

Figure 5.2.2 Constricted method of finger design

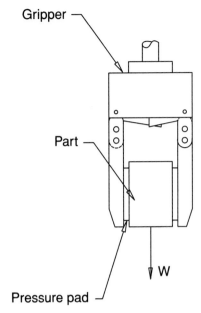

Figure 5.2.3 Force against a part must exceed the weight of the part.

friction method—if a force of sufficient magnitude is applied against the part in a direction parallel to the friction surfaces of the fingers, the part might slip out of the gripper. To resist this slippage, the gripper must be designed to exert a force that is greater than the weight of the part, as illustrated in Figure 5.2.3.

Mechanical grippers can have two different closing motions: angular or parallel. The action of the angular and parallel devices is illustrated in Figure 5.2.4. Some grippers are supplied with blank jaws that can be removed and machined into the configuration required for the application.

Grippers use both external and internal geometry features of the part to pick it up in an application. Another variation in mechanical grippers is the number of jaws or fingers used to grasp the part. Most applications use two-finger grippers. However, three-finger grippers and four-finger grippers are used when parts need to be centered by the gripping process, as shown in Figure 5.2.5. The gripper must be closed and opened by program commands as the robot moves through the production operation.

The robot controller supplies the electrical signals that result in the gripper's action. Most grippers are opened and closed with a pneumatic actuator. However, in limited applications, hydraulic or spring power is used. In some cases, grippers are spring-opened and power-closed less frequently; grip-

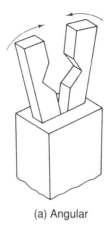
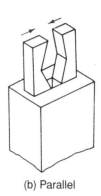

(a) Angular   (b) Parallel

Figure 5.2.4 Standard angular and parallel grippers

pers are spring-closed and power-opened. Each of the types is based on a specific application.

Electrical-powered jaws are also used in limited light applications with a solenoid or DC servomotor providing the opening and closing action. Servo drives will be used more frequently when grippers are developed to vary the applied pressure to the object as it is grasped. Tactile or touch sensing must be developed beyond its present state. New developments coming to market include sensor fusion and smart sensors, which are capable of microcomputer-based calibration, computation, and decision making. A key component in robot design is the end effector. A mechanical gripper comprises between 5 and 10 percent of the robot cost. If a specially designed gripper is required, sometimes the cost for design and fabrication can exceed 20 percent of the total robot cost. See Figure 5.2.6 for the classification of mechanical grippers.

5.3 GRIPPER FORCE ANALYSIS

As indicated earlier, the purpose of the gripper mechanism is to convert input power into the required motion and force to grasp and hold an object. Let us illustrate how to determine the magnitude of the required input power in order to obtain a given gripping force. The following force equations can be used to determine the required magnitude of the gripper force as a function of the given factors.

$$\mu\eta F_g = w \qquad \text{(Equation 5.3.1)}$$

$$\mu\eta F_p = wg \qquad \text{(Equation 5.3.2)}$$

(Continues on page 120)

118 Robot End Effectors

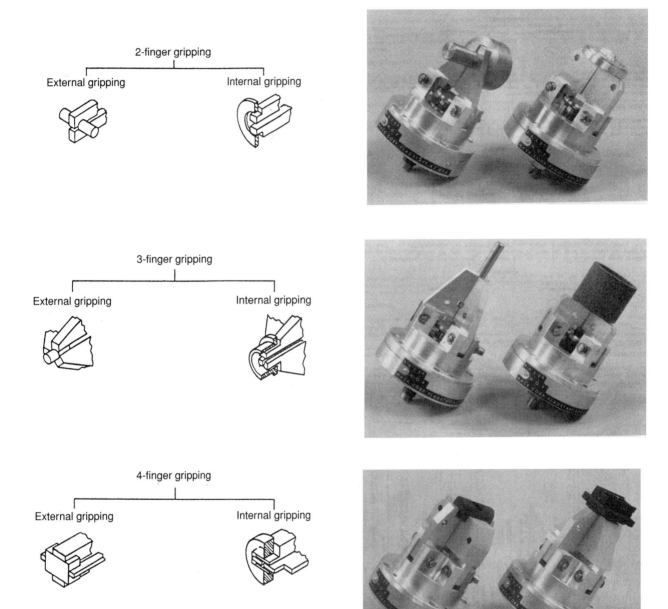

Figure 5.2.5 External and internal grippers with two, three, and four fingers. *(Courtesy of Mack Corporation)*

5.3 Gripper Force Analysis

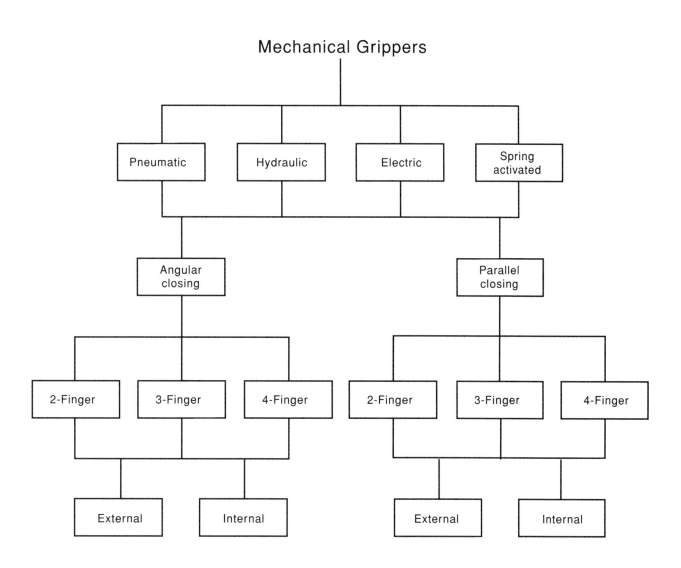

Figure 5.2.6 Classification scheme of mechanical grippers

where μ = coefficient of friction of the finger contacts surface against the part surface
η = number of contacting fingers
F = gripper force
w = weight of the part or object being gripped
g = factor of the combined effect of gravity and acceleration—given as follows:
  g = 3.0, if the acceleration force is applied in the same direction as the gravity force
  g = 2.0, if the acceleration is applied in a horizontal direction
  g = 1.0w, if the acceleration is applied in the opposite direction of the gravity force
SF = safety factor

Equation 5.3.1 covers the case in which weight alone is the force tending to cause the part to slip out of the gripper when the force of gravity is directly parallel to the contacting surface. Equation 5.3.2 covers the case when the force tending to pull the part out of the finger is greater than the weight of the object because the acceleration or deceleration of the part could exert a force that is twice the weight of the part.

Example 5.1 A cardboard carton weighing 10 pounds is held in a gripper using friction against two opposing fingers. The coefficient of friction is 0.25. The weight of the carton is directed parallel to the finger surfaces.

a. Determine the required gripper force for the condition given.
b. If SF = 1.5, what would be the value of the gripper force?

Solution

Find $F = ?$

$$\mu = 0.25 \qquad \eta = 2 \qquad w = 10 \text{lb} \qquad g = 3.0$$

a.
$$\mu \eta F = wg$$

$$F = \frac{wg}{\mu \eta} = \frac{(10)(3.0)}{0.25 \times 2} = 60 \text{lb}$$

The gripper must cause a force of 60 pounds to be exerted by the fingers against the carton surface.

b. If $SF = 1.5$,

$$F = 1.5 \times 60 = 90 \text{ lb}$$

The required gripper force would be 90 pounds. This safety factor would help to compensate for the potential problem of the carton being grasped at a position other than its center of mass.

Example 5.2 Determine the actual value of g factor in Example 5.1. Assume that the carton will experience a maximum acceleration of 40 ft/sec/sec in a vertical direction when it is lifted by the gripper fingers.

Solution

Find $g = ?$

The value of g would be 1.0 plus the quotient of the actual acceleration divided by gravity acceleration of 32.2 ft./sec/sec.

$$g = 1.0 + \frac{40}{32.2} = 1.0 + 1.24 = 2.24$$

In the following two examples, we assume that a friction-type grasping action is being used to hold the part, and so we will use the gripper force calculated in Example 5.1 as our starting point. We will demonstrate the gripper force analysis. (A detailed study of mechanism analysis is beyond the scope of this text, and the reader may refer to the references.)

Example 5.3 An angular motion gripper is used for holding the cardboard carton, as shown in Figure 5.3.1. The gripper force, calculated in Example 5.1, is 60 pounds. The gripper is to be activated by a piston device to apply an actuating force Fa. Determine the piston device force Fa to close the gripper.

Solution

The analysis would require that the moments about the pivot arms be summed and made equal to zero.

$$\Sigma M = 0$$

$$FL - Fa La = 0$$

$$(60 \text{lb})(12") - (Fa)(3") = 0$$

$$Fa = \frac{720}{3} = 240 \text{ lb}$$

The piston device would have to provide an actuating force of 240 pounds to close the gripper with a force against the carton of 60 pounds.

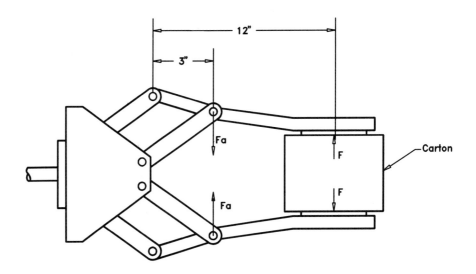

Figure 5.3.1 Pivot-type gripper used in Example 5.3

Example 5.4 Figure 5.3.2 shows the linkage mechanism and dimensions of a gripper used to handle a work part for a machining operation. The gripper force is determined to be 25 pounds. Determine the actuating force Fa applied to the plunger.

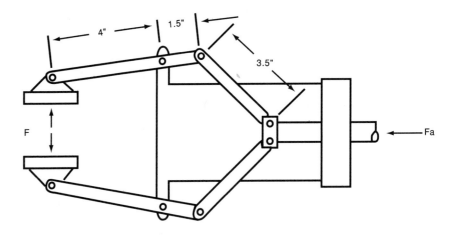

Figure 5.3.2 Gripper used in Example 5.4

5.3 Gripper Force Analysis 123

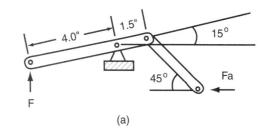

(a)

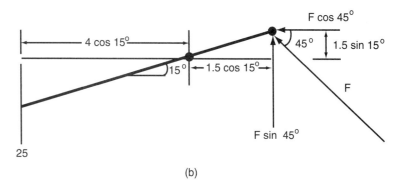

(b)

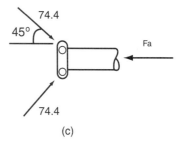

(c)

Figure 5.3.3 Linkage analysis of Example 5.4

Solution

Figure 5.3.3 (a) shows how the symmetry of the gripper can be used to advantage so that only one half of the mechanism needs to be considered. Figure 5.3.3 (b) shows how the moments might be summed about the pivot point for the finger line against which the 25-pound gripper force is applied. Therefore:

$$25(4 \cos 15°) = F \sin 45°(1.5 \cos 15°) + F \cos 45°(1.5 \sin 15°)$$

$$96.6 = (1.0246 + 0.2745)F = 1.2991F$$

$$F = 74.4 \text{ lb}$$

The actuating force applied to the plunger to deliver this force of 74.4 pounds to each finger is shown in Figure 5.3.3(c) and can be calculated as follows:

$$Fa = 2F \cos 45° = 2 \times 74.4 \times \cos 45°$$

$$Fa = 105.2 \text{ lb}$$

Some power input mechanism would be required to deliver this actuating force of 105.2 pounds to the gripper.

5.4 OTHER TYPES OF GRIPPERS

In addition to mechanical grippers, there are a variety of other devices that can be used to lift and hold objects. These devices are categorized according to the medium used for their gripping force, as follows:

1. Vacuum
2. Magnetic
3. Adhesive

Vacuum

Vacuum is used as the gripping force. The part or product is lifted by vacuum cups or by a vacuum surface.

The lifting power is a function of the degree of vacuum achieved and the size of the area on the part where the vacuum is applied.

Vacuum cups, also called suction cups, are made of neoprene or synthetic rubber. They are extremely lightweight and simple in construction. The number, size, and type of cups used will depend on the weight, size, shape, and type of material being handled. Usually they are between one and eight inches in diameter, round or oval in shape. The cups shown in Figure 5.4.1 are off-the-shelf items.

Multiple-cup vacuum grippers increase the contact surface area, as shown in Figure 5.4.2, and permit the size and weight of the workpiece to be increased.

The usual requirements of a workpiece to be handled by vacuum cups are to be smooth and in clean condition in order to form a satisfactory vacuum between the piece and the suction cup.

Vacuum grippers can be used on curved and contoured surfaces as well as on flat surfaces. They are ideal for handling fragile parts, such as glass, eggs, and sometimes flexible soft materials. In this last case, the suction cup would be made of a hard substance.

The flexibility of the vacuum cup provides the robot with a certain amount of compliance. However, to allow for surface unevenness, some vacuum cups are spring-loaded or mounted on a ball joint.

The lift capacity of the suction cup depends on the effective area of the cup and the negative air pressure between the cup and the object to be lifted. The relationship can be summarized in the following equation:

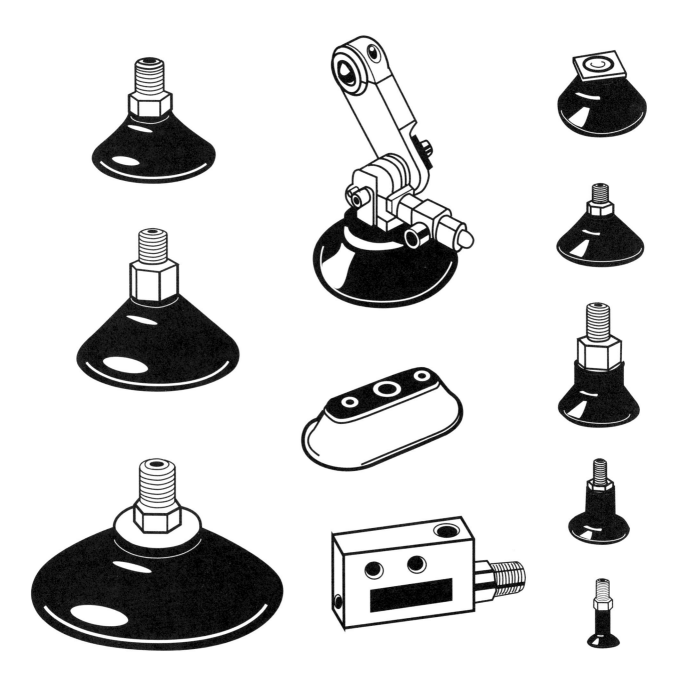

Figure 5.4.1 Commercial vacuum cups and venturi block

Figure 5.4.2 Multicup vacuum gripper *(Courtesy of Pacific Robotics Inc.)*

$$F = PA \qquad \text{(Equation 5.4.1)}$$

Where F = the force or lift capacity, lb
P = the negative pressure, lb/in.^{2}
A = the total effective area of the suction cup(s) used to create the vacuum, in.^{2}

The effective area of the cup is approximately equal to the area of the suction cup(s).

Example 5.5   A vacuum gripper is used to lift flat steel plates 0.25 x 24 x 36 inches. The gripper will utilize two suction cups, 5.0 inches in diameter each, and they will be located 18 inches apart for stability. A safety factor of 2 should be used to allow for acceleration of the plate. Determine the negative pressure required to lift the plates if the density of the steel is 0.28 lb/in.³

Solution

The weight of the plate would be:

$$w = 0.28 \times 0.25 \times 24 \times 36 = 60.48 \text{ lb}$$

This would be equal to the force F that must be applied by the two suction cups. The area of each suction cup would be:

$$A = 3.14 \left(\frac{5}{2}\right)^2 = 19.63 \text{ in.}^2$$

The area of the two cups would be

$$2 \times 19.63 = 39.26 \text{ in.}^2$$

From Equation 5.4.1:

$$P = \frac{w}{A} = 60.48/39.26 = 1.54 \text{ lb/in.}^2$$

Applying the safety factor of 2, we have:

$$P = 2 \times 1.54 \text{ lb/in.}^2 = 3.08 \text{ lb/in.}^2 \text{ negative pressure}$$

The advantages for using suction cup grippers are:

1. They require only one surface for grasping the part.
2. They apply a uniform pressure on the surface of the part.
3. They require a relatively lightweight gripper.
4. They are applicable to a variety of different materials.
5. They have a significantly low cost.

Vacuum surfaces are just an extension of the vacuum cup principle. In some material-handling applications the product to be lifted is not ridged enough for vacuum cups to be effective. To lift such material as cloth, paper, and plastic into place, a vacuum surface, as the one illustrated in Figure 5.4.3, is used. The vacuum gripper consists of a flat surface with tiny holes that forms one side of a vacuum chamber. Each hole in the vacuum surface provides a small lifting force so that the flexible cloth, paper, or plastic would be held into place against the vacuum surface from many points.

Vacuum grippers are usually venturi devices, applying Bernoulli's principle to create suction by using compressed air. The vacuum generator and venturi block (miniature vacuum pump) are two common devices used for this purpose. The vacuum generator is a piston-operated or vane-driven device powered by an electric motor, and it is capable of creating a relative high vacuum. The venturi on the other hand is a simple device, as shown in Figure 5.4.4, and can be operated

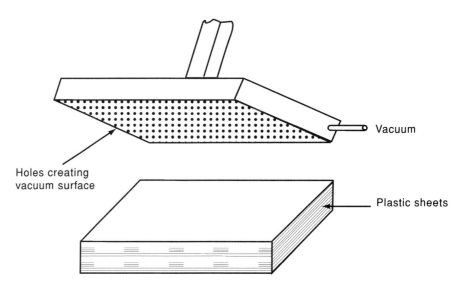

Figure 5.4.3 Vacuum surface

by means of shop air pressure. A single vacuum cup can produce 20+ inches of mercury vacuum from a 22 psi line. This enables the cup to support from 10 to 100 pounds of weight, depending on the sealing capabilities of the parts and the desired safety factor used.

Magnetic

Magnetic devices can handle ferromagnetic materials. The material can be lifted in the form of a sheet or plate with an electromagnet mounted on the robot tool plate. Figure 5.4.5 shows a single magnetic gripper and a dual magnetic gripper.

Magnetic grippers are similar in operation to vacuum grippers. However, instead of using vacuum to pick up the object, they employ a magnetic field created by an electromagnet or permanent magnet. Objects that have a flat, smooth, clean surface are the easiest to handle. The advantages of using magnetic grippers are:

1. Pickup times are very fast.
2. Part-size variations can be tolerated.
3. They are able to handle metal parts with holes.
4. They require only one surface for gripping.

The disadvantages with magnetic grippers include:

1. The residual magnetism remaining in the workpiece may cause problems in subsequent handling.

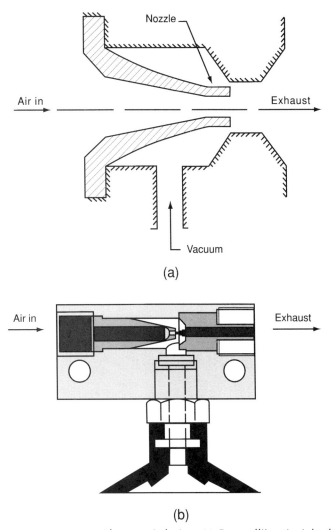

Figure 5.4.4 Vacuum cup with venturi device: (a) Bernoulli's principle that applies compressed air to form a vacuum; (b) assembly of suction cup and venturi device

2. The magnetic attraction tends to penetrate beyond the top layer in the stack, which can cause more than a single part to be lifted by the magnet.

Magnetic grippers can be divided into two categories, according to the type of magnets they use: electromagnets and permanent magnets. Electromagnetic grippers are easier to control but require DC power and an appropriate controller unit. Permanent magnets have the advantage of not requiring an external power

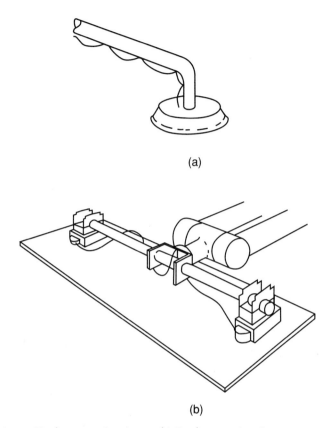

Figure 5.4.5 (a) Single magnetic gripper; (b) Dual magnetic gripper

source to operate the magnet. However, they require a stripping device for the part to be released at the end of the handling cycle. A stripping device is illustrated in Figure 5.4.6.

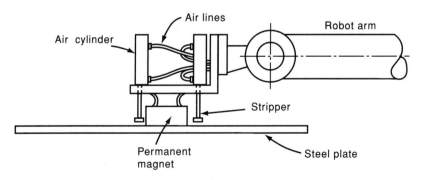

Figure 5.4.6 Stripper device with a permanent magnet gripper

Adhesive

Adhesive devices can be a very feasible means of handling fabrics and other lightweight materials. An adhesive substance is fed automatically to the robot wrist and performs the grasping action. The adhesive material is loaded in the form of a continuous ribbon into a feeding mechanism that is attached to the robot wrist and operates in a manner similar to a typewriter ribbon. The requirements of the items to be handled are that they must be gripped on one side only and that other forms of grasping, such as a vacuum or magnet, are not appropriate.

5.5 SPECIAL-PURPOSE GRIPPERS

Robot flexibility in various applications is responsible for the large variety of gripper devices. The development of off-the-shelf grippers to fill special applications is very limited. Often the user starts with an available basic design and then modifies the gripper to do the special job. In almost every case, special grippers must be fabricated to hold parts for specific jobs.

Research and development is being carried out with this objective of designing a universal gripper capable of grasping and handling a variety of objects with differing geometries. If such a universal device could be developed at a relatively low cost, it would save the time and expense of designing a special-purpose gripper for each new robot application.

Other miscellaneous devices can be used to grip parts or materials in robot applications.

Hook grippers can be used to handle containers of parts and to load and unload them from overhead conveyors. Obviously, the items must have some sort of handle to enable the hook to hold it.

Scoop and ladle grippers can be used to handle certain materials in liquid or powder form. A tool for ladling hot material, such as molded metal, is shown in Figure 5.5.1. One of this method's limitations is that the amount of material being scooped by the robot is sometimes difficult to control.

Collet grippers are used to pick and place cylindrical parts that are uniform in size. They obtain 360° of clamping contact with strong force for rapid part transfer. They are used for grinding and deburring operations. Collet grippers are available in round, square, or hex shapes. Figure 5.5.2 shows a round collet gripper.

Inflatable grippers have an inflatable diaphragm that expands to grasp the object. The inflatable diaphragm is fabricated out of rubber or other elastic material, which makes it appropriate for gripping fragile objects. The gripper applies a uniform grasping pressure against the surface of the object rather than a concentrated force typical of a mechanical gripper. Figure 5.5.3 shows an inflatable diaphragm grasping the inside diameter of a cup-shaped container.

Expandable grippers are similar to inflatable grippers but with a two- or three-finger design. Primarily, they are used to clamp an irregular-shaped workpiece. There are two types of expandable grippers: one that surrounds objects, gripping them from the outside, and one that grips hollow objects from the inside. In both cases, they make use of a hollow rubber envelope or other plastic material

132 Robot End Effectors

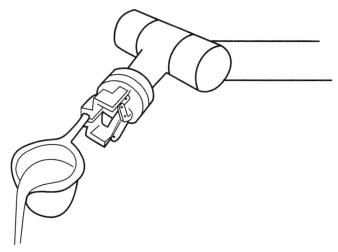

Figure 5.5.1 Ladle for pouring hot metal into a mold

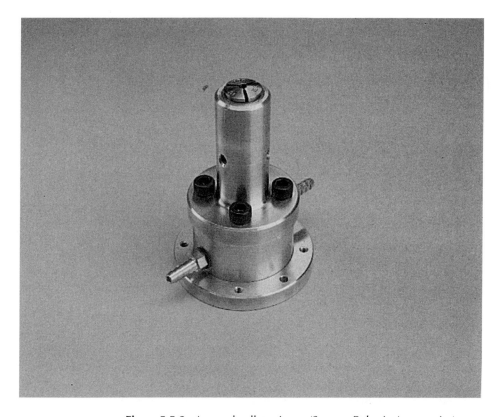

Figure 5.5.2 A round collet gripper *(Source: Robotic Accessories)*

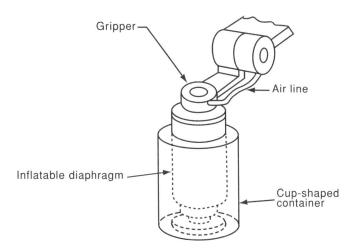

Figure 5.5.3 Inflatable gripper

that expands when pressurized. Expandable grippers are distributing even pressure on the part and are ideal for handling fragile parts or parts that vary a great deal in size.

In general, the grippers are classified in three categories:

1. Those that come in contact with only one face of the object to be lifted and use a method such as vacuum, magnetism, or adhesive action to capture the object.
2. Those that use two rigid fingers to grip an object. This type makes contact with the object at two specific points and may or may not deform the object.
3. Those that deform and attempt to increase the contact area between the gripper and object. This type includes multijointed fingers or a device operating on a principle similar to a balloon inflated inside or outside the object.

5.6 GRIPPER SELECTION AND DESIGN

The gripper selection capability is the most demanding process in any robot system to match the need for the production requirement. J. F. Engelberger, in *Robotics in Practice*, defines many of the factors that should be considered in assessing gripping requirements:

1. The part surface to be grasped must be reachable.
2. The size variation of the part must be accounted for because this might influence the accuracy of locating the part.
3. The gripper design must accommodate the change in size that occurs between part loading and unloading.

4. Consideration must be given to the potential problem of scratching and distorting the part during gripping.
5. If there is a choice between two different dimensions on the part, the larger dimension should be selected for grasping.
6. Gripper fingers can be designed to conform to the part shape by using resilient pads or self-aligning fingers.
7. The important factors that determine the required grasping force are:
   a. The weight of the object
   b. The speed and acceleration with which the robot arm moves, and the orientational relationship
   c. The physical constriction or friction that is used to hold the part
   d. The coefficient of friction between the object and the gripper fingers

5.7 PROCESS TOOLING

Process tooling is an end effector designed to perform work rather than to pick and place a work part. In a limited number of applications, the process tooling is a gripper that is designed to grasp and handle the tool. The reason for using a gripper in these applications is that there may be more than one tool to be used by the robot in the work cycle.

Process tooling refers to the general class of special end effectors that may be attached to the robot wrist.

A spot-welding gun can be attached to the robot wrist to place a series of welds on flat or curved surfaces. Generally, a three-degree-of-freedom wrist is required because of the dexterity required for maneuvering the gun. Gas-metal-arc-welding (GMAW) and Flux-core arc welding (FCAW) are the most commonly used methods for arc welding with robots. A welding gun can be attached to the robot wrist that carries the gas and bare wire for GMAW or cored electrode filled with flux for FCAW. The robot can position the welding gun for a single straight or curved run or use a weaving pattern for wider welds. Both methods are shown in Figures 5.7.1 and Figure 5.7.2. Spray-painting guns are also commonly used by industrial robots. In some cases, only two degrees of freedom may be required of the robot wrist for spray painting. The robot can spray parts with compound curved surfaces. Grinders, routers, wire brushing, or sanders are also easily attached to a robot wrist. Liquid cement applicators, heating torches, and waterjet cutting tools can also be incorporated in the robot wrist. A large class of assembly tools, such as drills, screwdrivers, and wrenches, can be used by the robot. In some cases, these tools are automatically interchangeable by the robot.

A variety of process tools are shown in Figure 5.7.3. These tools were developed by Unimation for the many applications of their industrial robots.

Multiple Tools

A single industrial robot can also handle several tools sequentially, with an automatic tool-changing operation programmed into the robot's memory. The tools can be different types or sizes, permitting multiple operations on the same work-

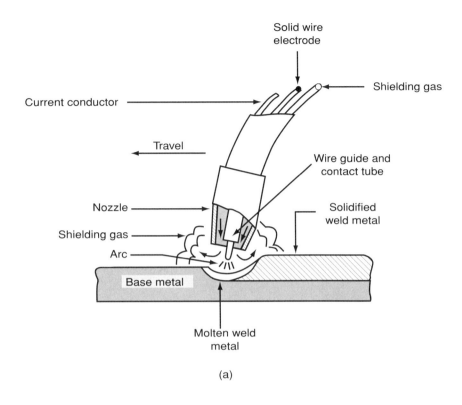

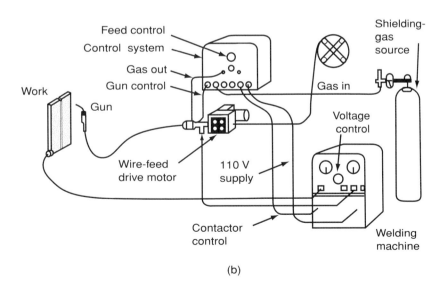

Figure 5.7.1 (a) Gas-metal-arc-welding process (GMAW); (b) Basic equipment used in gas-metal-arc-welding operations *(Source: American Welding Society)*

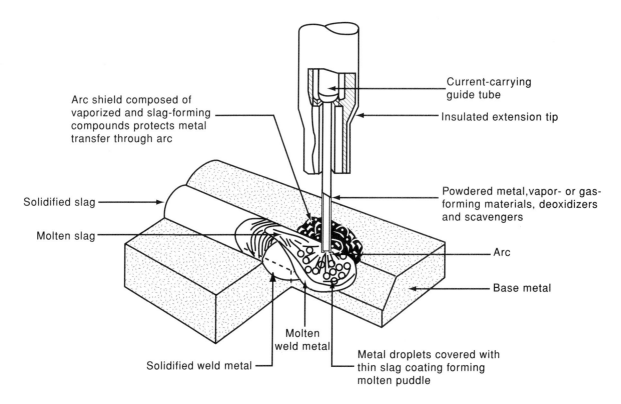

Figure 5.7.2 Schematic illustration of the Flux-core-arc-welding process (FCAW). The operation is similar to GMAW shown in Figure 5.7.1(b). *(Courtesy of the American Welding Society)*

place. To remove a tool, the robot lowers the tool into a cradle that retains the snap-in tool as the robot pulls its wrist away. The process is reversed to pick up another tool. Figure 5.7.4 shows a tool-changing operation.

5.8 COMPLIANCE

Compliance is a special end effector that is neither a gripper nor a process tool but rather a sensor or device that fits between the robot wrist and end effector for special assembly applications. In general, a compliant robot system is one that complies with externally generated forces to modify its motion for the purpose of alignment between mating parts.

If a robot uses a force sensor (piezoelectric, magnetic, or strain gauges) and modifies its control strategy based on that sensor's output, the term *active compliance* is used to describe the behavior. On the other hand, if the robot's gripper is constructed in such a way that the mechanical structure deforms to comply with those forces, the term *passive compliance* is used. Therefore, problems with mat-

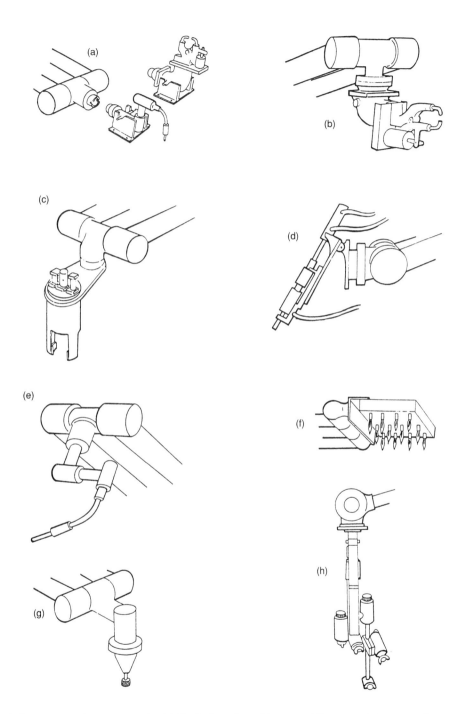

Figure 5.7.3 Various types of process tools: (a) Tool changing; (b) spotwelding gun; (c) pneumatic nut-runners, drills, and impact wrenches; (d) stud-welding head; (e) arc-welding torch; (f) heating torch; (g) routers, sanders, and grinders; (h) spray gun *(Source: Unimation-Westinghouse)*

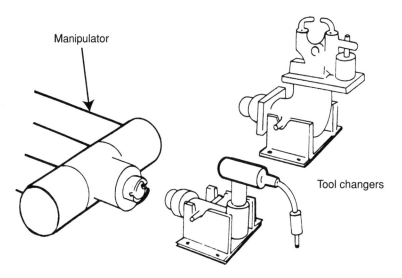

Figure 5.7.4 Tool changers decrease downtime and increase uptime for batchmode manufacturing, maintaining or repairing tools, and assembly applications *(Source: Unimation-Westinghouse)*

ing-part alignment in assembly and other applications are resolved using active and passive compliance techniques.

Active Compliance

The general task of inserting a pin into a hole represents three types of contact during the process: (a) The chamfer contact occurs when the pin is not perfectly aligned with the hole; (b) if the pin is not rigid it will rotate slightly and start to slide and make a contact along one side of the hole; and (c) if the misalignment is severe, the pin will make a two-point contact with the base of the pin and the far wall of the hole. Figure 5.8.1 shows how a misalignment of a pin into a hole results in an axial and lateral force, and a twisting moment by a contact force applied to the wrist sensor for correction. By moving in the correct direction with the compliance (Figure 5.8.2), the robot can reduce these forces on the pin.

Whitney (1983) has investigated the forces resulting from pin insertion in great detail and has arrived at a careful analysis of the forces acting on the pin being inserted into an unchamferred hole. See Figure 5.8.3.

Active compliance systems as indicated earlier measure the active force and torque when the robot performs the programmed task and often are called F/T-sensing systems. Force-sensing systems allow the robot to detect changes and variations in the workpiece or tooling during the operation and adapt the program to correct them. F/T sensing uses an adaptor placed between the gripper and the robot tool plate to measure the force and torque caused by contact between mating

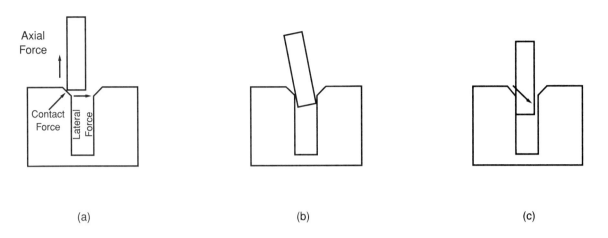

Figure 5.8.1 Compliance for inserting a pin into a chamfering hole: (a) Inserting with lateral error; (b) Inserting with rotational error; (c) Inserting with axial error

parts. Figure 5.8.4 shows a typical device. This device consists of two parallel plates that are separated by two rigid rods firmly attached to one plate but with a ball joint on the other. In addition, three elastic members are also placed between the plates that keep them separated and parallel. The rods and plates are arranged so that one plate is fixed and the other one has limited rotation and deflection.

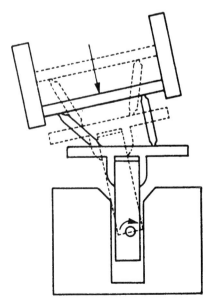

Figure 5.8.2 Compliance technique employed motion to reduce the force on the pin.

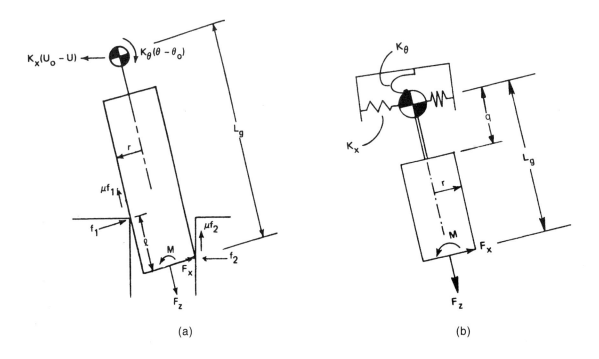

Figure 5.8.3 Active compliance: (a) forces acting on pin during two-point contact; (b) rigid peg supported compliantly by lateral springs K_x and angular spring K_θ at a distance q from peg's tip (*Source: ASME*)

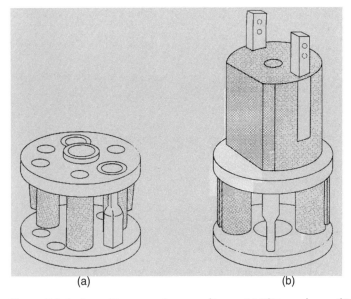

Figure 5.8.4 Force/Torque active compliance: (a) F/T transducer; (b) attached to the robot gripper

Other devices for assembly are also available. For example, engineers at the Kawasaki Laboratories in Japan can put together complex parts, such as motors and gearboxes, using high-precision feedback, cleverly designed grippers, and compliant fixtures.

Passive Compliance

Another approach to compliance is to allow the wrist to deform in such a way that the external forces are minimized. Passive compliance is using a spring-loaded wrist to provide the deformation. The concept of this principle applies to a Remote Center Compliance (RCC) device. This device was originally developed at the Charles Stark Draper Laboratories of Cambridge, Massachusetts, but now is available commercially in many forms by different manufacturers. The RCC device is a unique device that compensates for position errors due to machine inaccuracy, parts vibration, and fixturing tolerance. This minimizes the assembly forces and the possibility of parts jamming. Figure 5.8.5 shows how the device works.

The original RCC device consists of three plates: The center plate is connected to the top plate with four rods and to the bottom plate with four additional rods. In operation, four rods, one on each corner, are used for lateral compliance (only two rods are shown in Figure 5.8.5), and four angled rods, one on each corner, are used for rotational compliance (again, only two rods are shown). The flexible rods allow the plates to move relative to each other and provide a combination of lateral and rotational compliance; however, this device is rigid in the axial direction with no compliance provided.

Modern RCC devices, such as the model illustrated in Figure 5.8.6, consist of a set of six elastomertic shear pads sandwiched between two plates, and mechanical stops protect against overload movements in all directions.

The shear pads are stiff in the axial direction but highly compliant in the lateral and rotational directions. The upper plate is attached to the robot tool plate, and the lower plate is attached to the gripper. This device can also be furnished with a lockout feature. It is locked during movement and unlocked immediately before part insertions to allow the RCC to compensate for misalignment during assembly. This capability reduces assembly cycle time and increases the operation life of the shear pads.

In operation, the center of compliance is the point in space about which rotational and translation motion occurs. Positioning the center of compliance as the part-mating surface allows the part being inserted to translate laterally and rotate around the center of compliance, which reduces assembly forces and the possibility of parts jamming. This compensation for lateral and rotational misalignment reduces wear on the gripper as well as the need for high-accuracy machines and fixturing. The lockout device is illustrated in Figure 5.8.7.

Another form of passive compliance is found in some SCARA configuration robots. The Yamaha robot in Figure 5.8.8 uses SCARA technology to provide a variable tool movement for insertion compliance at programmed points. In oper-

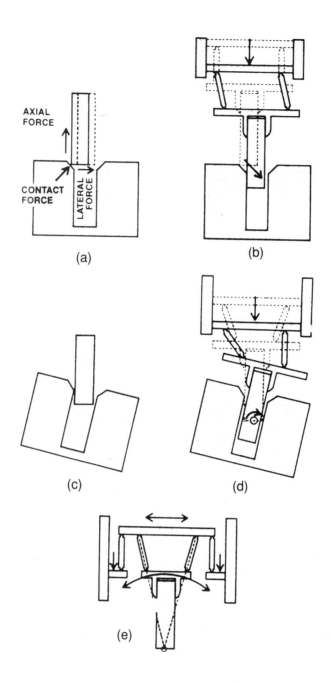

Figure 5.8.5 Operation of Remote Center Compliance (RCC) device: (a) lateral error; (b) axial error; (c) rotational error; (d) easily inserted by rotating about the compliance center; (e) combining the two modes of freedom to a useful compliant device

5.8 Compliance   143

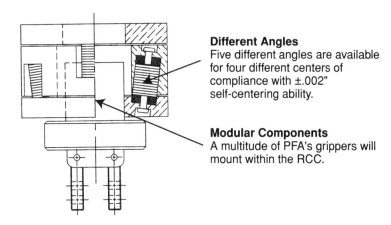

Different Angles
Five different angles are available for four different centers of compliance with ±.002" self-centering ability.

Modular Components
A multitude of PFA's grippers will mount within the RCC.

Figure 5.8.6 Automated assembly compliance device model ASP-85 *(Courtesy of PFA Inc.)*

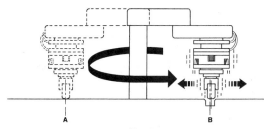

Without Lock Out System
Effects of inertia cause residual oscillation. The net result is a slower cycle time.

Features
- Reduces cycle times
- Permits rapid accelerations/decelerations
- Integrates with modified AST-100 Accommodator RCCs
- Prevents X, Y, Z, and rotational travel during transition
- Allows a high degree of repeatability

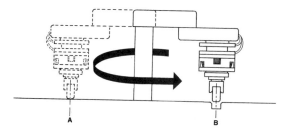

With Lock Out System
ALS System rigidly locks the RCC and tooling in place during acceleration, transport and deceleration. The net result is a faster cycle time.

Figure 5.8.7 RCC Accommodator with PFA lockout system *(Courtesy of PFA Inc.)*

5.8 Compliance

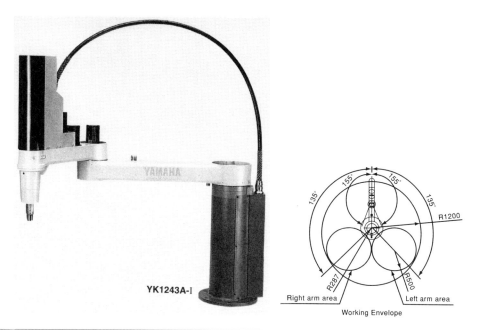

	Robot Type	YK541A-I	YK641-I	YK741-I	YK841-I	YK1041-I	YK1043A-I	YK1243-I
	Number of Servo Motor Axes	4 Axes					4 Axes	
Axis Types	X Axis Arm Length	250mm	350mm	350mm	450mm	550mm	500mm	700mm
	Max. Angle of Movement	±110°					±135°	
	Y Axis Arm Length	250mm		350mm		450mm	500mm	
	Max. Angle of Movement	±145°		±140°			±145°	±155°
	Z Axis Drive Method	Servo Motor					Servo Motor	
	Stroke	100, 200, 300mm		200, 400mm			200, 400mm	
	R Axis Drive Method	Servo Motor					Servo Motor	
	Max. Angle of Movement	±180°					±180°	
Motor	X Axis	300W		400W			400W	
	Y Axis	100W		200W			400W	
	Z Axis	100W		200W			400W	
	R Axis	60W		100W			200W	
Max. Speed	X, Y Resultant	3.5m/sec	4.2m/sec	5.0m/sec	5.7m/sec	4.7m/sec	3.35m/sec	3.76m/sec
	Z Axis	500mm/sec					250mm/sec	
	R Axis	432°/sec					180°/sec	
	Standard Cycle Time	0.9 sec		0.95sec		1.0sec		
	Repeatability	±0.03mm					±0.05mm	±0.05mm
	Max. Payload	10kg			20kg		50kg	
	Weight	34kg	35kg	78kg	79kg	84kg	88kg	90kg
	Travel Limits	1. Soft Limit 2. Mechanical Limit (X, Y axis)						
	Robot Cable	3.5m					3.5m	
	R Axis Allowable Moment of Inertia*	1.2kg • cm • sec^2		3.2kg • cm • sec^2			25kg • cm • sec^2	

*There are limits to the acceleration parameter settings.

Figure 5.8.8 Yamaha SCARA Industrial robot model YK1243A-1 shown. Other models with their parameters are listed in the table. *(Courtesy of Yamaha Robotics Inc.)*

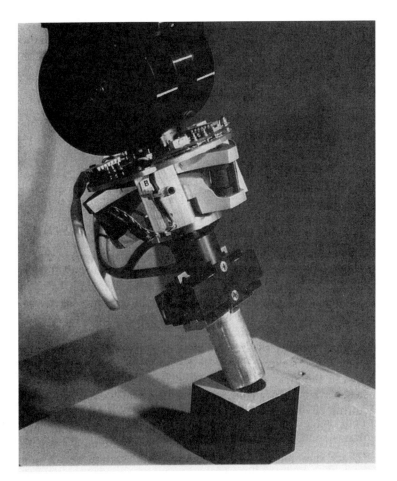

Figure 5.8.9 Chamferless insertion of a pin in a hole using an IRCC device *(Source: Charles Stark Draper Laboratories, Inc.)*

ation, when the tool reaches the programmed point, the controller frees the servo system so that the gripper can move freely over a selected distance in the x and y axes. If the gripper attempts to move beyond the selected range, the servo system stiffens and the gripper position is maintained.

In addition to the SCARA arms that initiated the selective compliance technique, other arms like the Seiko RT-300 have language commands that permit any of the axes to have compliance.

Also, by employing different types of displacement sensors on an RCC device, it is possible to obtain electrical signals proportional to various torques and forces that result in any misalignment between two parts. This information can then be utilized in a force or torque feedback scheme. Such sensors have been developed at the MIT Draper Laboratories, recently, and are called Instrumented Remote Center Compliance (IRCC). Figures 5.8.9 and 5.8.10 show such devices.

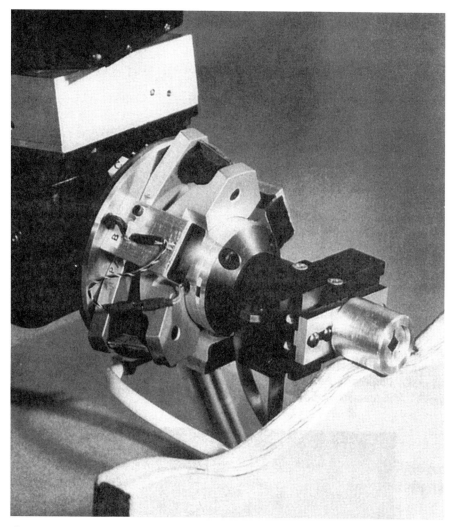

Figure 5.8.10 Edge-following rotary cutting tool using an IRCC *(Source: Charles Stark Draper Laboratories, Inc.)*

5.9 SUMMARY

The objective of this chapter was to introduce the different types of end effectors that are available today for use on robot arms. An end effector is a device mounted on the tool plate of the robot's wrist to perform a particular task. There are two major categories of end effectors: grippers and process tooling. Grippers are used to grasp, hold, and transfer objects. Process tooling is designed to perform work on an object rather than merely grasp it.

Grippers can be classified as:

a. Single grippers with two, three, or four fingers or jaws
b. Multiple grippers (two or more grippers mounted on the same robot's wrist)
c. Exterior and interior surface grippers
d. According to the method of gripping mechanism: vacuum, magnetic, adhesive, special-purpose grippers
e. According to input power of gripping mechanism: pneumatic, hydraulic, electric, mechanical (spring activated)
f. According to the closing motion of the jaws: angular closing, parallel closing

Process tooling can be classified as tool and fixture.

As a tool, it can be attached to the robot wrist to perform various operations, such as spot welding, arc welding, spray painting, drilling, routing, wire brushing, grinding, heating torches, water jet cutting, and as a cement applicator for assembly.

As a fixture, it is sometimes used to hold a set of interchangeable tools that are to be manipulated during the work cycle. An example of tool changing is shown in Figure 5.7.4.

Compliance is a device mounted between the robot wrist and end effector that allows for a certain amount of give-and-play in the motions of the end effector in various directions during assembly applications. Active compliance is the device using force sensors to detect the behavior of the parts to be assembled. Passive compliance is the device using mechanical structure of deformation to comply with the behavior of the parts.

5.10 REVIEW QUESTIONS

5.1 What is an end effector and what function does it serve?
5.2 What are the two major categories of end effectors?
5.3 In what ways do end effectors differ from the human hand?
5.4 What are the five characteristics that an end effector must satisfy?
5.5. What are the three classifications of an end effector?
5.6 Describe the classification and characteristics of mechanical grippers.
5.7 Describe the classification and characteristics of process tooling.
5.8 Describe vacuum, magnetic, and adhesive grippers.
5.9 Describe special-purpose grippers.
5.10 Describe the three categories of grippers.
5.11 What should you consider when selecting a gripper?
5.12 What is compliance?
5.13 Describe active and passive compliance.
5.14 What is an RCC and an IRCC device?
5.15 What other compliance techniques are available?

5.11 PROBLEMS

5.1 A part weighing 8 pounds is to be held by a gripper using friction against two opposing fingers. The coefficient of friction between the fingers and the part surface is 0.3. The g factor to be used in force calculations should be 3.0. Compute the required gripper force.

5.2 Solve Problem 5.1, this time using a safety factor of 1.5 in the calculations.

5.3 Solve Problem 5.1, using a maximum acceleration of 56 ft/sec/sec and a safety factor of 1.5. Find the required gripper force.

5.4 A part weighing 15 pounds is to be grasped by a mechanical gripper using friction between two opposing fingers. The coefficient of static friction is 0.35 and the coefficient of dynamic friction is 0.20. The direction of the acceleration force is parallel to the contacting surfaces of the gripper fingers. Which value of coefficient of friction is appropriate to use in the force calculations? Why? Compute the required gripper force by assuming a g factor of 2.0.

5.5. For the information given in the mechanical gripper design of Figure 5.11.1, determine the required actuating force if the gripper force is to be 30 pounds.

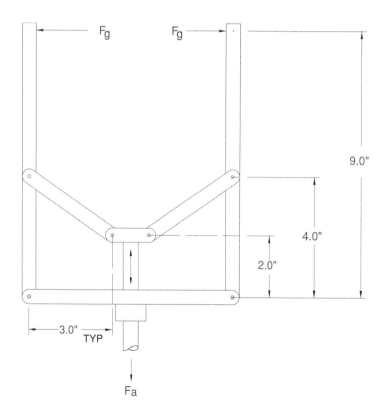

Figure 5.11.1 Mechanical gripper for Problem 5.5

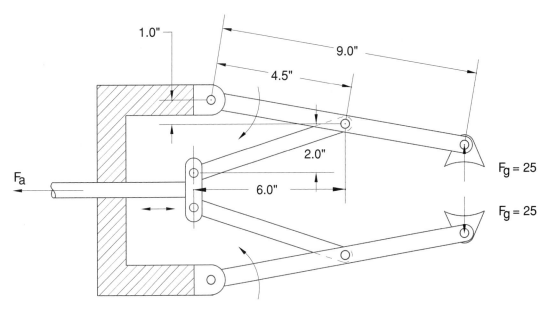

Figure 5.11.2 Mechanical gripper for Problem 5.6

5.6 For the information given in the mechanical gripper design in Figure 5.11.2, calculate the required actuating force *Fa*.

5.7 A vacuum gripper is to be designed to handle flat plate glass in an automobile windshield plant. Each plate weighs 35 pounds. A single suction cup will be used, and the diameter of the suction cup is 5.0 in. Determine the negative pressure required (below atmospheric pressure) to lift each plate. Use a safety factor of 2 in your calculations.

5.8 A piston is to be designed to exert an actuation force of 150 pounds on its extension stroke. The inside diameter of the piston is 2.0 in., and the ram diameter is 0.375 in. What shop air pressure will be required to provide this actuation force? Use a safety factor of 1.5.

5.12 REFERENCES

Chae, A., C. Atkeson, and J. Hollerbach. *Model-Based Control of a Robot Manipulator*. Cambridge, MA: MIT Press, 1988.

Chen, F. Y. "Gripping Mechanisms for Industrial Robots." *Mechanism and Machine Theory* 17, no. 5 (1982): 299–311.

Engelberger, J. F. *Robotics in Practice*. Cambridge, MA: MIT Press, 1993.

Kankaanranta, R. and H. Koiro. Dynamics and Simulation of Compliance Motion of a Manipulator. *IEEE Trans. Robot. Autom.* 4 (April 1988): 163–73.

Lewis, F. L., C. T. Abdallah, and D. M. Dawson. *Control of Robot Manipulators*. New York: Macmillan Publishing Co., 1993.

Mason, M. T. and J. K. Salisbury. *Robot Hands and the Mechanics of Manipulation.* Cambridge, MA: MIT Press, 1983.

Russell, R. A. *Robot Tactile Sensing.* Englewood Cliffs, NJ: Prentice Hall, 1990.

Schilling, R. J. *Fundamentals of Robotics Analysis and Control.* Englewood Cliffs, NJ: Prentice Hall, 1990.

Spong, M. and M. Vidyasagar. *Robot Dynamics and Control.* New York: John Wiley, 1989.

Whitney, D. E. Quasi-static assembly of compliantly supported rigid parts. *Journal of Dynamic Systems, Measurement, and Control* (March 1982) 65–77; also in *Robot Motion, Planning and Control*, pp. 439–471. Brady, Hollerbach, Johnson, Lozano-Pérez, and Mason, (eds.) M.I.T. Press, 1983.

CHAPTER 6

Sensors

6.0 OBJECTIVES

After studying this chapter, the reader should:

1. Be acquainted with robot sensors
2. Recognize sensor classification
3. Understand microswitches
4. Be able to define solid-state switches
5. Be familiar with proximity sensors
6. Be aware of photoelectric sensors
7. Be informed of rotary position sensors
8. Comprehend usage and selection of sensors
9. Understand signal processing
10. Realize sensor and control integration

6.1 ROBOT SENSORS

A robot sensor is a device or transducer that detects information about the robot and its surroundings, and transmits it to the robot's controller.

A sensor produces a signal for purposes of determining or measuring a property, such as position, force, torque, pressure, temperature, humidity, speed, acceleration, and vibration. Traditionally, sensors, actuators, and switches have been used to set limits on the performance of machines. Sensor technology is essential to data acquisition, monitoring, communication, and computer control of machines and systems.

Because sensors convert one quantity to another, they are also often referred to as transducers, meaning transfer. Analog sensors produce a signal, such as voltage, that is proportional to the measured quantity. Digital sensors have numeric or digital outputs that can be directly transferred to computers. **Analog-to-digital** converters (A/D) are available to interface analog sensors with computers.

Robot systems should be able to sense, evaluate, make decisions, and interact with their environment. By perfection of sensors, robots will be able to achieve their full potential.

Sensors can help robots to:

1. Detect positions and orientation of parts
2. Ensure consistent product quality
3. Discover variations of shape and dimensions of parts

4. Identify unknown obstacles
5. Determine and analyze system malfunctions

6.2 SENSOR CLASSIFICATION

Sensors that are of interest in manufacturing may be classified, generally, as:

a. Mechanical—for measuring quantities such as position, shape, velocity, force, torque, pressure, vibration, strain, and mass
b. Electrical—for measuring voltage, current, change, and conductivity
c. Magnetic—for measuring magnetic field, flux, and permeability
d. Thermal—for measuring temperature, flux, conductivity, and specific heat
e. Others—such as acoustic, proximity, chemical, photoelectric, radiation, lasers, optical systems (fiber optics and light-emitting diodes), tactile, voice and visual sensing

Sensors also are grouped into three main categories: function performed; location and type of detection; physical activation.

Function Performed

According to their function performed, sensors can be categorized under manipulation or acquisition.

Those that allow the robot to interact with its surroundings, like **tactile** and force sensors attached to the manipulator are grouped under manipulation. Tactile sensors are devices that indicate contact between themselves and another solid object. Tactile sensing devices can be divided into two classes: touch sensors and force sensors. Touch sensors provide a binary output signal that indicates whether or not contact has been made with the object. Force sensors (also sometimes called stress sensors) indicate not only that contact has been made with the object but also the magnitude of the contact force between the two objects.

Those sensors that let the robot know its own present state are grouped under acquisition. These devices measure the distance to the nearest object within a zone of information-collection space. An example of a point-measuring device may be found in the distance-measuring ultrasound devices developed by Polaroid, as shown in Figure 6.2.1.

Location and Type of Detection

According to their location and type of detection, sensors are categorized as internal, external, or interlock.

Internal sensors use feedback information internally to ascertain their present condition. The first complex sensor used by industrial robots is known as **haptic perception**. This is the robotic equivalent of the human sense of kinesthesia,

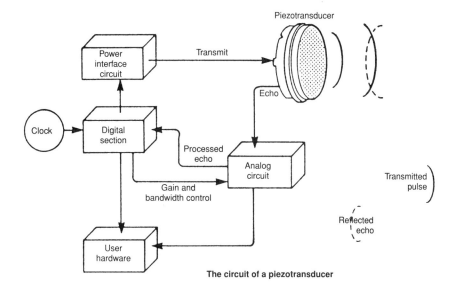

Figure 6.2.1 Piezotransducer is an ultrasonic sensor suitable for robotics, security systems, industrial controls, automotive, level measurement, and energy conservation applications. *(Courtesy of Polaroid Corp.)*

156 Sensors

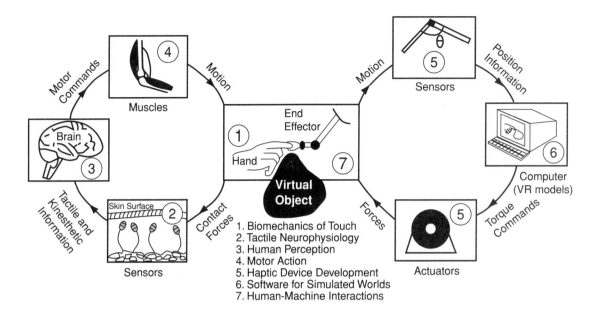

Figure 6.2.2 Haptic perception—human and machine haptics enhance interactions in virtual reality and teleoperator systems, ordered according to the number in the functional block diagram. *(Courtesy of Dr. M. Srinivasan, MIT Laboratory for Human and Machine Haptics).*

which is sensing information that comes from the joints and muscles without locating them visually. Figure 6.2.2 illustrates such a sensor.

Presently, an industrial robot's internal sensors use mechanical, electrical, electronic, and hydraulic devices to obtain feedback information. These devices are called closed-loop and servo-controlled systems.

One example is a shaft encoder that can be used to detect very fine rotational movement. Another example is the direct-readout encoders or absolute-readout encoders that can give the degrees of rotation from 0° to 360° directly. These readings are expressed in terms of switch openings or closings and thus are already in a digital form that can easily be entered into the robot's electronic memory and interpreted by the robot's controller.

Figure 6.2.3 shows a simplified direct-readout shaft encoder. Each ring of the disk has twice as many divisions on it as does the ring immediately inside it, but only one division mark accounts for the value changes at a time. This produces a count value with an accuracy of plus or minus one division.

The pulse-reading shaft encoder or incremental encoder is less expensive and emits a pulse for each increment of shaft rotation. External counter circuits keep track of the number of pulses and the present position of the shaft. Figure 6.2.4 shows a pulse-reading encoder with a single set of windows and two sensors. Other

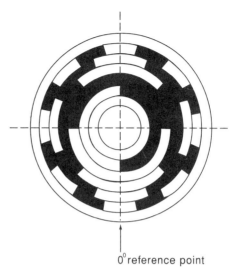

Figure 6.2.3 Direct-readout shaft encoder

inexpensive but less accurate internal sensors are strain gauges, potentiometers, and synchro system. (See Figure 6.2.5.)

External sensors are, as the name implies, physically mounted on the robot or on process equipment in the robot cell so that they can be visually seen.

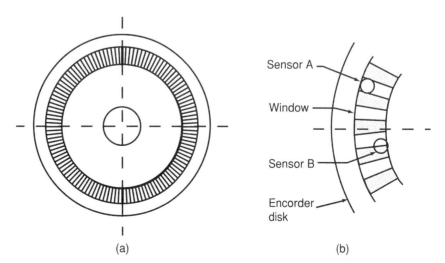

Figure 6.2.4 Pulse-reading shaft encoder: (a) encoder disk; (b) sensors—the direction of rotation can be determined by using two sensors 90° from each other and half a window apart.

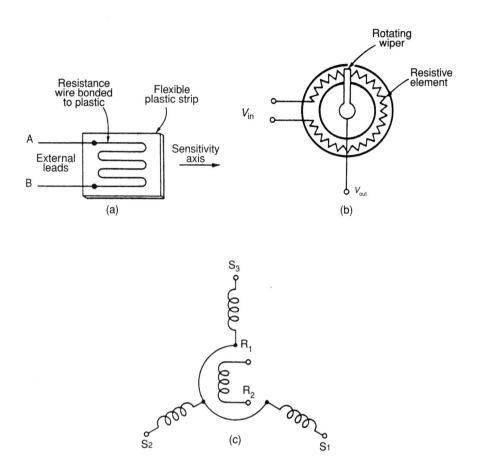

Figure 6.2.5 Inexpensive internal sensors: (a) bonded-wire strain gauge; (b) circular wirewound potentiometer; and (c) two circuits synchro.

The four most common external sensors are the microswitches, the simple touch or tactile sensors, the photoelectric devices, and the proximity sensors.

The senses of hearing, speech, vision, and touch for robots are still in the experimental stages. Development of the sense of voice recognition (hearing) is mainly intended to make it easier for humans to give commands to the robots when training them to do a specific task. Speech synthesis or voice will enable a robot to communicate warnings to humans when something is wrong. Vision will make self-orientation and proper grasping of parts easier for the robot, because vision can be used to correct small errors of positioning or alignment. The tactile sense will help a robot tell when it is gripping a part and what the orientation of the part is.

Interlock sensors are devices that do not allow an operation to be performed until certain conditions exist. They are very old devices that are used to protect the unauthorized person from harmful conditions. Some interlock devices may sound an alarm or stop the motion without switching its power off in an emergency.

An interlock sensor may also be used to protect a robot. Examples include the airflow in the robot's controller that can reach the level before power is applied to the controller's system; hydraulic pressure that can reach some minimum level before hydraulic actuators are allowed to move; the temperature of the hydraulic fluid, which can remain below a certain level if the hydraulic system is to be used. In all these cases, the robot must receive an electrical interlock signal indicating that it has finished its cycle and is open before the robot can be allowed to move to the next operation. Interlock sensors are electromechanical or completely mechanical. They also can be viewed as internal sensors as well as external sensors, because they can be used for measuring the state of the device as well as warning of an intruder.

Physical Activation

According to their physical activation, sensors are categorized as contact or noncontact.

Contact sensors, as the name implies, must physically touch an object before the sensor is activated. Contact sensors include microswitches and all tactile-sensing devices. In each category, the sensors can be in direct or indirect action with digital or analog output signals.

Noncontact sensors measure the condition of an object without physically touching the part. Most frequently, noncontact sensors are those that measure a point of response and others that give a spatial array of measurements at neighboring points of information. A vision system is the most common example of a device that measures spatial information.

By point of response, these sensors may be further divided into sensors that measure proximity and sensors that measure the presence of an object, like the photoelectric devices.

Further subdivision of the noncontact sensors includes devices that measure spectral range, such as infrared, visible, and X ray. Such sensors may be used for inspection tasks, part identification, and other applications and are based on electrical fields, ultrasonic, and radiation. The classifications of sensors available for use on an intelligent robot are shown diagrammatically in Figure 6.2.6. Microswitches, solid-state switches, proximity, photoelectric, and rotary position sensors are discussed in more detail in the following sections. (Vision systems are discussed in Chapter 7.) In reality there are two main classifications of sensors based on their output: discrete sensors and complex sensors.

Discrete sensors are also known as simple or digital sensors whose output has only two states—on/off, yes/no, high/low, or 1 or 0—in which case it is frequently

160 Sensors

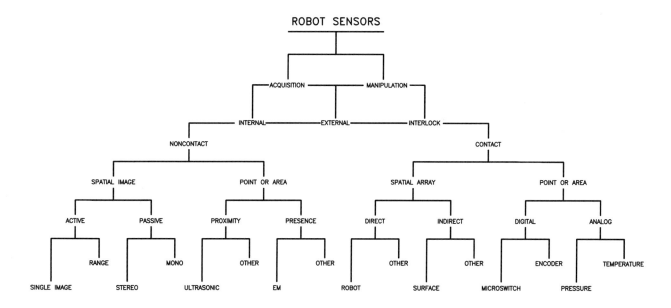

Figure 6.2.6 Classification of sensors available for an intelligent robot

referred to as a binary device. Simple sensor interface is the process in which a device transmits signals from a simple (digital) sensor to a controller.

Complex sensors' output has more than two states and can be of any value within the range of values it measures. Complex sensors would include analog sensors and serial communication sensors. Complex sensor interface is the process that transmits a complex (**analog**) sensor's signals to a controller after converting them to digital signals or otherwise conditioning them.

Both types of these sensors used in interface with most manufacturing systems can be grouped also in three categories: contact sensors, noncontact sensors, and process control sensors.

A simple and complex sensor system block diagram is shown in Figure 6.2.7.

6.3 MICROSWITCHES

Microswitches have been used in flexible automation for many years and are reliable and easy devices to interface in an automatic robot cell.

Microswitches are electromechanical devices actuated either by some part or motion of a robot or machine to alter the electrical circuit. This event is indicated by means of opening or closing the switch contacts. Because of the importance and widespread applications of these switches in their varying classifications, a general discussion of switches and switch contacts follows.

6.3 Microswitches

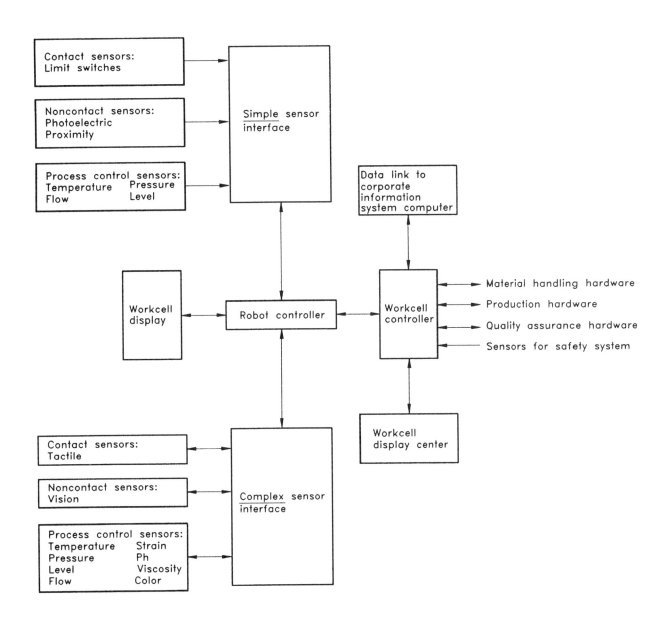

Figure 6.2.7 Sensor system block diagram

Mechanical Contact Switches

In order for a switch to provide meaningful control data, it must pass an electrical current with a virtual short circuit in the ON state, and break that current flow in the OFF state. The location at which current is interrupted in a mechanical switch is known as the *contacts*.

The basic problem with mechanical contact switches is that current, once flowing through the closed contact pair, has a tendency to keep flowing when the contacts are opened. The result is an arc, or spark, jumping from one contact to the other as the contacts separate. The greater the amount of current flowing through the contacts, and the higher the voltage across the open contact pair, the greater the tendency to arc.

As the contacts are opened, arcing across the air gap results in spark erosion of the contact surfaces as well as contact oxidation. When the contacts are closed again, the arc-eroded contact surfaces offer slightly higher resistance than they did on the previous closing.

It is not difficult to understand, then, that all mechanical switch contacts will fail sooner or later by virtue of the degradation inherent in the switching operation itself. To maximize contact life, switch designers attempt to minimize contact resistance (thus heating) by using oxidation-resistant metals (gold, platinum) and by providing contact pressure in the ON state to squeeze the contact interface, increasing the contact area and breaking through any oxide film.

Switching Resistive and Inductive Loads

There are two basic types of electrical loads that must be switched by machine-control systems: resistive and inductive. Resistive loads include lights and heaters; inductive loads include motors and solenoids.

Contacts carrying DC current to purely resistive loads have a tendency to arc when the circuit is broken, which is proportional to the open-circuit voltage and the amount of current being carried through the circuit.

The problem with switching inductive loads carrying DC power is that an abrupt opening of the circuit results in very high voltage surge, which not only aggravates arcing but also is potentially damaging to any other electronic components in the same circuit.

Another problem associated with switching motors, relays, and solenoids is that they exhibit high inrush currents upon start-up. These initial high currents can result in contact welding, as discussed earlier, and so need sufficient spring force to break the contacts.

Contact arcing can be suppressed magnetically by placing a coil in close proximity to the opening contacts with a magnetic field that opposes that of the arc itself. Contact arcing in DC inductive load circuits can also be suppressed by inserting resistors, capacitors, and/or diodes in parallel with the contacts and the inductive load. These components allow the inductive power surge to be dissipated at circuit locations other than across the opening contacts.

Control Arrangements

The state of a switch's contacts in the unactuated mode determines whether a switch is normally open (**NO**) or normally closed (**NC**) with respect to a single circuit. An NO contact pair will close (pass current) or *make* upon switch actuation. An NC contact pair will open (stop current flow) or *break* upon switch actuation.

The number of poles associated with a switch is the number of independent circuits that the switch makes or breaks at one actuator position.

The throw of the switch is the number of circuits that can be switched by a single pole. For example, a single-throw switch might make a circuit in one switch position and break that same circuit in another position. A double-throw switch might have circuit A closed and circuit B open in one switch position, and vice versa in the other switch position.

Switch break is the number of contact pairs that are opened or closed in each independent circuit as the switch moves from one position to the other. Figure 6.3.1 shows schematic diagrams of contact arrangements depicting NO, NC, poles, throws, and breaks.

Figure 6.3.2 shows the ANSI contact symbols for a number of switch contact arrangements. The form is identified by a letter, as shown in the figure. The number of poles, or the number of separate circuits that can be switched, is identified by a number preceding the form letter. Thus, a 4A switch would be a four-pole NO switch. If it were a relay, then the relay coil switch would control four pairs of NO contacts. A 2C relay would have two pairs of transfer contacts controlled by the relay coil. Transfer contacts are double-throw contacts because they are normally closed for circuit A (NO for circuit B) in the unactuated state, and vice versa in the actuated state. A form 2C relay could also be called a double-pole, double-throw switch.

Types of Microswitches

There are five types of microswitches used in workcells:

1. Manual switches
2. Limit switches
3. Impulse limit switches
4. Reed switches
5. Pressure switches

Manual switches are manually actuated and classified as momentary or maintained contact types. Momentary contact switches include push-button and toggle switches, and this type of action is widely used to start and stop machine operations. Maintained contact switches include push-button, toggle, sliding, and rotary types. The contacts are transferred by the actuator and remain transferred after the operator's hand is removed. Maintained contact manual switches often have integral indicator lights to show whether they are in the OFF or ON state.

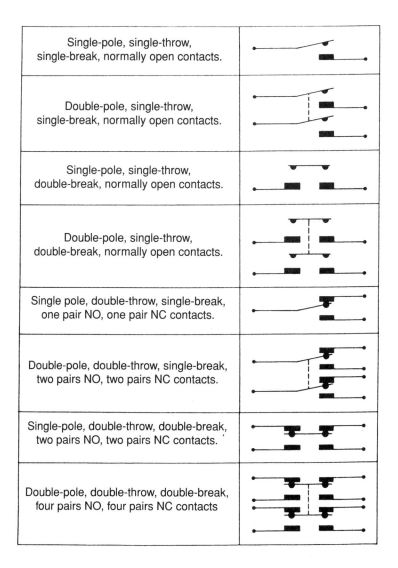

Figure 6.3.1 Electromagnetic contact arrangements

Limit switches are perhaps the most common linear position sensors used in robots and work cells. A **limit switch** is designed to be mechanically actuated when a machine member or object reaches a particular position. The machine member physically contacts the limit switch actuator and switches the contacts, normally one to four poles. The limit switch can be used to control machine operation sequencing, or to provide machine protection functions by shutting down the machine when machine members or workers are at a prohibited location.

Form	Description	Symbol
A	Make or SPSTNO	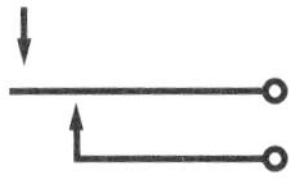
B	Break or SPSTNC	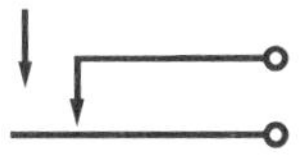
C	Break, make or SPDT (B-M) or transfer	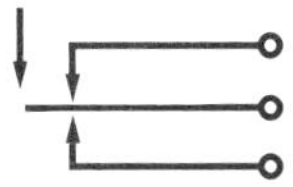
D	Make, break or make-before-break, or SPDT (M-B) or "Continually transfer"	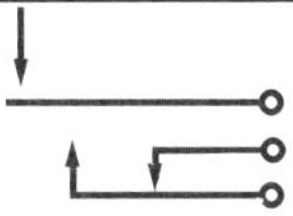
E	Break, make, break or break-make-before-break or SPDT (B-M-B)	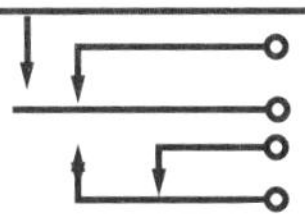
F	Make, make SPST (M-M)	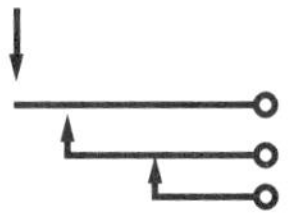
G	Break, break or SPST (B-B)	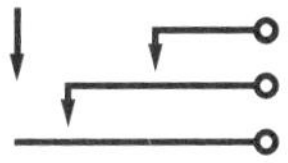
H	Break, break, make or SPDT (B-B-M)	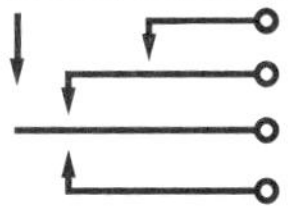
I	Make, break, make or SPDT (M-B-M)	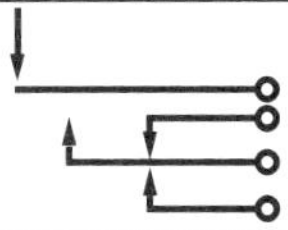
J	Make, make, break or SPDT (M-M-B)	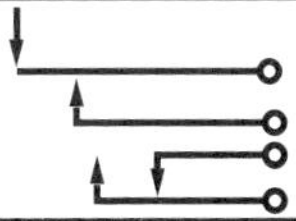

Form	Description	Symbol
K	Single pole, double throw, center off or SPDTNO	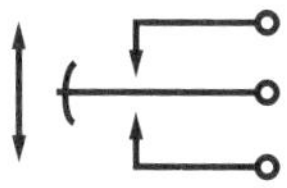
L	Break, make, make or SPDT (B-M-M)	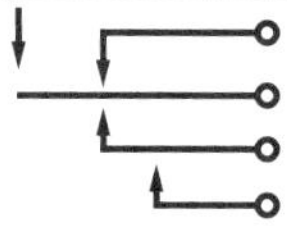
M	Single pole, double throw, closed neutral SP DT NC	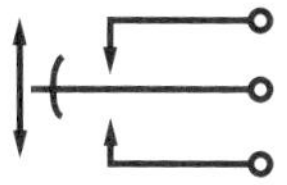
U	Double make, contact on arm SP ST NO DM	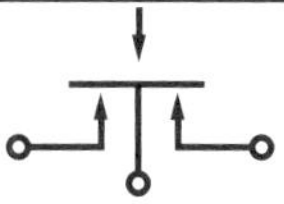
V	Double break, contact on arm SP ST NC DB	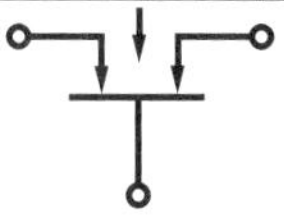
W	Double break, double make, contact on arm. ST DT NC-NO (DB-DM)	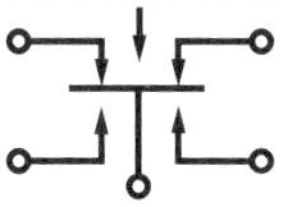
X	Double make or SP ST NO DM	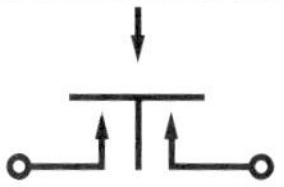
Y	Double break or SP ST NC DB	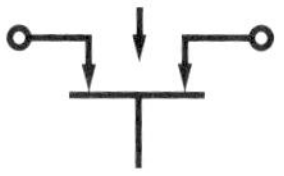
Z	Double break, double make SP DT NC-NO (DB-DM)	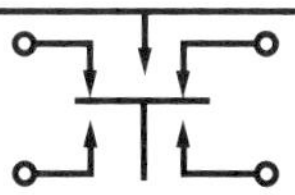

Figure 6.3.2 ANSI symbols for electromechanical contacts

166 Sensors

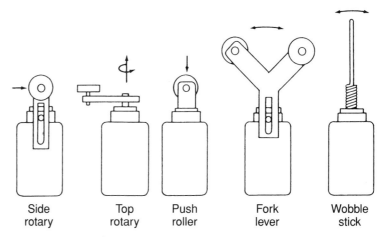

Figure 6.3.3 Limit switch actuators

Industrial limit switches are usually enclosed in cases in order to protect the switch from dust, water, and human abuse. Standard duty enclosed limit switches are rugged; however, heavy-duty limit switches are available as oil-tight, corrosion-resistant, and generally more rugged enclosures. Heavy-duty limit switches should undergo 20 million cycles, without failure, under moderately severe industrial usage.

Table 6.1 lists the common NEMA type enclosures for limit switches, and the conditions against which they must protect switches.

Figure 6.3.3 shows the common limit switch actuators used in workcells. Actuator types are selected primarily on the basis of where the limit switch might be located with respect to the controlled machine member. It is important to install the limit switch in a location and position such that:

1. The moving member will not destroy the limit switch.
2. The limit switch is accessible for maintenance.
3. The limit switch is protected from accidental actuation (such as an overhanging workpiece, etc.).
4. Chips, moisture, grease, or oil accumulation does not take place on the actuator.

Figure 6.3.4 illustrates proper cam design for mechanically actuated limit switches. Improper cam design can result in greatly reduced switch life.

The primary reason for premature limit switch failure is caused by repeated high impacts of cams, particularly at high speeds. The leading edge of the cam must contact the actuator roller at the proper angle.

Just as it is important to avoid excessive impact of cam or actuator, so is it important to avoid rapid spring back as the cam overrides the actuating roller of

Table 6.1 NEMA switch enclosures

Non-Hazardous Locations

Type 1	For indoor use primarily to provide a degree of protection against contact with the enclosed equipment.
Type 3	For outdoor use primarily to provide a degree of protection against windblown dust, rain, sleet, and external ice formation.
Type 3R	For outdoor use primarily to provide a degree of protection against falling rain, sleet, and external ice formation.
Type 4	For indoor or outdoor use primarily to provide a degree of protection against windblown dust and rain, splashing water, and hose-directed water.
Type 4X	For indoor or outdoor use primarily to provide a degree of protection against corrosion, windblown dust and rain, splashing water, and hose-directed water.
Type 6	For indoor or outdoor use primarily to provide a degree of protection against the entry of water during occasional temporary submersion at a limited depth.
Type 12	For indoor use primarily to provide a degree of protection against dust, falling dirt, and dripping noncorrosive liquids.
Type 13	For indoor use primarily to provide a degree of protection against dust, spraying of water, oil, and noncorrosive coolant.

Hazardous Locations

Type 7	For use indoors in locations classified as Class I, Groups B, C, or D by the National Electrical Code
	Group B - Atmospheres containing hydrogen or manufactured gas.
	Group C - Atmospheres containing diethyl ether, ethylene, or cyclopropane.
	Group D - Atmospheres containing gasoline, hexane, butane, naphtha, propane, acetone, toluene, or isoprene.
Type 9	For use in indoor locations classified as Class II
	Group E - Atmospheres containing metal dust.
	Group F - Atmospheres containing black carbon, coal dust, or coke dust.
	Group G - Atmospheres containing flour, starch, or grain dust.

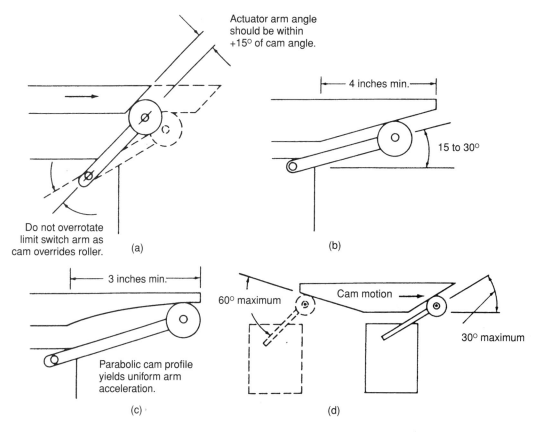

Figure 6.3.4 Limit switch cam design: (a) for speeds less than 50 fpm; (b) for speeds between 50 and 200 fpm; (c) for speeds greater than 200 fpm; (d) constant acceleration cam design.

a switch. Under no circumstances should the lever of a switch be allowed to snap back. Figure 6.3.4 (d) shows the trailing edge of the cam not exceeding 60° between the level arm and the cam surface. Note that the use of an override cam will result in the actuation of the switch on the return stroke unless a one-way lever is used. See Figure 6.3.4 (a) and (b).

Limit switches can come with momentary contacts (standard signal) or impulse contacts (spontaneous signal).

Impulse limit switches produce a switching signal only while the plunger travels inward. The contacts switch back when the end of the plunger travel is reached and remain OFF during return travel of the plunger. This switching action overcomes some control system design problems where an actuator sitting on a limit switch will not allow subsequent operations to take place.

The duration of the switching impulse is inversely proportional to actuation speed. Generally, 50 ft/min. is the maximum speed for reliable operation. Posi-

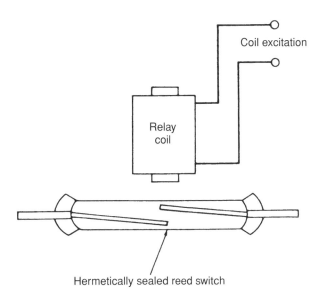

Figure 6.3.5 Normally open sealed reed switch

tioning of the impulse switch, relative to its actuator, is critical because the actuator-plunger travel must be enough to complete the switching pulse, but not so much as to cause overtravel damage.

Reed switches consist of two or more thin metal strips (reeds) enclosed in an evacuated hermetically sealed glass capsule. The reeds overlap and can be closed by moving a magnet close to the reeds, and opened by moving the magnet away. Alternatively, the reeds can be closed by energizing a nearby coil (reed relay). Figure 6.3.5 shows an **NO reed switch**. A coil wound around the switch creates a reed relay, and when energized, the reeds snap together and remain closed until the coil is de-energized. To create an NC relay contact, the reeds are biased into the closed position with a permanent magnet. A reverse polarity coil cancels the permanent magnetic field when energized, and the reeds snap open. Reed switches are not generally used to switch large inductive loads directly (i.e., motors) but often are used to switch power to another control source.

Pressure switches are used in robot cells as sequence control devices and as machine protection devices. A pressure switch is used in a machine control system in the same way as a position sensing limit switch is used. Snap-acting contacts open (or close) when a preset pressure is sensed by the pressure switch.

The pressure switch is not unlike the limit switch in construction; indeed, some pressure switches employ a basic plunger-actuated limit switch in their design. What must be added to a limit switch to make it a pressure switch is a pressure-sensitive subsystem that outputs an actuating stroke (proper force and distance) to open and/or close snap-acting contacts. The key to designing a useful pressure-sensing subsystem is to make the pressure switch contacts adjustable.

Pressure switches are available in most NEMA-type enclosures for protection from various environments. The switch assembly has one or two NPT ports to which the pressure being sensed is connected and can be used with liquid or gas media. If switching is to occur at a given differential pressure, then fluid lines at the two pressures (high and low) are connected to the switch. If switching is to occur at a set gage pressure (set point), then only one fluid line is connected to the switch.

Pressure switches use either pistons, bellows, or diaphragms as the pressure-to-stroke switch actuator. Bellows and diaphragms are most common because their static seals can prevent fluid leakage very reliably. The sensing element (diaphragm or bellows) is usually biased with an adjustable spring load so that the pressure at which the switch actuates can be adjusted.

Use of pressure switches can often simplify component placement problems in designing a workcell. Figure 6.3.6 shows a cylinder with pressure switches sensing the cylinder inlet and outlet pressures. This arrangement can replace two-position sensing-limit switches and the attendant problems produced by switch placement and actuator design. The system of Figure 6.3.6 can also be used as a **load-sensing** control scheme, a feature not possible with limit switches. This feature allows pressure in the cylinder to rise to the point where the applied force equals the maximum permissible force so that the switch can close, and by virtue of a properly designed control system, retracts immediately for the next machine cycle.

6.4 SOLID-STATE SWITCHES

Solid-state switches consist of multiple layers of semiconductor material constructed in a way that the output terminals are nonconducting (open) until relatively low power is applied to the input of switch terminals. The output terminals then become conducting (closed) without the inherent unreliability of moving contacts. Each solid-state switch is equivalent to a single-pole, single-throw (SPST) mechanical switch, except that actuation is accomplished electrically rather than mechanically. Solid-state switching takes place in the microsecond-to-nanosecond range, as opposed to milliseconds required for mechanical contact switching.

Semiconductor switches do have a finite "contact" resistance, even in the conducting or ON mode. A 40-amp switch might typically dissipate 50 watts, so it is important that these devices be properly heat sinked and cooled. Unlike metallic contacts, semiconductor switches are very temperature sensitive, both in terms of poor performance at elevated temperatures and failure tendency due to thermal cycling (i.e., fatigue due to expansion and contraction).

Solid-state switches are also highly intolerant of excessive voltage and current. It is, therefore, essential that switching circuits be protected from high inrush currents and voltage surges encountered when switching OFF inductive loads. It

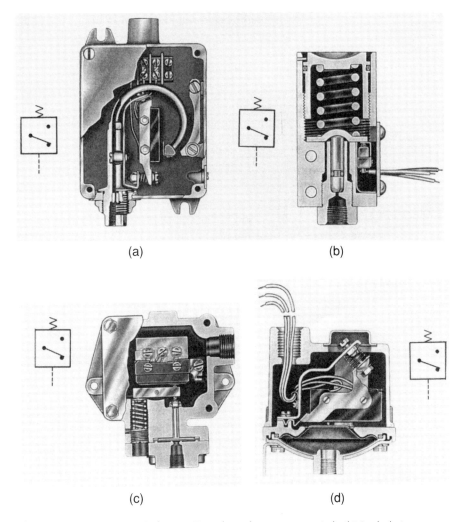

Figure 6.3.6 Pressure switches: (a) Bourdon tube pressure switch; (b) Sealed piston pressure switch; (c) Dia-Seal piston pressure switch; and (d) Diaphragm pressure switch *(Courtesy of Barksdale, Inc.)*

is not uncommon to step up from solid-state control circuitry using the solid-state switch or relay to activate an electromechanical relay, which in turn switches the load.

Solid-state switches are now widely used as integral output switches for sensors that are not mechanically actuated as are standard limit switches. Major examples are proximity sensors, photoelectric sensors, and timers. These sensors can be applied in control systems exactly as their electromechanical counterparts might be, as long as the switched loads do not exceed the output rating of the device.

6.5 PROXIMITY SENSORS

Proximity sensors are linear position sensors that can be applied in automatic machinery in the same manner as limit switches, but with one important distinction: Physical contact between the switch and a machine element is not required to open or close the switch. The target or machine element approaches the proximity switch sensing area in either a head-on mode [Figure 6.5.1(a)] or a slide-by mode [Figure 6.5.1(b)].

The proximity sensor is not a simple device. It generally requires an external source of excitation power, an analog sensor that responds to the proximity of a material, a circuit to transform the analog proximity signal to a repeatable switch signal, and an output switch, usually solid-state. In spite of the complexity, proximity switches are available in completely self-contained packages ranging in size from 3/8 inch in diameter to 3 inches in diameter. The same NEMA enclosure options available for limit switches apply to proximity sensors.

The most widely used proximity sensors are of the inductive type. The sensor or probe projects a high-frequency (200,000 Hz) electromagnetic field in the space in front of the sensor. When a conductive target enters the field, eddy currents induce the field and the probe activates the device.

Inductive-type proximity sensors are limited to use with conductive (usually metallic) targets. Sensing ranges vary from 0 to 1 inch. Switching capability and speed depend upon the type of switch output. Because of the high-frequency fields

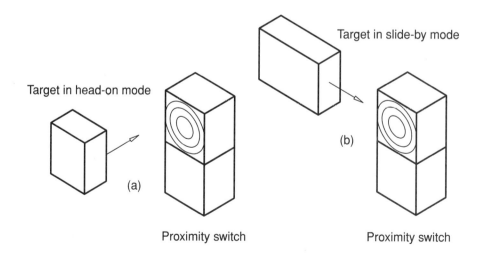

Figure 6.5.1 (a) Proximity switch mounted in head-on mode; (b) proximity switch mounted in slide-by mode

involved, a sensor must be shielded first from metallic machine elements, and second, from nearby sensors.

Other proximity sensors are available commercially and utilize different operating principles.

Capacitive-type proximity sensors obtain their switching signal based on the air gap between the sensor and nonconductive materials.

Magnetic reed switch sensing schemes are used for proximity sensing by placing a permanent magnet on the traveling target. When the magnet comes into proximity to the stationary switch, the reeds close.

Hall effect sensors are used where the output of a voltage is proportional to the magnetic field. A magnet mounted on the target creates a Hall voltage as it nears the sensor. This signal is amplified, and when it reaches a threshold value, solid-state switch outputs are actuated.

Ultrasonic proximity sensors operate on the principle of reflected ultrasound. The time between signal output and return of the reflected signal is proportional to distance between the sensor and the target. Developed initially as automatic focusing devices for cameras, these sensors operate over the range of 1–20 feet. Accuracy of switching is not as repeatable as that of the previously discussed sensors.

Ring-type proximity sensors are available that actuate (switching signal) when objects pass through the ring. Coin counters or small-parts counters are examples of ring proximity switch applications.

Proximity probes of all types are usually sealed. Cylindrical or square devices have leads protruding from the nonactive end. Heavy-duty proximity sensors are often square assemblies, much like mechanical limit switches but without the actuator.

6.6 PHOTOELECTRIC SENSORS

Photoelectric sensors are noncontact position-sensing devices that the output responds to:

1. Interruption of a light beam by an object or part
2. Reflection of a light beam back to its source by an object or part passing in front of the projected beam

The four types of photoelectric sensors are defined as follows and are shown in Figure 6.6.1:

1. Light source and receiver
2. Retroreflector
3. Diffuse scanner
4. Reflect scanner

The light source and receiver type consists of a light source and photoreceiver for a direct scan or through a scan photoelectric control system.

174 Sensors

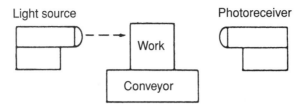

As work moves between light source and receiver, the light beam is interrupted, resulting in a switch closure (or opening).
(a)

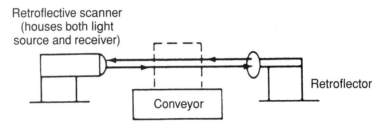

Light beam from source to retroflector and back to scanner is interrupted by object, initiating a control signal.
(b)

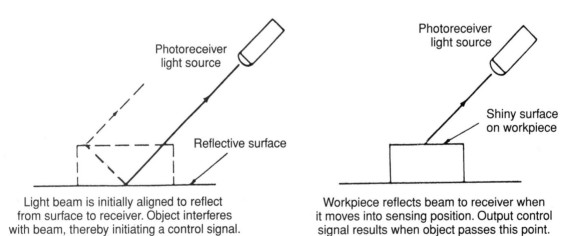

Light beam is initially aligned to reflect from surface to receiver. Object interferes with beam, thereby initiating a control signal.
(c)

Workpiece reflects beam to receiver when it moves into sensing position. Output control signal results when object passes this point.
(d)

Figure 6.6.1 Photoelectric sensors: (a) light source and receiver type; (b) retroflector type; (c) diffuse scanner type; (d) reflect scanner type

When the light path between source and receiver is interrupted, an output in the form of a switch closure is obtained. Output contacts of Forms A, B, and C are widely available with ratings of up to 10 amps. AC devices can switch circuit loads such as relays, solenoids, or small motors directly. DC devices are usually used for logic-level switching; the output switches are solid-state and switch DC currents on the order of 150 ma.

Scanning arrangements can be used with source/receiver separation distances of over 100 feet, although workcell applications rarely call for such a range.

The retroreflector type is shown in Figure 6.6.1(b), where both light source and photoreceiver are packaged in the same housing, known as a scanner. The retroreflective scanner can switch when the light beam from a fixed reflector is interrupted, or from a passing workpiece or machine element. This type of scanner has limited distance between target and light source.

The diffuse scanner type is shown in Figure 6.6.1(c), where both light source and photoreceiver are built into the same housing. A workpiece reflects the beam to the receiver when it moves into the sensing position. An output control signal results when the object passes this point.

The reflect scanner type is shown in Figure 6.6.1(d), where the sensor works like a combination of the retroreflector and diffuse scanner types. The light source and receiver are both located in the same enclosure, and the beam uses the part's surface to reflect light back to the receiver. The light reflected must be a definitive beam because diffused light alone will not be sufficient to activate the receiver. The angle of the light leaving the sensor is adjusted to improve sensitivity and establish a specific distance at detection.

The light source for most industrial photoelectric control systems is the light-emitting diode (LED), which generates light in the infrared (IR) spectrum. Most industrial sensors use silicon photodetectors, which are sensitive to IR light and are economical.

The photoreceiver detects light from the source, amplifies it, and switches its output contacts when a change in nominal conditions occurs. A receiver can be energized on light or energized on dark depending on its intended application.

Miniature photoelectric sources and receivers are available in which the light is transmitted to the source point and received through fiber-optic cables. This allows use of photoelectrics for small parts in very tight machinery spaces.

Photoelectric devices should always be mounted so as to avoid or be protected from dirty environments. The devices should also be selected with excess gain G appropriate to the environment. The excess gain for clean air is 1.0, that is, no special allowance needs to be made. The excess gain requirement increases to 2.0 for low contamination, to 10.0 for moderate contamination, and to 25.0 for high contamination.

Excess gain charts are available in manufacturers' catalogs; these charts plot excess gain versus working distance (between source and receiver). The closer the working distance, the higher the excess gain characteristic of the photoelectric device. Thus, if a source was located in an area of low contamination ($G = 2.0$), and the receiver in an area of moderate contamination ($G = 10.0$), the total excess gain for the source/receiver pair should be 2.0 x 10.0 = 20.0. The designer would then make sure that the working distance was less than that required for an excess gain of 20.0 for the particular devices used.

6.7 ROTARY POSITION SENSORS

In most cases in robots, a primary interest is to control the position of the arm. There is a large variety of devices available for sensing position. However, the most popular angular-position sensors are the following devices:

a. Encoders
b. Synchros
c. Resolvers
d. Potentiometers

Encoders

The three general types of angular-position encoders using digital output are: tachometer, incremental optical, and absolute optical.

The tachometer encoder is a single-output device that generates pulses in response to sensor rotation. This device senses only increments of motion and not direction. Figure 6.7.1 illustrates a rotary pulse generator that outputs a signal every time a magnet passes by. The angular resolution of this device in degrees is $360/n$, where n is the number of magnets. If the device is energized or sampled for a fixed-time interval, the number of output pulses per unit time represents angular velocity.

The incremental optical encoder is illustrated in Figure 6.7.2. It consists of a glass disk marked with alternating transparent and opaque stripes aligned radially. A phototransmitter (a light source) is located on one side of the disk, and a photoreceiver is on the other. As the disk rotates, the light beam is alternately completed and broken. The output from the photoreceiver is a pulse train whose frequency is proportional to the speed of rotation of the disk. In a typical encoder, there are two sets of phototransmitters and receivers aligned 90° out of phase. This phasing provides direction information; that is, if signal A leads signal B by 90°, the encoder disk is rotating in one direction; but if B leads A, then it is going in the other direction. By counting the pulses and by adding or subtracting based on the sign, it is possible to use the encoder to provide position information with respect to a known starting location.

The absolute optical encoder employs the same basic construction as incremental optical encoders except that there are more tracks of stripes and a corresponding number of receivers and transmitters. Usually, the stripes are arranged to provide a binary number proportional to the shaft angle. The first track might have two stripes, the second four, the third eight, and so on. In this way the angle can be read directly from the encoder without any necessary counting. Figure 6.7.3 illustrates an absolute optical encoder. The resolution of an absolute optical

6.7 Rotary Position Sensors

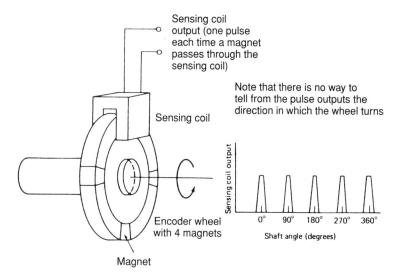

Figure 6.7.1 Magnetic tachometer encoder

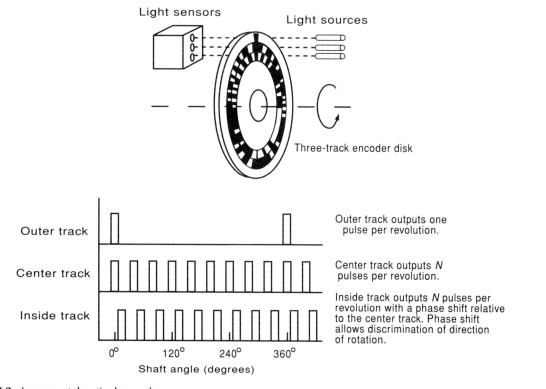

Figure 6.7.2 Incremental optical encoder

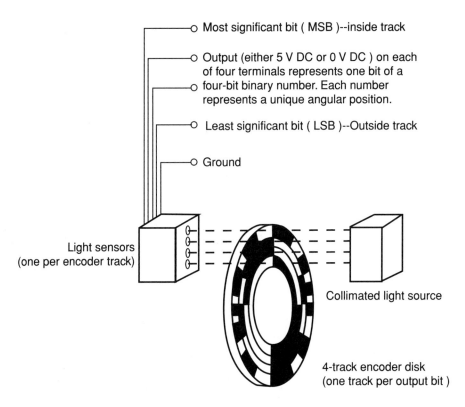

Figure 6.7.3 Absolute optical encoder

encoder is dependent on the number of tracks and is given by 2^n, where n is the number of tracks on the disk.

Example 6.1 What is the resolution, in degrees, of an encoder with 10 tracks?

Solution

The number of increments per revolution is $2^{10} = 1{,}024$ increments/rev. The angular width of each control increment is therefore:

$$360°/\,2^{10} = 360°/\,1{,}024 = 0.3515°$$

The output of an absolute encoder or of an incremental encoder and counter combination is represented by:

$$\text{out}(t) = K_e \theta(t)$$

where *out* is a number,

K_e is the number of pulses per radian, and

θ is the shaft angle, expressed in radians.

Example 6.2 What is the output value of an absolute encoder if the shaft angle is 1 rad and the encoder has 8 tracks?

Solution

The resolution is $2^8 = 256$ parts/rev. There are 2π rad/rev. Therefore, the output is: $256 / 2\pi = 41$.

Synchros

A synchro is a rotating transformer in which a single-pole rotor winding is excited by AC voltage (60 Hz or 400 Hz). AC rotor current induces voltage in three stator coil windings that are wound 120° apart. The synchro is no different in appearance than a small AC motor.

Figure 6.7.4 shows the basic construction of a synchro and its circuit representation. If a voltage excitation of $A \sin wt$ is applied to the synchro rotor, then the output that will appear at the various stator terminals will be:

$$V_{13} = A \sin wt \sin \theta$$

$$V_{32} = A \sin wt \sin (\theta + 120°)$$

$$V_{21} = A \sin wt \sin (\theta + 240°)$$

where θ is the shaft angle of the synchro.

One unique feature of a synchro is that two identical devices may be connected remotely, stator terminals to stator terminals and rotor terminals to rotor terminals to excitation voltage source, to create a "torque chain." As the input shaft is turned on the first synchro (called the torque transmitter), the shaft on the other synchro (called the torque receiver) turns through an equal angle.

Torque synchros are accurate to ±1° and were once widely used to transmit shaft position to remote indicators in aircraft or ship navigation instruments. The amount of torque that can be transmitted is limited by the amount of current that can flow through the synchro windings, and is relatively small.

Control synchros are used in a control chain as are torque synchros, but instead of exciting the output synchro rotor, it is held fixed and the output voltage measured. The output voltage varies from maximum to zero to maximum as the shaft of the input synchro (control transmitter) is rotated through 360°.

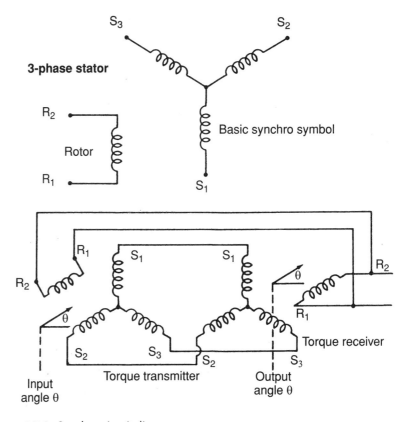

Figure 6.7.4 Synchro circuit diagram

Resolvers

Resolvers are a form of synchro, in which there are only two stator windings, placed 90° mechanically out of phase with each other. If the rotor is excited with a voltage $A \sin wt$, then the outputs on stator terminals will be:

$$V_1 = A \sin wt \sin \theta$$

$$V_2 = A \sin wt \cos \theta$$

where θ is the angle of the rotor with respect to the stator.

This signal may be used directly, or it may be converted into a digital representation using a device known as a "resolver-to-digital" converter. Because a resolver is essentially a rotating transformer, it is important to remember that an AC signal must be used for excitation. If a DC signal were used, there would be no output signal.

Example 6.3 At time t the excitation voltage to a resolver is 24 V. The shaft angle is 90°. What is the output signal from the resolver?

Solution

$$V_1 = (24 \text{ V})(\sin 90°) = 24 \text{ V}$$

$$V_2 = (24 \text{ V})(\cos 90°) = 0 \text{ V}$$

Example 6.4 At time t the excitation voltage to a resolver is 24 V and $V_1 = 17$ and $V_2 = -17$ V. What is the angle?

Solution

$$\arcsin (17/24) = 45° \text{ or } 135°$$

$$\arccos (17/24) = 135° \text{ or } 225°$$

The shaft angle must be 135°.

Potentiometers

Potentiometers are analog devices whose output voltage is proportional to the position of a wiper. Figure 6.7.5 illustrates a typical potentiometer (pot). A voltage is applied across the resistive element. The voltage between the wiper and ground is proportional to the ratio of the resistance on one side of the wiper to the total resistance of the resistive element. Essentially, the pot acts as a voltage divider network. That is, the voltage across the resistive element is divided into two parts by a wiper. Measuring this voltage gives the position of the wiper. The function of the potentiometer can be represented by the following function:

$$V_0(t) = K_p \theta(t) \qquad \text{(Equation 7.6.1)}$$

where $V_0(t)$ is the output voltage,

K_p is the voltage constant of the pot in volts per radian (or volts per inch in the case of a linear pot), and

$\theta(t)$ is the position of the pot in radians (or inches).

Because a pot requires an excitation voltage in order to calculate V_0, we can use:

$$V_0 = V_{ex} (\theta_{act} / \theta_{tot}) \qquad \text{(Equation 7.6.2)}$$

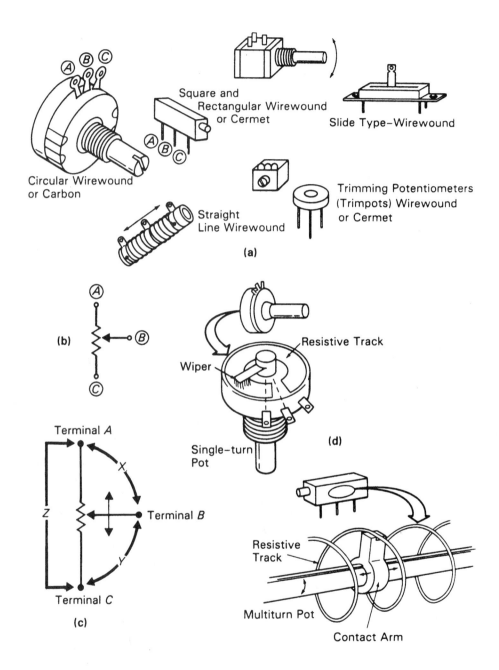

Figure 6.7.5 Potentiometer: (a) Physical units; (b) schematic symbol; (c) operation; (d) construction

where V_{ex} is the excitation voltage,

θ_{tot} is the total travel available of the wiper, and

θ_{act} is the actual position of the wiper.

Example 6.5 Find the output voltage of a potentiometer with the following characteristics. Also determine the K_p. The excitation voltage is 12 V; total wiper travel is 320°; and the wiper position is 64°.

Solution

$K_p = V_{ex} / \theta_{tot}$, which is 12V / 320° = 0.375 V/deg. The output voltage is:

$$(64°)(0.0375 \text{ V / deg}) = 2.4 \text{ V}$$

Table 6.2 summarizes the types of rotary position sensors used in robotics and automation.

6.8 USAGE AND SELECTION OF SENSORS

The major uses of sensors in industrial robots and other automated manufacturing systems can be divided further into four basic categories:

1. Safety monitoring
2. Control interlocking
3. Quality control inspection
4. Positions and related information

Safety monitoring is one of the important applications of sensor technology in automated manufacturing operations that concerns the protection of human workers who work in the vicinity of the robot or other equipment.

Control interlocking in robots is used to coordinate and verify the sequence of activities of the different pieces of equipment in the workcell and to verify before proceeding with the next element of the cycle.

Table 6.2 Rotary position sensors

Device	Output	Accuracy
Multipole resolver	analog (voltage)	7 arc seconds
Absolute optical encoder	digital pulse	23 arc seconds
Standard resolver	analog (voltage)	7 arc minutes
Potentiometer	analog (resistance)	7 arc minutes
Incremental optical encoder	digital pulse	11 arc minutes
Contact encoder	digital pulse	26 arc minutes

Quality control inspection is used to determine a variety of part quality characteristics. This generally involves the checking of parts, assemblies, or products for conformance to certain criteria specified by the design engineering department.

Position and related information sensors are used to determine the positions and other related information about various objects in the robot cell: such as work parts, fixtures, people, equipment, and so on. This kind of data will determine work-part identification, random position and orientation of parts, and accuracy on requirements by feedback supporting data.

The selection of a sensor for a particular application depends on factors such as the quantity to be measured or sensed, the sensor's interaction with other components in the system, expected service life, level of sophistication, difficulties associated with the sensor's use, power source, and cost. Another important consideration is the environment in which the sensors are to be used. Rugged sensors are being developed to withstand extremes of temperature, shock and vibration, humidity, corrosion, dust and various contaminants, fluids, electromagnetic radiation, and other interferences.

6.9 SIGNAL PROCESSING

There are three basic elements that constitute the interaction of a modern robot:

1. Controller—A mix of hardware and software that controls the positioning and motion of the robot, as dictated by the teach pendant or computer
2. Sensors—Monitor the robot's surroundings and behavior and provide the information directly, either to the controller or the computer
3. Decision making—Generally, software that provides the robot with decision-making capability (primarily made by the programmer)

To see these elements in a proper order, we consider the relatively standard matter with which a sensor signal is processed in order to produce a corresponding response.

A schematic processing loop is shown in Figure 6.9.1. Follow the signal as it proceeds around the loop in terms of its transition from one numbered station to the next:

$1 \to 2$ The force sensor has been calibrated so that a force F_0 corresponds to an output voltage V_0. Because this signal is generally rather weak, it needs to be amplified.

$2 \to 3$ The amplified signal voltage is then fed into an analog-to-digital (A/D) converter, replacing the analog (continuous) signal with discrete signal levels. Generally, such analog-to-digital converters may include an encoder, which assigns a binary code corresponding to each signal level.

$3 \to 4$ This binary signal now moves to a controller, a piece of hardware that basically keeps track of where a signal comes from and where it

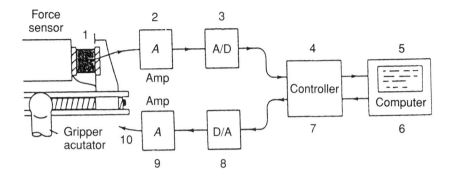

Figure 6.9.1 Sensor-to-actuator signal processing

is going. Here, the controller will note that it received a signal of a certain level, say from force sensor 1 on the gripper.

4 → 5   The controller will then pass this binary signal to the computer, with an identifying code indicating that the signal comes from force sensor 1. The computer, in turn, through its software, may now display a statement of the form "force sensor 1 indicates a force F_0." That is, it recognizes where the signaled force level exists and displays this to an operator or operating program.

5 → 6   The computer now processes this information by comparing F_0 to some required force F_1; for example, with $F_0 < F_1$ implying an order to continue closing the gripper.

6 → 7   Hence the computer now sends a signal to the controller, indicating a certain coded binary voltage level for the gripper motor and an identifying code for the motor.

7 → 8   The controller again identifies where the signal is to go and sends out a binary signal to the motor. If the motor runs on an analog signal, then there is a decoder and a digital-to-analog (D/A) converter, which eventually provides the specified analog voltage level.

8 → 9   Because computer and controller voltages may run between 0 V and 5 V, this may not be enough to drive the motor, so that amplification is again necessary.

9 → 10   The motor now turns the amount indicated by the voltage level. If it is a stepper motor, the signal voltage may simply be enough to have it step through a specified number of segments, or we might have some other device indicating the amount of rotation of the motor. Finally, of course, in the absence of these, the preceding cycle may be repeated as often as needed to reach the required force level $F_0 =$

F_1, at which time the motor is provided with just enough voltage (torque) to maintain this level.

This type of cycle is repeated for every sensor that the robot has, all of the information being sent in time with the pulses of some universal clock. In each case, the incoming signal must be interpreted, a response prescribed, and a signal fed back, indicating that the desired response has been carried out. Computational power clearly is of the essence in this context, for microprocessor to mainframe. The level of intelligence in connection with the resolution of the sensors determines the sophistication of the robotic response.

6.10 SENSORS AND CONTROL INTEGRATION

Because a variety of sensors may be required for an intelligent robot, the problem of integrating the sensory information with the stored information to develop a control strategy is also important. In some cases, a single computer may be powerful enough to control the robot. In more-complex systems, a hierarchical distributed computer can be used by a robot or flexible manufacturing system. An executive controller may be used to implement the overall strategy. It communicates to a series of dedicated processors that control the robot functions and receive input from the sensor systems. Sublevels in the hierarchy may also be used for related tasks. A large central microprocessor with high-level language capabilities, connected to smaller microprocessors on a common bus, provides an implementation method for hierarchical control. The software strategy may then be contained in the master controller, and high-speed actions are controlled by the distributed microprocessors.

The distributed sensor and control system is similar in many ways to the central nervous system in the human. Many actions are controlled by neural networks in the spinal cord without conscious control. These local reflexes and autonomous functions are vital to human survival and are important in discovering how they may be simulated in a robot. The study of these mechanisms in robots may ultimately lead to a greater understanding of how we function as humans.

From the preceding, the seven desirable features of sensors can be identified as the following:

1. Accuracy—means no systematic positive or negative errors in the measurement.
2. Precision—means no random variability in the measurement.
3. Operating range—means that the entire **operating range** is accurate and precise.
4. Speed of response—means that it should be capable of responding in minimum time.
5. Calibration—means that it should be easy to calibrate.
6. Reliability—means that it should possess a high reliability without failures.

7. Cost and ease of operation—means that the costs to purchase, install, and operate should be as low as possible.

6.11 SUMMARY

In this chapter, we have presented the topic of non-vision-based robot sensors. A sensor is a device that detects information about the robot and its surroundings and transmits it to a computer or to its controller for further actions.

Sensors are classified according to:

1. Measuring quantity values—such as mechanical, electrical, magnetic, thermal, and other sensors
2. Function performed—such as manipulation and acquisition sensors
3. Location and type of detection—such as internal, external, and interlock sensors
4. Physical activation—such as contact and noncontact sensors

The four major uses of sensors in robots and automated systems are:

1. Safety monitoring, which concerns the protection of human workers
2. Control interlocking, which coordinates and verifies the sequence of activities
3. Quality control inspection, which conforms with the criteria of design
4. Positions and related information, which determines the accuracy and reliability of parts

All non-vision sensors are grouped into a contact or noncontact category, as shown in Figure 6.2.6.

Contact sensors include microswitches and all tactile-sensing devices. Noncontact sensors include proximity sensors, photoelectric, and rotary position encoders.

Microswitches are electromechanical devices activated either by some part or motion of a robot or machine to alter the electrical circuit. Because of the widespread application of these switches and their configuration, a general discussion of contact arrangements and switching loads is included in this chapter.

Proximity sensors can be applied in robots and automatic machinery in the same manner as limit switches, but without physical contact of the part. The device detects the presence of a part within a specified range of the device.

The sensors are available in a variety of types and sizes. The most common types of proximity sensors are:

    a. Inductive (conductive targets)
    b. Capacitive (nonconductive targets)
    c. Magnetic
    d. Hall effect

e. Ultrasonic
f. Ring-type

Output circuits DC and AC, in normally open and normally closed configurations, are available.

Proximity probes of all types are usually sealed, cylindrical, or square devices with leads protruding from the nonactive end. Acoustic devices can also be used as proximity sensors.

Ultrasonic frequencies (above 20,000 Hz) are often used in such devices because the sound is beyond the range of human hearing. This kind of device can also be used as a range sensor.

Photoelectric sensors detect the presence of a part when a transmitted light or reflected beam is broken to a receiver. The four types of photoelectric sensors commonly used in the industry are:

1. Light source and receiver
2. Retroreflector
3. Diffuse scanner
4. Reflect scanner

The operational characteristics and the excess gain charts for the four types of photoelectric devices are available from manufacturers' catalogs. The closer the working distance, the higher the excess gain characteristic. Therefore, the primary difference of the four types is range or distance to the detected part. A variety of packages, shapes, and output configurations are available. Rotary position encoders are devices used to convert position data to electrical signals. As more systems become controlled by computers and related devices, the use of digital position encoders increases. Encoders are available as three basic types: tachometer, incremental, and absolute. The last two types are optical encoders. Tachometers are generally used to provide velocity information to the controller. This can be used for performing velocity control of a device or, in many cases, to improve the stability of the workcell system and its response to disturbances. Potentiometers, synchros, and resolvers are analog angular-position sensors with an output signal similar to encoder outputs. They are capable of turning at relatively high speed for use in servo positioning systems.

6.12 REVIEW QUESTIONS

6.1 What is the basic function of sensors in a workcell?
6.2 What are the five basic reasons sensors are used in a workcell?
6.3 How are sensors classified for an intelligent robot?
6.4 What purpose do acquisition and manipulation sensors serve in a robot?
6.5 What purpose do internal, external, and interlock sensors serve in a robot?
6.6 What is a haptic perception sensor and how does it work?
6.7 What purpose do contact and nonconduct sensors serve in a robot?
6.8 What is the function of a microswitch?

6.9 Name the most-common types of microswitches.
6.10 What are solid-state switches and where are they used?
6.11 Describe the six different types of proximity sensors.
6.12 What is an acoustic and ultrasonic sensor?
6.13 Describe the four different types of photoelectric sensors.
6.14 What is the function of a rotary position encoder?
6.15 What are the three basic types of encoders?
6.16 What is the distinction between tactile sensing and simple touch sensing?
6.17 Describe how tactile sensors operate.
6.18 What are the three basic elements that constitute the interactions of a modern robot?
6.19 What are the four major uses of sensors in industrial robots and automated manufacturing systems?

6.13 PROBLEMS

6.1 A potentiometer is to be used as the feedback device to indicate the position of the output link of a rotational robot joint. The excitation voltage of the potentiometer is 5.0 V, and the total wiper travel of the potentiometer is 300°. The wiper arm is directly connected to the rotational joint. Determine:
   a. The voltage constant of the potentiometer, K_p
   b. The resulting output voltage of the potentiometer if the wiper position is 38°
   c. The corresponding angular position of the wiper and the output link if the output voltage of the potentiometer is 3.75 V

6.2 A resolver is used to indicate angular position of a rotational wrist joint. The excitation voltage to the resolver is 24 V. The resolver is directly connected to the wrist joint. At a certain moment, the two pairs of stator terminals have V_{S1} of 10.0 V and V_{S2} of 21.82. Determine the angle of the rotational joint.

6.3 At time t the excitation voltage to a synchro is 36 V, and the shaft angle is 60°. What is the output signal from the synchro at the three stator terminals?

6.4 What is the resolution of an absolute optical encoder that has six tracks? nine tracks? twelve tracks?

6.5 For an absolute optical encoder with ten tracks, determine:
   a. The value of the encoder constant K_e
   b. Its output value if the shaft angle of the encoder were 0.73 rad

6.6 A DC tachometer is to be used as the velocity feedback device on a twisting joint. The joint actuator is capable at a maximum velocity of 0.75 rad/s, and the tachometer constant is 8.0 V/rad/s. Determine:
   a. The maximum output voltage of the tachometer if it rotates with twice the angular velocity of the joint
   b. The output voltage of the DC tachometer if the joint rotates at a speed of 25°/s

6.7 A linear array of light sensors is to be used to determine the distance x in the setup illustrated in Figure 6.13.1. If the surface of the object is parallel

190 Sensors

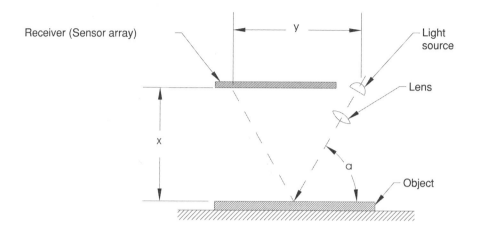

Figure 6.13.1 Scheme of linear sensor array for Problem 6.7

to the sensing array, the angle a of the light source is 60°, and the reflector position of the object reaches $y = 100$ mm, determine the value of x. [**Hint:** $x = 0.5\ y\ \tan(a)$]

6.8 With reference to Problem 6.7 above (Figure 6.13.1), discuss what will be the effect on the accuracy of the measurement if the surface of the object is not parallel to the linear sensor array.

6.9 Draw a classification scheme that includes all the sensors described in this chapter.

6.10 Design a decision chart to select the best sensor for a given manufacturing application.

6.14 REFERENCES

Annaswamy, A. M., and M. A. Srinivasan. "The Role of Compliant Fingerpads in Grasping and Manipulation: Identification and Control." The IMA Volumes in Mathematics and its Applications, *Volume 104: Essays in Mathematical Robotics*. J. Bailleul, S. Sastry, and H. J. Sussman, Eds. New York: Springer-Verlag, 1998

Beauregard, G. L., and M. A. Srinivasan. "Sensorimotor Interactions in the Haptic Perception of Virtual Objects." The Engineering Foundation Conference on Biomechanics and Neural Control of Movement, Mount Ceferling, OH, June 1996.

Dandekar, K., and M. A. Srinivasan. *Role of Mechanics in Tactile Sensing of Shape*. RLE TR-604, MIT, 1996.

Diesing, H., and K. V. Schmucker. "Nonvisual Sensors for Industrial Robots." Dissertation submitted to the Academy of Science of the German Democratic Institute of Automation, Berlin, 1990.

Gulati, R. J., and M. A. Srinivasan. *Determination of Mechanical Properties of the Human Fingerpad in Vivo, Using a Tactile Stimulator.* RLE TR-605, MIT, 1996.

Karason, S. P., and Srinivasan, M. A. "Passive Human Grasp Control of an Active Instrumented Object." Proceedings of the ASME Dynamic Systems and Control Division, DSC-Vol. 57-1, pp. 641–47, ASME, November 1995.

Lee, M. H. *Intelligent Robotics.* New York: Halsted Press, 1989.

Lewis, F. L., C. T. Abdallah, and D. M. Dawson. *Control of Robot Manipulators.* New York: Macmillan Publishing Co., 1993.

Russell, R. A. *Robot Tactile Sensing.* Englewood Cliffs, NJ: Prentice Hall, 1990.

Srinivasan, M. A., and Dandekar, K. "An Investigation of the Mechanics of Tactile Sense Using Two Dimensional Models of the Primate Fingertip." *Journal of Biomechanical Engineering* 118 (1996): 48–55.

CHAPTER 7

Vision

7.0 OBJECTIVES

After studying this chapter, the reader should:

1. Be acquainted with the basic systems of Visual Sensing
2. Understand the operation and the five functions of machine vision
3. Recognize machine vision applications
4. Be familiar with other optical methods

7.1 VISUAL SENSING

Visual sensing (machine vision or computer vision) is the capability to "see" conferred on a robot by a vision system. A **vision system** is a device that can collect data and form an image that can be interpreted by a robot computer to determine an appropriate position or to "see" an object.

Therefore, machine vision can be defined as the acquisition of image data, followed by the processing, analysis, and interpretation of these data by machine or computer for some useful application. Machine vision is a rapidly growing technology, with its principal applications in health care (diagnosis of a disease, using computed tomography images), industry, agriculture, and other fields. Machine vision systems have been used for quality control of products ranging from miniature parts to turbine blades, and from submicron structures on wafers to auto-body panels.

In this chapter, we examine how machine vision works and its industrial applications in quality-control inspection and other areas.

There are two basic systems of machine vision: linear array and matrix arrays.

Linear array is sensing only one dimension, such as the presence of an object or some feature on its surface.

Matrix arrays are classified as either two-dimensional or three-dimensional. Two-dimensional systems view the scene as a two-dimensional image, which is quite adequate for applications involving objects lying in one plane. Examples include dimensional measuring and gaging, verifying the presence of components, and checking features on a flat surface.

Three-dimensional systems are used for applications requiring a three-dimensional analysis of the scene, where contours or shapes are involved—for example, capable of detecting a properly inserted printed circuit or a properly made solder joint. When used in automated inspection systems, they can also detect cracks and flaws. The majority of current industrial applications are two-dimensional systems, and therefore the following discussion focuses mostly on this type of technology. Several applications of machine vision in manufacturing are illustrated in Figure 7.1.1.

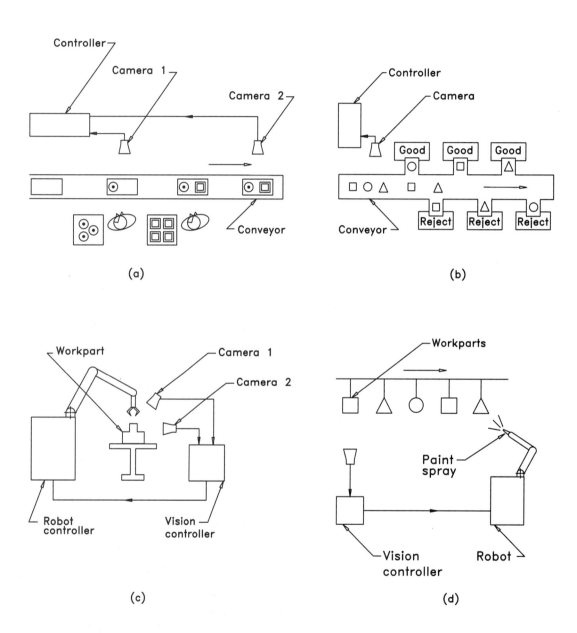

Figure 7.1.1 Examples of machine-vision application: (a) On-line inspection of parts; (b) Identification of parts with various shapes; (c) Use of cameras to provide positional input to a robot relative to the workpiece; (d) With input of a camera provides printing to parts of different shapes.

7.2 MACHINE VISION

The goal of a **machine vision** system is to create a model of the real world from images. A machine vision system recovers useful information about a scene from its two-dimensional projections. Because images are two-dimensional projections of the three-dimensional world, the information is not directly available and must be recovered. The recovery requires the inversion of a many-to-one mapping, knowledge about the objects in the scene, and their projection geometry. This knowledge can be obtained by image interpretation of the machine-vision system.

The operation of a machine-vision system can be divided into the following five steps:

1. Image acquisition
2. Image digitization
3. Image processing
4. Image analysis
5. Image interpretation

The function of the five steps and their relationship to each other are illustrated schematically in Figure 7.2.1.

Image Acquisition

It is important to understand the relationship between the geometry of **image** formation and its representation in the **computer**. There is a bridge from the mathematical notation used to develop machine vision **algorithms** to the algorithmic notation used in programs.

A **pixel** is a sample of the image intensity quantized to an integer value, as illustrated in Figures 7.2.2 and 7.2.3.

An image is a two-dimensional array of pixels. The row and column indices (i, j) of a pixel are integer values that specify the row and column in the **array** of pixel values. Pixel $(0,0)$ is located at the top left corner of the image. The index i points down, and j points to the right. This index notation corresponds closely to the array **syntax** used in computer programs. The positions of points in the image plane have x and y coordinates. The y coordinate corresponds to the vertical direction, and the x coordinate corresponds to the horizontal direction. The y axis points up, and the x axis points to the right. The directions corresponding to the two indices i and j in the pixel index (i, j) are the reverse of the directions corresponding to the respective coordinates in the position (x, y).

The x and y coordinates are real numbers stored as floating-point numbers in the computer. Image plane coordinates (x, y) can be computed from pixel coordinates (i, j) of an $m \cdot n$ pixel array as follows:

$$x = j - (m - 1) / 2 \qquad \text{(Equation 7.2.1)}$$

$$y = - (i - (n - 1) / 2) \qquad \text{(Equation 7.2.2)}$$

196 Vision

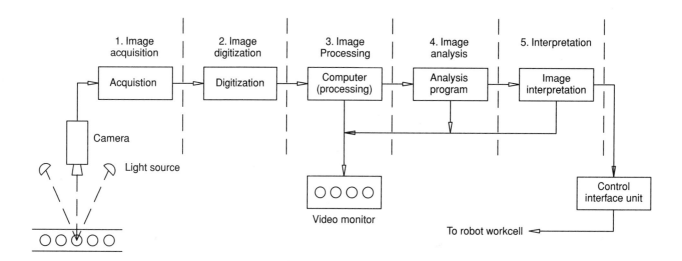

Figure 7.2.1 The five basic functions of a machine vision system

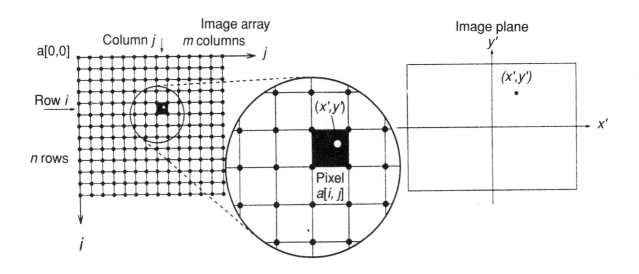

Figure 7.2.2 Relationship between image plane candidates and image array indices

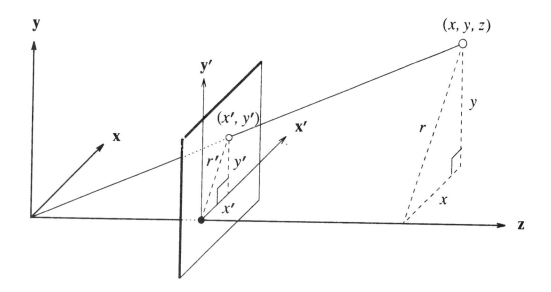

Figure 7.2.3 Illustration shows the line of sight that is used to calculate the projected point from the object point.

The origin of the image plane coordinate system corresponds to the center of the image array. In an imaging system, each pixel occupies some finite area on the image plane.

Machine vision algorithms that depend on the exact shape of the pixel footprint will not be covered in this text, so we may assume for concreteness that the pixels partition the image plane into equal-sized squares. Positions in the image plane can be represented to fractions of a pixel. The coordinates (x_{ij}, y_{ij}) of the pixel with indices (I, j) are the location of the center of the pixel in the coordinate system. Because we are concerned only with the location of the center at which the image sample was taken, the pixel may be further abstracted to a point in the image plane. The array of pixels in the computer program corresponds to the grid of image plane locations at which the samples were obtained.

Diagrammatically, the images can be shown as a square tessellation of the rectangular region, with each square shaded to indicate the image intensity of that pixel. The image can be modeled as a square grid of samples of the image intensity, represented in the computer as an array of pixel values. The camera and **digitizing electronics** are designed to ensure that this assumption is satisfied. Some variations, such as different spacing between the rows and columns in the grid, distortions due to lens imperfections, and errors in the construction of the camera, can

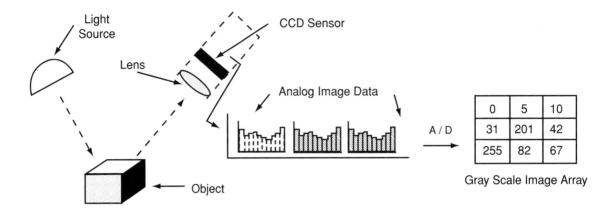

Figure 7.2.4 The image acquisition process

be removed through calibration without changing the algorithms that process the image. A schematic of the process leading from the light source to an algebraic image array is shown in Figure 7.2.4.

In summary, a pixel is both a gray value, which is a quantized sample of the continuous image intensity, and an image location specified by the row and column indices in the image array. The image array is obtained by sampling the image intensity at points on a rectangular grid. Points in the image plane, specified with coordinates x and y, may lie between the grid locations at which pixels were sampled.

Image Digitization

Image digitization is accomplished by using a video camera and a digitizing system to store the image data for subsequent analysis. The camera is focused on the subject of interest and an image is obtained (as previously explained) by dividing the viewing area into a matrix of discrete picture elements (pixels), in which each element has a value that is proportional to the light intensity of that portion of the scene. The intensity value from each pixel is converted into its equivalent digital value by an analog-to-digital converter (ADC). The operation of viewing a scene consisting of a simple object that divides the scene into a matrix of picture elements and intensity values corresponding to the image is shown in Figure 7.2.5.

The figure illustrates the probable image obtained from the simplest type of vision system, a binary vision system. In binary vision, the light intensity of each pixel is ultimately reduced to either of two values, white or black, depending on whether the intensity exceeds a given threshold level. A more sophisticated vision

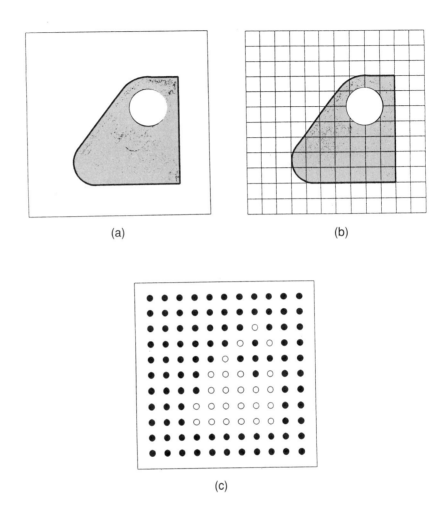

Figure 7.2.5 Illustration shows the image divided into a matrix of picture elements: (a) the scene; (b) matrix of picture elements superimposed on the scene; (c) picture element intensity values for the scene.

system is capable of distinguishing and storing different shades of gray in the image. This is called a gray-scale system. This type of system can determine not only an object's outline and area characteristics, but also its surface characteristics, such as texture. Gray-scale vision systems typically use 4, 6, or 8 bits of memory. Other vision systems can also recognize color. Each bit corresponds to 2^8 or 256 intensity levels, which is generally more levels than the video camera can really distinguish, and more than the human eye can discern.

Each set of pixel values is called a frame. Each frame, consisting of the set of digitized pixel values, is stored in a computer memory device called a frame buffer. The process of reading all the pixel values in a frame is performed with a frequency of thirty times per second (twenty-five times per second for European vision systems).

Resolution of a vision system is its ability to sense fine details and features in the image. This depends on the number of pixels used. Common pixel arrays include 256×256 and 512×512 picture elements. The more pixels in the vision system, the higher its resolution. However, system cost increases as pixel count increases. Also, the time required to read the picture elements and process the data increases with the number of pixels.

Two types of cameras are used in machine vision applications: vidicon cameras (similar to television) and solid-state cameras.

Vidicon cameras operate by focusing the image onto a photoconductive surface and scanning the surface with an electron beam to obtain the relative pixel values. Different areas on the photoconductive surface have different voltage levels corresponding to the light intensities striking the areas. The electron beam follows a well-defined scanning pattern, in effect dividing the surface into a large number of horizontal lines and reading the lines from top to bottom. Each line is in turn divided into a series of points. The number of points on each line, multiplied by the number of lines, gives the dimensions of the pixel matrix, as shown in Figure 7.2.5(b). During the scanning process, the electron beam reads the voltage level of each pixel.

Solid-state cameras operate by focusing the image onto a two-dimensional array of very small, finely spaced photosensitive elements. The photosensitive elements form the matrix of pixels shown in Figure 7.2.5(c). An electrical charge is generated by each element according to the intensity of light striking the element. The charge is accumulated in a storage device consisting of an array of storage elements corresponding one-to-one with the photosensitive picture elements. These charge values are read sequentially in the data-processing and analysis function of machine vision. The solid-state camera has several advantages over the vidicon camera in industrial applications. It is physically smaller and more rugged, and the image produced is more stable.

The vidicon camera suffers from distortion that occurs in the image of a fast-moving object because of the time lapse associated with the scanning electron beam as it reads the pixel levels on the photoconductive surface. The relative advantages of solid-state cameras have resulted in their growing dominance in machine vision systems. The most popular types of solid-state cameras used are charge-coupled-devices (CCD), charge-injection-devices (CID), and photodiode arrays.

Three-dimensional vision can be achieved by using two cameras for moving and fast-response vision. A single camera can produce the effect of three-dimensional vision if it is shifted between taking pictures. Because the camera must move, however, this is a slower method of achieving three-dimensional vision.

Infrared and ultraviolet cameras work similarly to a normal television camera, except that they are sensitive to different frequencies of light.

No matter what type of vision sensor a machine uses, the resolution of the vision system—its ability to sense fine details and features in the image—is dependent on the number of picture elements used. The more pixels designed into the vision system, the higher its resolution. However, the cost of the camera increases as the number of pixels is increased. The following example illustrates this principle.

Example 7.1 A certain video camera has a 256×256–pixel matrix. Each pixel must be converted from an analog signal to the corresponding digital signal by an A/D. The A/D conversion process takes 0.1 µs (0.1×10^{-6} s) to complete. Assuming that there is no lost time in moving between pixels, how long will it take to collect the image data for one frame, and is this time compatible with processing at the rate of 30 frames/s?

Solution:

There are 65,536 (256×256) pixels to be scanned and converted. The total time to complete the analog-to-digital conversion process is:

$$(65{,}536 \text{ pixels})(0.1 \times 10^{-6} \text{ s}) = 0.00655 \text{ s}$$

At a rate of 30 frames/s, the processing time for each frame is $1/30 = 0.0333$ s. This is significantly larger than the 0.00655 s required to perform analog-to-digital conversion. There would be a certain time loss between pixels, but this would be relatively small in comparison to the A/D process.

Another important aspect of machine vision is lighting. The scene viewed by the vision camera must be well illuminated, and the illumination must be constant over time. There are numerous ways to illuminate the scene. One way to classify the possibilities is according to the location of the light source: front lighting, back lighting, and side lighting. The three light source locations are shown in Figure 7.2.6.

In front lighting, the light source is located on the same side of the object as the camera. This produces a reflected light from the object that allows inspection of surface features. In back lighting, the light source is placed behind the object being viewed by the camera. This creates a dark silhouette of the object, which contrasts sharply with the light background, to inspect part dimensions and distinguish outlines. Side lighting causes irregularities in an otherwise smooth plane surface to cast shadows that can be identified by the vision system. This can inspect defects and flaws in the surface of an object.

The basic types of lighting devices used in machine vision may be grouped into the following categories:

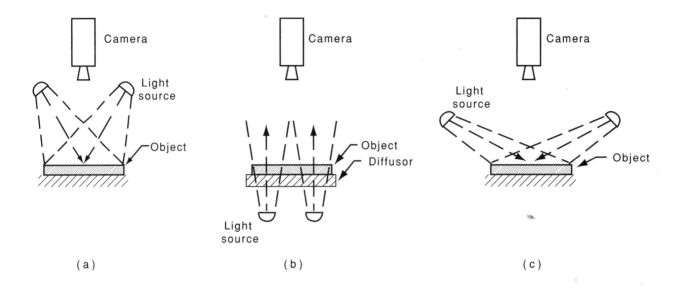

Figure 7.2.6 Lighting illumination: (a) front lighting; (b) back lighting; (c) side lighting.

a. Diffuse surface devices—for example, fluorescent lamps and light tables.
b. Condenser projectors—transform an expanding light source into a condensing light source.
c. Flood or spot projectors—floodlights and spotlights are used to illuminate surface areas.
d. Collimators—provide a parallel beam of light on the subject.
e. Images—for example, slide projectors and optical enlargers, which form an image of the target at the object plane.

Various illumination techniques have been developed to use these lighting devices. Many of these techniques are presented in Table 7.1. The purpose of these techniques is to direct the path of light from the lighting device to the camera so as to display the subject in a suitable manner to the camera.

As shown in Table 7.1, there are two basic illumination techniques used in machine vision: front lighting and back lighting. Front lighting simply means that the light source is on the same side of the scene as the camera. In back lighting, the light source is directed at the camera and is located behind the objects of interest. Back lighting is suitable for applications in which a silhouette of the object is sufficient for recognition or where there is a need to obtain relevant measurements.

Table 7.1 Illumination Techniques

Technique	Function
Front light source	
1. Front illumination	Area flooded such that surface is defining feature of image.
2. Specular illumination (dark field)	Used for surface defect recognition (dark background)
3. Specular illumination (light field)	Used for surface recognition; camera in-line with reflected rays (light background)
4. Front imager	Structured light applications; imaged light superimposed on object surface—light beam displaced as function of thickness.
Back light source	
1. Rear illumination (light field)	Uses surface diffusor to silhouette features; used in parts inspection and basic measurements.
2. Rear illumination (condenser)	Produces high-contrast images; useful for high magnification application.
3. Rear illumination (collimator)	Produces parallel light ray source such that features of object do not lie in same plane.
4. Rear offset illumination	Useful to produce feature highlights when feature is in transparent medium.

Image Processing

The third function in the operation of a machine vision system is **image processing**. As indicated by the preceding example, the amount of data that must be processed is significant. A microprocessor processes the image, usually in less than one second. The image is measured, and the measurements are digitized (image recognition), as was explained in the previous section, by an analog-to-digital converter (A/D). The digital signal represents light intensity values over the entire image. These values are stored in memory, allowing the digital image to be analyzed and interpreted.

Image processing is a well-developed field. Processing techniques usually transform images into other images—such as image enhancement, image compression, and correcting blurred or out-of-focus images. On the other hand, machine vision algorithms take images as inputs but produce other types of outputs, such as representations for the object contours in an image. Image processing algorithms are useful in this stage to enhance particular information and suppress noise.

There are two parts to the image formation process:

1. The geometry of image formation, which determines where in the image plane the projection of a point in the scene will be located
2. The physics of light, which determines the brightness of a point in the image plane as a function of scene illumination and surface properties

Computers also generate images from geometry primitives such as lines, circles, and free-form surfaces that play a significant role in visualization and virtual reality.

Pattern recognition classifies numerical and symbolic data. Many statistical and syntactical techniques have been developed for classification of patterns. Techniques from pattern recognition play an important role in machine vision for recognizing objects.

Image analysis

Image analysis involves gathering the information from image processing and analyzing its data within the time required to complete the conversion. A number of techniques have been developed for analyzing the image data in a machine vision system. One category of techniques is called segmentation. Segmentation techniques are intended to define and separate regions of interest within the image. Two of the common segmentation techniques are thresholding and edge detection.

Thresholding involves the conversion of each pixel intensity level into a binary value, representing either white or black. This is done by comparing the intensity value of each pixel with a defined threshold value. If the pixel value is greater than the threshold, it is given the binary bit value of white, say 1. If less than the defined threshold, it's given the bit value of black, say 0. Reducing the image to binary form by means of thresholding usually simplifies the subsequent problem of defining and identifying objects in the image.

Edge detection is concerned with determining the location of boundaries between an object and its surroundings in an image. This is accomplished by identifying the contrast in light intensity that exists between adjacent pixels at the borders of the object, as explained in the preceding section.

Another set of techniques in image processing that normally follows segmentation is that of feature extraction. Most machine vision systems define an object in the image by means of the object's features. Some features of an object include the area of the object; length, width, or diameter of the object; perimeter, center of gravity, and aspect ratio. Feature extraction methods are designed to determine these features based on the object's area and its boundaries (using thresholding, edge detection, and other segmentation techniques). For example, the area of the object can be determined by counting the number of white (or black) pixels that make up the object. The object's length can be found by measuring the distance (in terms of pixels) between the two extreme opposite edges of the part.

Image interpretation

For any given application, an interpretation of the image must be accomplished based on the extracted features. The **image interpretation** function is usually concerned with recognizing the object, a task called object or pattern recognition. The objective in this task is to identify the object in the image by comparing it to predefined models or standard values. Two commonly used interpretation techniques are template matching and feature weighing.

Template matching is the name given to various methods that attempt to compare one or more features of an image with the corresponding features of a model or template stored in computer memory. The most basic template-matching technique is one that compares the image, pixel by pixel, with a corresponding computer model. Within certain statistical tolerances, the computer determines whether or not the image matches the template. One of the technical difficulties of this method is the problem of aligning the part in the same position in front of the camera, to allow the comparison to be made without complications.

Feature weighing is a technique in which several features are combined into a single measure by assigning a weight to each feature according to its relative importance in identifying the object. The score of the object in the image is compared with the score of an ideal object residing in computer memory to achieve proper identification.

Another future technique for interpretation of images will be artificial intelligence. Artificial intelligence is concerned with designing systems that are comparable with studying computational aspects of intelligence. Artificial intelligence is used to analyze scenes by computing a symbolic representation of the scene contents after the images have been processed to obtain features. Artificial intelligence may be viewed as having three stages: perception, cognition, and action. Perception translates signals from the world into symbols, cognition manipulates symbols into signals, and action translates signals into images that effect changes in the world. Many techniques from artificial intelligence play important roles in all aspects of computer vision. In fact, computer vision is often considered a subfield of artificial intelligence.

Design and analysis of neural networks has become a very active field in the last decade, as explained by Kosko (1992). Neural networks are being increasingly applied to solve some machine vision problems. However, because this field is in its infancy, there are no established practical techniques for machine vision yet.

Commercial Robot Vision Systems

Three different types of vision systems for robot control and inspection are available commercially:

1. The type that deals with two-dimensional binary images, or silhouette imagery

2. The type that uses gray-scale imagery, from which more discrimination is possible but for which longer computation times may be required
3. The type that uses some sort of structured light and stereo triangulation to determine the three-dimensional surface of objects.

Each approach requires a lighting system for illuminating the objects, a camera for gathering the image, and a storage device for recording the image during processing.

7.3 MACHINE VISION APPLICATIONS

The reason for interpreting the vision system is to achieve some practical objective in robot application. Machine vision systems are being used increasingly in manufacturing and robot automation to perform various tasks and can be divided into six distinct categories:

1. Inspection
2. Part identification
3. Part orientation
4. Part location
5. Visual guidance and control
6. Safety monitoring

Inspection is the biggest category for quality control of parts. According to M. P. Groover (1996) it is estimated that inspection constitutes as much as 90 percent of machine vision.

Machine vision installations in industry perform a variety of automated inspection tasks, most of which are either on-line/in-process or on-line/postprocess. The applications are almost always in mass production, where the time required to program and set up the vision system can be spread over many thousands of units. One hundred percent inspection is done almost exclusively. Typical industrial inspection tasks include the following:

- Dimensional measurement or gauging is used for accuracy and geometrical integrity. The parts are measured or gauged by the camera, and the dimensions are calculated and compared with a computer-stored model to determine the size values.
- Verification of product is used in flexible automated assembly systems to verify the presence of components in an assembled product.
- Verification of holes is checked by the vision system for missing holes, location, and number of holes in a part.
- Identification of flaws, defects, or printed label is checked on the surface of a part, which often reveals a change in reflecting light. The vision system can identify the deviation from an ideal model of the surface.

In addition to optical methods, various nonoptical techniques are used in inspection. These include sensor techniques, based on electrical fields, radiation, and ultrasonics.

Part identification is used commercially in many applications where the vision system stores data for different parts in active memory and uses it to recognize or distinguish between parts so that some action can be taken as they enter the workcell. The applications include counting, part sorting, and character recognition.

Part orientation is used in parts to be gripped in a specified manner by the end effector. The vision system supplies the information and data to drive the gripper into the correct orientation for pick-or-place action.

Part location is used to locate randomly placed parts on x, y axes. The vision system measures the x and y distances and identifies the center of the camera to coincide with the center of the randomly placed part.

Visual guidance and control involves applications in which a vision system is teamed with a robot or other devices in the robot cell to control the movement of a machine. Examples of these applications include seam tracking in continuous arc welding, part positioning and/or reorientation, and bin picking.

Safety monitoring involves applications of machine vision in which the vision system is used to monitor the operation of a production cell. Its purpose is to detect irregularities that indicate a condition that is hazardous to equipment or people working in the cell. It can also be used to detect human intruders who might be at risk by wandering into the robot cell.

7.4 OTHER OPTICAL METHODS

Machine vision is a rapidly growing technology, perhaps because it is similar to one of the most important human senses. Its potential for applications in industry is substantial. However, there are other optical sensing techniques that can provide equal interest.

Scanning Laser Systems

Lasers have been used in a number of industrial processing and measuring applications. High-power **laser** beams are used for heat treatment, welding, cutting, hole piercing, scribing, and marking applications. Low-power lasers are utilized in various measuring, gauging, and inspection situations.

The scanning laser device falls into the low-power category. Industrial systems that use low-power lasers generally consist of four basic parts: (a) the laser light source; (b) an optical system to direct and structure the light; (c) an optical system to collect the image or the light after it has interacted with the object; and (d) a detection system. The detection system may be a human observer or an optoelectronic device interfaced with an electronic display unit or computer.

Devices used to **scan** laser beams include mirrors and prisms. These devices are generally oscillated sinusoidally or rotated with constant angular velocity. A

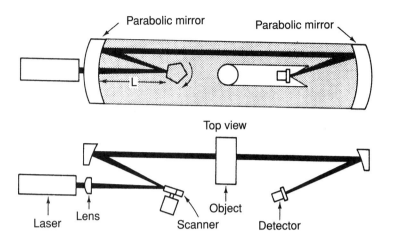

Figure 7.4.1 Simplified diagram of laser system designed for gauging or inspection.

simplified diagram of a laser system designed for gauging or inspection is shown in Figure 7.4.1.

This **system** uses an off-axis parabolic mirror to eliminate spherical aberration. A converging lens is used to focus the laser beam on a multisided mirror scanner placed at the focal point of the parabolic mirror. The beam is collimated after reflection from the parabolic mirror. By using an off-axis parabolic mirror, the rotating mirror can be placed so that it will not interfere with the reflected beam. The detector consists of a photodiode placed at the focal point of a second off-axis parabolic mirror. As the laser beam scans across an object placed between the scanner and detector, the beam is blocked and interrupted by the object.

This period can be timed with great accuracy and related to the size of the object in the path of the laser beam. The scanning laser beam device can complete its gauging or inspection in a very short time cycle. A microprocessor-based system counts the time interruption of the scanning laser beam as its sweeps past the object, makes the conversion from time to a linear dimension, and signals other equipment to make adjustments in the manufacturing process and/or activates a sortation device on the production line. Applications of the scanning laser technique include alignment, surface inspection, and some machining and grinding operations (see Figure 7.4.2 and Figure 7.4.3).

Diode-Array Camera Systems

Many optical inspection devices found in industry have a camera lens to focus the image of an object on a one- or two-dimensional diode array. White-light

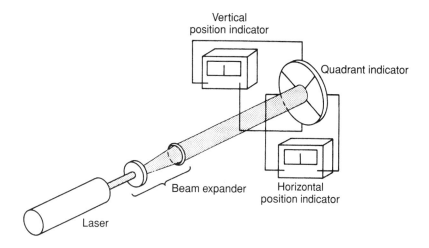

Figure 7.4.2 Simplified diagram of a laser alignment system

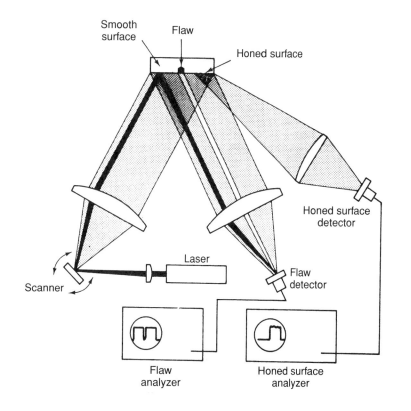

Figure 7.4.3 Laser scanner system designed to detect defects

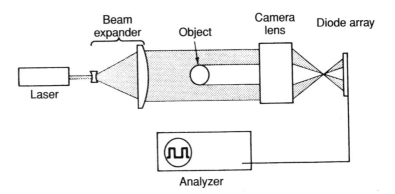

Figure 7.4.4 Typical optical gauge or optical inspection system designed by using a diode array camera

illumination is usually preferred in these applications. Laser illumination is often used when the light must be highly structured or directional or both. Unless otherwise stated, we assume here that a laser is being used as a light source. As in all applications, backlighting should be used whenever possible and reflected light used when necessary.

A typical diode-array camera system using laser light to backlight an object is illustrated in Figure 7.4.4. A beam spreader collimates and increases the diameter of the laser beam. This laser light backlights an object that is to be inspected or gauged. A camera lens images the sharp edges of the object on a diode array. The output of the diode array can be displayed on an oscilloscope as illustrated in Figure 7.4.4. Those diodes that are not illuminated by the image produce little or no signal, and those diodes that are illuminated produce positive pulses as the diodes are scanned in sequence. By counting the number of diodes illuminated, or those not illuminated, or both, along with the magnification for the lens system, the dimensions of the object can be determined. If one-dimensional measurements are sufficient, then a linear diode array can be used.

Two-dimensional measurements can be made by using a two-dimensional diode array. If a system of this type is used for gauging, then the lens must be designed so that a linear relationship exists between the object and image.

When using a diode-array camera system, it is convenient to predict, mathematically, the magnification and the location of the image for a given object distance. The camera lens must be treated as a thick lens. We have found the Newtonian form of the lens equations particularly useful for this purpose. The Newtonian form of the lens equation is:

$$xx^1 = F^2 \qquad \text{(Equation 7.4.1)}$$

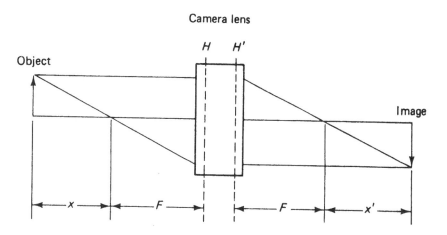

Figure 7.4.5 Notation for Newtonian form of lens equation.

where x and x^1 are measured relative to the primary and secondary focal planes, as shown in Figure 7.4.5. The focal planes are easily found by passing a collimated laser beam through the camera lens in both directions and finding where the beam focuses. The focal length of the lens can be obtained from the specifications on the lens. Magnification using the Newtonian form is

$$m = (y^1/y) = -(x^1/F) = -(F/x) \qquad \text{(Equation 7.4.2)}$$

The resolution with which a measurement of the image on the diode array can be made is plus or minus the distance between two adjacent photodiodes. Thus the theoretical resolution with which a measurement of the object can be determined is plus or minus the distance between adjacent photodiodes divided by the magnification.

Industrial parts can be inspected by frontlighting the parts if sufficient contrast is present so that the image can be interpreted by the diode array. Figure 7.4.6 shows a simplified diagram of an optical inspection system designed to detect the presence or absence of threads. If the threads are present, the output of the linear diode array is a sinusoidal waveform, as illustrated. If threads are not present, the output for each diode tends to be the same and will produce a DC signal. This system could also be used to determine the pitch of the threads and to detect stripped threads. Rotating the object about the axis of the hole allows a complete inspection of the threaded hole to be made.

Laser illumination is undesirable because of diffraction effects and speckle due to the coherent nature of laser light. When light of high irradiance must be directed from long distances into small confined areas, a laser is often required.

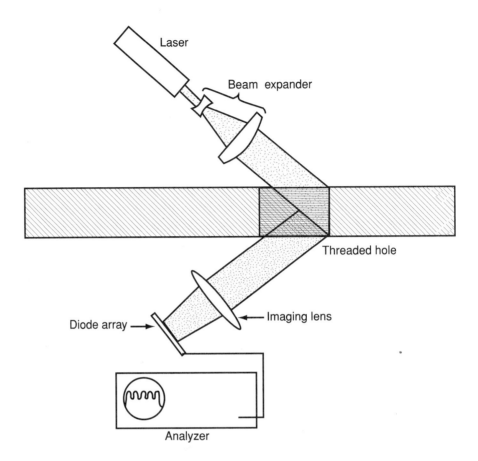

Figure 7.4.6 Optical thread checker

Optical Triangulation Techniques

Optical triangulation provides a noncontact method of determining the displacement of a diffused surface. Figure 7.4.7 is a simplified diagram of a laser-based system that is successfully used in many industrial applications. A low-power HeNe or diode laser projects a spot or line of light on a diffusive surface. A portion of the light is scattered from the surface and is imaged by a converging lens on a linear diode array or linear position detector. If the diffusive surface has a component of displacement parallel to the light incident on it from the laser, the spot of light on the surface will have a component of displacement parallel and perpendicular to the axis of the detector lens. The component of displacement perpendicular to the axis causes a corresponding displacement of the image on a

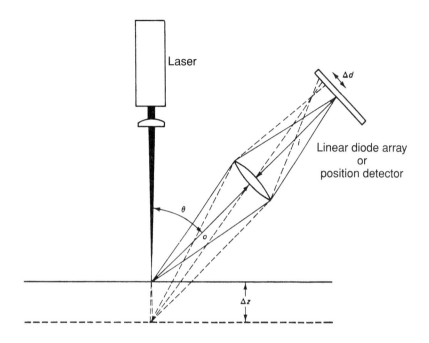

Figure 7.4.7 Diagram of laser-based optical triangulation system

detector. The displacement of the image on the detector can be used to determine the displacement of the diffusive surface.

Many triangulation systems are built with the detector perpendicular to the axis of the detector lens, as shown in Figure 7.4.7. The displacement Δd of the image on the detector in terms of the displacement of the diffusive surface Δz, parallel to the incident laser beam, is approximately:

$$\Delta d = \Delta z \, m \, \sin \theta \qquad \text{(Equation 7.4.3)}$$

where $m = i/o$ is the magnification and θ is the angle between a line normal to the surface and the light scattered to the imaging lens. The displacement Δd and distance o and i can be determined by optical measurement or by Equation 7.4.3.

7.5 SUMMARY

In this chapter, we have intended to provide a practical introduction to vision systems. Vision techniques are being applied in many areas ranging from medical imaging to remote sensing, industrial to document processing, and nanotechnology to multimedia databases.

Vision system or machine vision is a device that can collect data and form an image that can be interpreted by a robot computer for some useful movement in the workcell to determine an appropriate position or to see an object.

Vision in industry is used to identify and inspect parts, determine position orientation and verification for three-dimensional parts, and is involved in visual guidance control, safety monitoring, and other applications.

Vision systems are classified as being either linear array or matrix arrays. In linear array, only one dimension is sensed. Matrix arrays are classified as either two-dimensional or three-dimensional.

The basic components of a vision system include a lighting system, one or more cameras, camera controller, interface equipment, video image monitors, and programming devices. Three types of vision sensors are used in machine vision applications:

1. Solid-state camera
2. Vidicon camera
3. Laser camera (transmitter/receiver)

Solid-state cameras are smaller and more rugged. The most popular type is the charged-coupled-device (CCD).

Vidicon cameras are the type used in television, and its output is analog. If processing is to be done on a signal, it has to be digitized using the gray-scale levels. Vidicon cameras, which use filaments to emit the electron beams, have a shorter lifetime than solid-state cameras.

Laser cameras are the type of sensor used in video systems as a source and detector. Such systems use a helium-neon laser source whose power output is usually no higher than a milliwatt.

Other sensor systems used in robot vision include diode-array camera systems, optical triangulation techniques, X rays, and the ultrasonic sensors.

Three-dimensional vision can be achieved by using two cameras for moving and fast-response vision.

The operation of a machine vision system can be divided into five functions:

1. Image acquisition
2. Image digitization
3. Image processing
4. Image analysis
5. Interpretation

Machine vision applications in automated manufacturing with robots are numerous. Some of the distinct categories are:

1. Inspection
2. Part identification
3. Part orientation

4. Part location
5. Visual guidance and control
6. Safety monitoring

7.6 REVIEW QUESTIONS

7.1 What is a vision system?
7.2 How can vision systems be classified? Explain the categories.
7.3 What are the three types of sensors used in vision systems?
7.4 Describe the operation of a vidicon and a solid-state camera.
7.5 What are the basic components of a vision system?
7.6 Describe the three light-source locations used in vision applications.
7.7 How can three-dimensional vision be achieved?
7.8 What are the five functions of a machine vision system?
7.9 What is image acquisition?
7.10 What is image digitization?
7.11 What is image processing?
7.12 What is image analysis and how can it be obtained?
7.13 What is interpretation of image and what are the two techniques used?
7.14 What are the six most distinct categories of machine vision applications?
7.15 Explain in detail each category of question 7.14.
7.16 What other optical methods are used in machine vision?
7.17 Describe how the low-power scanning laser systems work and what their industrial applications are.
7.18 Describe how the diode-array camera systems work and what their industrial applications are.
7.19 Describe how the optical triangulation techniques work and what their industrial applications are.
7.20 Why can't we use artificial intelligence (AI) in vision systems?
7.21 What are the three main components in AI that can help vision systems?

7.7 PROBLEMS

7.1 A CCD solid-state camera has a 256×256–pixel matrix. The analog-to-digital converter takes $0.25\ \mu s$ (0.25×10^{-6} s) to convert the analog charge signal for each pixel into the corresponding digital signal. If there is no time loss in switching between pixels, determine the following:
   a. How much time is required to collect the image data for one frame?
   b. Is the time determined in *a* compatible with the processing rate of 30 frames/s?

7.2 The pixel count on the photoconductive surface of a vidicon camera is 512×512. Each pixel is converted from an analog voltage signal to the corresponding digital signal by an A/D. The conversion process takes $0.1\mu s$ (0.1×10^{-6} s) to complete. In addition to the A/D process, it takes $2\ \mu s$ to move from one horizontal line of pixels to the one below.

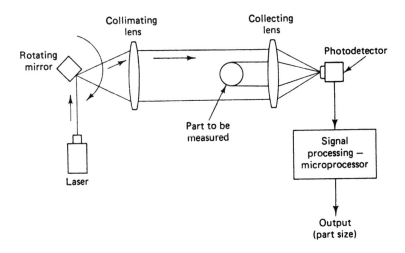

Figure 7.7.1 Scanning laser device for Problem 7.3

    a. How long will it take to collect and convert the image data for the one below?
    b. Can this be done thirty times per second?

7.3 A scanning laser device, similar to the one shown in Figure 7.7.1, is to be used to measure the diameter of shafts that are ground in a centerless grinding operation. The part has a diameter of 0.475 in. with a tolerance of ±0.002 in. The four-sided mirror of the scanning laser beam device rotates at 250 rev/min. The collimating lens focuses 30° of the sweep of the mirror into a swath that is 1.000 in. wide. Assume that the light beam moves at a constant speed across this swath. The photodetector is 100 ns (100×10^{-9} s). This resolution should be equivalent to no more than 10 percent of the tolerance band (0.004 in.).
    a. Determine the interruption time of the scanning laser beam for a part whose diameter is equal to the nominal size.
    b. How much of a difference in interruption time is associated with the tolerance of ±0.002 in.?
    c. Is the resolution of the photodetector and timing circuitry sufficient to achieve the 10 percent rule on the tolerance band?

7.4 Triangulation computations are to be used to determine the distance of parts moving on a conveyor. The setup of the optical measuring is illustrated in Figure 7.7.2. The angle between the beam and the surface of the part is 25°. Suppose that for one given part passing on the conveyor, the baseline distance is 6.55 in., as measured by the linear photo-

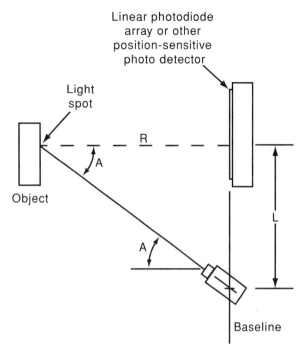

Figure 7.7.2 Optical triangulation system for Problem 7.4

sensitive detection system. What is the distance of this part from the baseline?

7.5 A planar sheet of light is focused on a surface in the manner illustrated in Figure 7.7.3. The angle that the light sheet makes with the nominal surface of the part is 30°. At this angle, the intersection line that the light sheet makes with the surface deviates 0.055 in. from a nominal straight line when viewed by a vision camera from a position normal to the surface. How much of a change in surface elevation is this equivalent to?

7.6 The beam scan angle θ for an oscillating mirror scanner is given by:

$$\theta = \theta_{max} \sin 2\pi ft$$

and its displacement by:

$$y = 2 F\tan(\theta/2)$$

where f is the oscillating frequency
θ_{max} is the angular displacement amplitude of the mirror
F is the focal length (See Figure 7.7.4.)

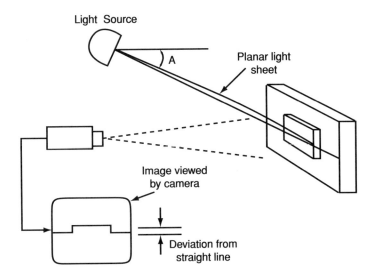

Figure 7.7.3 Variation of the triangular principle for Problem 7.5

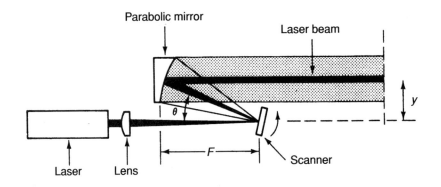

Figure 7.7.4 Scanner system with a parabolic mirror for Problem 7.6

Determine: The beam scan angle θ and plot the beam displacement and velocity for $0 \le t \le 120/\mu s$ if the lens focal length $F = 20$cm, $f = 2000$Hz, and $\theta_{max} = 12°$.

7.7 A diode-array camera gauge is configured as shown in Figure 7.4.4. A 1024 linear array with photodiodes placed on 25μs centers is used as a detector. Using the Newtonian form of the lens equation, determine:
for $x = 62.5$mm and $x^1 = 48.4$mm
  a. The focal length of the lens
  b. The magnification

7.8 With the data from Problem 7.7 determine:
  a. The number of diodes that would not be illuminated if the gauge is used to measure the diameter of .025 in. rod
  b. The theoretical resolution of the gauge

7.8 REFERENCES

Besl, P. J., and N. D. McKay. "A Method for Registration of 3-D Shapes." IEEE trans. *Pattern Analysis and Machine Analysis* 14, no. 2 (February 1992): 239–256.

Bhanu, B., and C. Ho. "CAD-based 3-D Object Recognition for Robot Vision." *Computer*, 20, no. 8 (1987): 19–36.

Fairchild, W. S. *CCD Databook: Solid State Imaging Technology.* Milipital, CA: Fairchild Weston Systems Inc., 1987.

Faugeras, O. *Three-dimensional Computer Vision, a Geometric View-point.* Cambridge, MA: MIT Press, 1993.

Gonzalez, R., and R. Woods, eds. *Digital Image Processing.* Reading, MA: Addison-Wesley, 1992.

Groover, M. P. *Fundamentals of Modern Manufacturing.* Upper Saddle River, NJ: Prentice Hall, 1996.

Haralick, R. M., and L. G. Shapiro. *Computer and Robot Vision.* Reading, MA: Addison-Wesley, 1992.

Harris, N. C., E. M. Hemmerling, and A. J. Hallmann. *Physics Principles and Applications.* New York: McGraw-Hill, 1989.

Jain, R. C., and A. K. Jain. *Analysis and Interpretation of Range Images.* New York: Springer-Verlag, 1990.

Kasturi, R., and R. C. Jain. *Computer Vision Principles.* Los Alamitos, CA: IEEE Computer Society Press, 1991.

Kohl, T. R. *Vision Tutor: Lab Guide.* Amherst, MA: Amerix Artificial Intelligence Inc., 1992.

Kosko, B. *Neural Networks and Fuzzy Systems.* Englewood Cliffs, NJ: Prentice Hall, 1992.

Nalwa, V. S. *A Guided Tour of Computer Vision.* Reading, MA: Addison-Wesley, 1993.

O'Gorman, L., and R. Kasturi. *Document Image Analysis.* Los Alamitos, CA: IEEE Computer Society Press, 1995.

Pratt, W. K. *Digital Image Processing.* New York: John Wiley, 1991.

Schalkoff, R. J. *Digital Image Processing and Computer Vision.* New York: John Wiley, 1989.

Tuceryan, M., and A. K. Jain. *Handbook of Pattern Recognition and Computer Vision.* World Scientific Publishing Co., 1993, pp. 235–276.

Wechsler, P., et al. *Computational Vision.* Boston, MA: Academic Press, 1990.

CHAPTER 8

Control Systems

8.0 OBJECTIVES

After studying this chapter, the reader should:

1. Be acquainted with the Control System correlation
2. Be able to define control system requirements
3. Be familiar with Programmable Logic Controller
4. Understand PLC programming terminals
5. Have knowledge of Proportional-Integral-Derivative control
6. Recognize the basics of Computer Numerical Control
7. Apprehend the principles of the microprocessor
8. Recognize the benefits of the Universal Robot Controller
9. Realize the importance of interfacing
10. Perceive the significance of workcell control

8.1 CONTROL SYSTEM CORRELATION

Control systems are an integral part of a workcell or large complex of equipment and have the primary purpose of controlling, monitoring, analyzing, or measuring a process or other equipment.

Most robot control units today are based on a **microcomputer** system, which can sense, evaluate, make decisions, and interact with their environment. The parts of a controller, types of motion, and path control systems used in robots have been discussed in Chapters 2 and 3.

When a robot executes a stored program for instructions that defines the sequence of motions and positions in the work cycle, the program may also include instructions for other functions, such as interacting with external equipment, responding to sensors, and processing data. The activities of such coordination must be established by means of control-system requirements and workcell control.

In this chapter, we present the following topics that apply to the control systems:

- Control-system requirements
- Programmable logic controller (PLC)
- PLC programming terminals
- Proportional-Integral-Derivative (PID) control
- Computer numerical control (CNC)
- Microprocessor unit (MPU)

- Universal robot controller (URC)
- Interfacing
- Workcell control

8.2 CONTROL SYSTEM REQUIREMENTS

Control systems began with hard-wired panels of logic, motor starters, fluid actuators, and solenoid valves. Servo control systems used vacuum tubes, and variable-speed drives consisted of DC generators driving DC motors. In the 1950s numerical control machines appeared, and machine control entered the world of digital control. These NC machines were programmed off-line and used punched paper tape for storage of the program. Programmable controllers appeared about 1970, replacing hard-wired relay logic with reprogrammable computer logic.

Improvements in semiconductor switching elements and memory systems set the stage for the microprocessor in the 1980s. The microprocessor generated the changes that are finally bringing process control and machine control under a single umbrella of unified control system technology.

The microprocessor gave us the means to move the control loop back to the plant floor and the ability to communicate with this distributed intelligence. We can have the short loop advantages of distributed control center and the fully informed operator of a centralized control system. Figure 8.2.1 shows a fully integrated, distributed control system for a hypothetical manufacturing plant that has both process control and machine control systems. The path that crosses the top and runs down the left side represents the communication network. The terms *data highway* and *local area network* (LAN) are used for different approaches to the communication network.

All of the control loops are closed on the plant floor. The local control units (LCUs) contain I/O modules that condition the input signals and controller modules that carry out the control algorithms.

Local control units can handle continuous processes, batch processes, and robotic workcells. In addition, other units, such as programmable logic controllers (PLCs), can be interfaced with the communications network. Local operators, central control room operators, plant maintenance, plant engineering, and plant management all have access to all the information via color display and keyboards.

A typical block diagram configuration of a control system for a robot joint is shown in Figure 8.2.2. The input command is the defined position to which the joint is directed to move. The output variable is the actual position of the joint. As indicated previously, most of the computational functions of robot joints are controlled and carried out by a microprocessor. Therefore, the purpose of the controller is to compare the actual output of the actuator with the input command and to provide a control signal that will reduce the error possibility to close to zero.

8.2 Control System Requirements   223

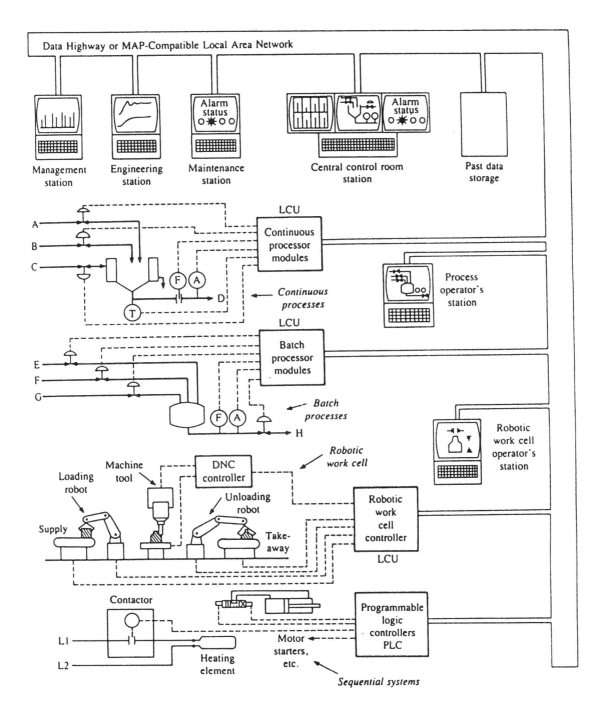

Figure 8.2.1 Integrated, distributed control system for a hypothetical manufacturing plant *(Source: Bateson, Control System Technology)*

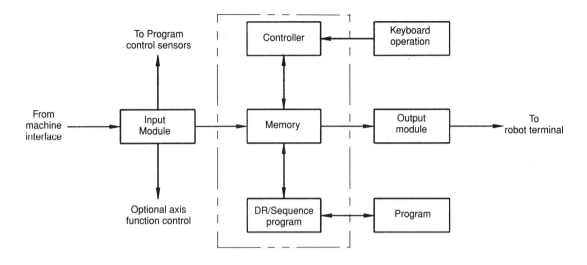

Figure 8.2.2 Basic robot control system

8.3 PROGRAMMABLE LOGIC CONTROLLER (PLC)

A programmable logic controller is a user-friendly computer that carries out control functions of many types and levels of complexity. It can operate any system that has output devices that turn on and off, as well as any setup with variable outputs. A PLC can be operated on the input side by on-off devices or by variable input devices. It can be programmed, controlled, and operated by a person who is unskilled due to its easy operational and powerful functions.

NEMA Standard ICS3-304-1978, defined a programmable logic controller as follows:

> A digital-operated electronic apparatus that uses a programmable memory for the internal storage of instructions for implementing specific functions, such as logic, sequencing, timing, counting, and arithmetic, to control through digital or analog input/output modules, various types of machines or processes.

The advantages of the PLC make it the heart of modern industry and the brains of the leading edge in control systems. Programmable logic controllers were originally designed to replace relay logic and to provide flexibility in control based on programming and executing instructions.

The first concept of PLC was conceived by a group of engineers from the Hydramatic division of General Motors in 1968. In the mid-1970s program entry, reprogramming, and reloading made a major step forward for the programmable logic controller. In 1978, the introduction of the microprocessing **chip** increased computer power and, therefore, all types of automation, robots, and PLC underwent many improvements. PLC programs became more understandable and the cost of the systems more affordable.

8.3 Programmable Logic Controller (PLC)

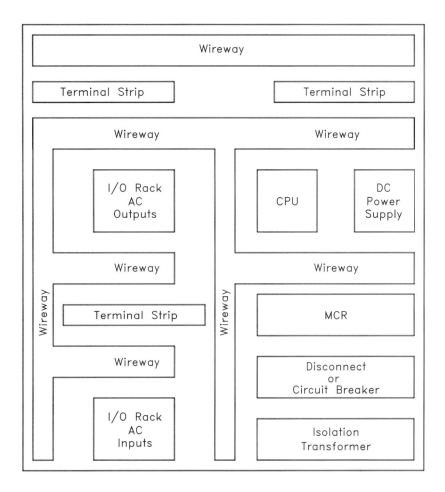

Figure 8.3.1 PLC system layout

The rapidly growing market for PLCs combined with the wide variety of PLCs available indicate that all other previous systems used for industrial control will eventually become obsolete. There are presently over fifty different manufacturers of programmable logic controllers in the United States.

Figure 8.3.1 shows a typical PLC system layout. The termination points for the hard-wired field equipment would be connected at the terminal strips. An emergency disconnect switch and isolation transformer are mounted inside the enclosure. A hard-wired master control relay (MCR) is also used to interrupt power to the I/O rack in the event of a system failure, but it will still allow power to the CPU. The main and auxiliary power supplies are shown in the layout. Tandem power supplies are often used in situations where downtime is critical. If one power supply fails, the auxiliary can be either manually or automatically switched into the circuit to maintain continuous levels of voltage and current at the CPU.

The following are some advantages derived from PLCs:

1. PLCs are generally easy to program and install.
2. The speed with which internal timers operate is much faster than conventional time-delay relay systems.
3. Access to PLCs is restricted by hardware features, such as keylocks, and by software features, such as passwords.
4. Problem-solving with PLCs is a major advantage over any other type of control system.
5. PLCs can be designed with communications capabilities that allow them to converse with local and remote computer systems or to provide human interfaces.
6. PLCs are extremely reliable and easier-to-maintain control devices and can be obtained in forms that can survive and function in harsh conditions.
7. PLCs require less floor space than relay logic controls.

All programmable logic controllers have input and output interfaces, memory, a central processing unit (CPU), a programming device, and a power supply. These components are shown in Figure 8.3.2, and their functions are as follows:

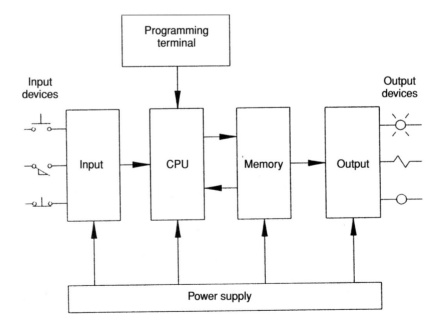

Figure 8.3.2 Block diagram of a programmable logic controller

1. The input interface provides a connection to the machine or process being controlled. The principal function of this interface is to receive and convert field signals into a form that can be used by the CPU. This conversion involves changing contact closures, current signals, analog voltages, and so on, into simple voltage levels that can be understood by the CPU. The input interface is modular and can be expanded by adding more modules to allow more inputs when the control task increases. The number of possible inputs is usually limited by the CPU and the size of memory.
2. The output interface performs the opposite function of the input interface. It takes signals from the CPU and translates them into forms that are appropriate to produce control actions by external devices such as solenoids, motor starters, and so on. The output interface is also modular in nature, so that additional output functions can be incorporated when required.
3. Memory and CPU provide the main intelligence of a PLC. Fundamental operating information is stored in **memory** as a pattern of electrical charges (bits) that is organized into basic working groups called words. Each word stored in memory is a piece of data, an instruction, or a part of an **instruction**. Data comes from the input section and is based on the stored program. The CPU performs logical decisions, drives the outputs, and continually refers to the program stored in memory for instructions concerning its next action for reference data. It also uses memory to store outside data for future use or for intermediate action when some sort of decision-making operation is involved.
4. A programming device, or programming terminal, allows a user to enter instructions into memory in the form of a program. The programming device produces the pattern of electric signals that correspond to the symbols, letters, or numbers entered by the user. Programming devices vary widely in complexity, ease of use, and cost. They can be small, hand-held units, or they may contain large CRT screens allowing better overviews of programs. Many PLCs can now be programmed, using a teach pendant or personal computer (PC). By special PLC software, a PC can function as a programming terminal, allowing program entry and editing directly from the PC's keyboard.
5. The power supply provides all the voltage levels needed to operate the PLC. The power supply converts 120V or 240V AC into the DC voltage required by the CPU, memory, and Input/Output modules.

A **program** that is written by a user and stored in a PLC's memory is a representation of the actions required to produce the correct output control signals for a given process condition. Such a program includes sections that allow process data to be brought into the controller memory, sections that represent decision-making, and sections that deal with converting a decision into physical output action.

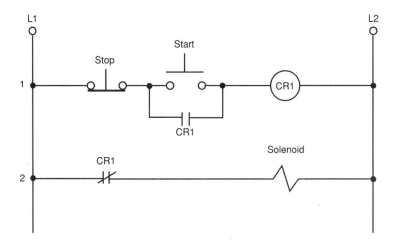

Figure 8.3.3 Line diagram using relay logic

Programming languages have many forms. Like all languages, the **language** of a PLC has grammar, syntax, and vocabulary that allow the user to write a program stating what the CPU is required to do. Each PLC manufacturer uses a slightly different language to do this. Most PLC languages are based on ladder logic, which is an advanced form of relay logic. Ladder logic is a term used to describe the format of a schematic diagram to be entered into a PLC. Figure 8.3.3 shows a typical relay logic diagram with a Stop/Start station and a Control Relay (CR) operating a solenoid. In this circuit, when CR1 is OFF, the solenoid is energized.

If the same diagram were to be drawn in ladder logic, it would appear as shown in Figure 8.3.4. In this circuit, the input and output devices are now represented by numbers. These numbers are referred to as addresses. The addresses shown in Figure 8.3.4 are typical of an Allen-Bradley PLC.

The basic points regarding ladder logic diagrams are as follows:

1. All devices that represent resistive or inductive loads to the circuit are shown at the right of the diagram.
2. All devices that represent or make or break electrical contacts are shown at the left of the diagram.
3. Devices connected in parallel with other devices are often called branches. Each complete horizontal line of a ladder diagram is typically referred to as a rung.

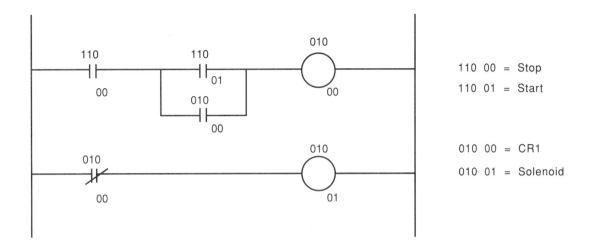

Figure 8.3.4 Ladder diagram using ladder logic

4. Electrical devices are shown in their normal conditions. An NC contact would be shown as normally closed, or represented as XIO (examine if open), and an NO contact would appear as a normally open device or represented as XIC (examine if closed). All contacts associated with a device will change state when the device is energized.
5. Devices that perform a stop function are usually connected in series; devices that perform a start function are connected in parallel.
6. The two parallel vertical lines where all rungs are connected provide power to the components.

Writing a Ladder Logic Diagram

In many cases, it is possible to prepare a ladder logic diagram directly from the narrative description of a control event sequence. In general, the following suggestions apply to writing ladder logic diagrams:

1. Define the process to be controlled.
2. Make a sketch of the operational process.
3. Create a written sequence of operations by listing the steps in as much detail as possible.
4. Add the sensors needed to carry out the control sequence on the sketch.
5. Add manual controls needed for the process setup or for operational checks.

6. Consider the safety of the operating personnel and make additions and adjustments as needed.
7. Add master stop swiches required for safe shutdown.
8. Create a ladder logic diagram that will be used as a basis for the PLC program.
9. Consider the "what if's" where the process sequence may go astray.

The ladder logic program in a computer takes the place of the external wiring required for a process regulation and can be seen as a control circuit within a PLC. Basically, there are two types of logic control circuits: a state combination control circuit and a sequential control circuit. The feature distinguishing state combination circuits and sequential control circuits is whether the control depends solely on a combination of the inputs or includes a sequence of the event. Control circuits that are solely determined by input states are referred to as state combination circuits, whereas those that are determined by both state combination and sequence of occurring events are called sequential control circuits. In essence, two types of ladder programs exist: one is developed from a state combination of logic control circuits, and the other is developed from a sequential logic control circuit. The focus of the following example is to provide a procedure that will demonstrate how to develop state combination circuits for a PLC.

Example 8.1 This example has three inputs and two outputs. The task is to design a control circuit to implement the control logic in such a way that when two of the three input switches turn on, light one (L1) will be On; and, when only one of the three input switches turns on, or none of the input switches turns on, light two (L2) will be On.

Solution

The control task can be defined completely by the following eight steps:

Step 1: Identify inputs and outputs. The task is to give each input and output device a specified name, so that it can be conveniently represented later.

Three inputs:

    PB1: The first push button
    PB2: The second push button
    PB3: The third push button

Two outputs:

    L1: The first light
    L2: The second light

Row	PB1/A	PB2/B	PB3/C
1	0	0	0
2	0	0	1
3	0	1	0
4	0	1	1
5	1	0	0
6	1	0	1
7	1	1	0
8	1	1	1

Figure 8.3.5 Input state table

Step 2: Build an input state table. The key to designing a state combination circuit is to construct a state table. The input state table contains all possible combinations of the input state. A two-input system, for example, has a total of 4 (or 2^2) combinations of input states. Likewise, a four-input system would have a total of 16 (or 2^4) possible combinations of input states. In this example, a three-input system will have a total of 8 (or 2^3) possible combinations of input states. Figure 8.3.5 shows the input state table.

Step 3: Determine the output states. The output state of each possible input combination is determined by the control task. In this case, L1 will be On when two of the three inputs have "1" input. Thus, for output X/L1, rows 4, 6, and 7 have "1" value and the other rows have "0" value. L2 will be On when just one of the three inputs has "1" or all of the three inputs have "0" input. Next, for output Y/L2, rows 1, 2, 3, and 5 have "1" value and the other rows have "0" value. Figure 8.3.6 depicts a completed state table that includes all logic relationships of the control task. The relationships between inputs and outputs are clearly indicated in the state table, which is then used to derive a Boolean equation.

Step 4: Derive the Boolean equations. The rows of the state table that have "1" output value are used to develop the Boolean equations. Each of these rows represents a subset of the Boolean equations. A complete equation is the sum of these subsets for each output. In this example, there are two outputs, X and Y. The Boolean equation of the X output is computed first.

There are three rows with "1" output, and the three subsets are:

$$\text{ROW4: L4} = \bar{A} \cdot B \cdot C$$

$$\text{ROW6: L6} = A \cdot \bar{B} \cdot C$$

Row	Input			Output	
	PB1/A	PB2/B	PB3/C	X/L1	Y/L2
1	0	0	0	0	1
2	0	0	1	0	1
3	0	1	0	0	1
4	0	1	1	1	0
5	1	0	0	0	1
6	1	0	1	1	0
7	1	1	0	1	0
8	1	1	1	0	0

Figure 8.3.6 Completed state table

$$\text{ROW7: } L7 = A \cdot B \cdot \overline{C}$$

Therefore, the completed Boolean equation for the X output becomes:

$$X = L4 + L6 + L7 = \overline{A} \cdot B \cdot C + A \cdot \overline{B} \cdot C + A \cdot B \cdot \overline{C}$$

Using the same method as above, the Y Boolean equation can be derived:

$$Y = \overline{A} \cdot \overline{B} \cdot \overline{C} + \overline{A} \cdot \overline{B} \cdot C + \overline{A} \cdot B \cdot \overline{C} + A \cdot \overline{B} \cdot \overline{C}$$

Step 5: Simplify the Boolean equation. In most cases, the Boolean equations derived from the state table can be simplified with Boolean laws or the Karnaugh map method. In the X output equation, no further simplification can be achieved

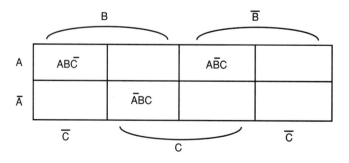

Figure 8.3.7 A Karnaugh map of the output X

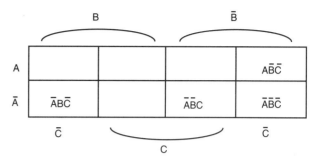

Figure 8.3.8 A Karnaugh map of the output Y

because there are no adjacent pairs in the Karnaugh map. The Karnaugh map for the output X is illustrated in Figure 8.3.7.

In the Karnaugh map of the Y output as shown in Figure 8.3.8, there are three adjacent pairs in the group that can be used to simplify the logical equation:

1. $\overline{A} \cdot \overline{B} \cdot \overline{C}$ and $\overline{A} \cdot B \cdot \overline{C}$
2. $\overline{A} \cdot \overline{B} \cdot \overline{C}$ and $\overline{A} \cdot \overline{B} \cdot C$
3. $\overline{A} \cdot \overline{B} \cdot \overline{C}$ and $A \cdot \overline{B} \cdot \overline{C}$

A common part, $(\overline{A} \cdot \overline{B} \cdot \overline{C})$, can be seen in the three pairs so that only one of the pairs can be selected to simplify the Boolean equation. However, any one of them can be selected. The third pair is used to simplify the Boolean equation of Y in this example. The simplified output can be figured by using Boolean laws; however, using a Karnaugh map is a more direct method.

$$Y = \overline{A} \cdot \overline{B} \cdot \overline{C} + A \cdot \overline{B} \cdot \overline{C} + \overline{A} \cdot B \cdot \overline{C}$$
$$= (\overline{A} + A) \cdot \overline{B} \cdot \overline{C} + \overline{A} \cdot (\overline{B} \cdot C + B \cdot \overline{C})$$
$$= \overline{B} \cdot \overline{C} + \overline{A} \cdot (\overline{B} \cdot C + B \cdot \overline{C})$$

Step 6: Construct the ladder diagram and the PLC circuit. In a Boolean equation, there are three logic operators: (1) the "·" is the "and" operator; (2) the "+" is the "or" operator; and (3) the "–" is the "not" operator. If the input has a "not" operator in the Boolean equation, a normally closed relay should be selected to represent it; otherwise, a normally open relay should be selected to represent the input in the ladder diagram. If two input parts are connected by the "or" operator in the Boolean equation, these two parts should be connected so they are parallel in the ladder diagram. If two parts are connected by the "and" operator in the Boolean equation, these two parts should be connected serially in the ladder diagram.

234 Control Systems

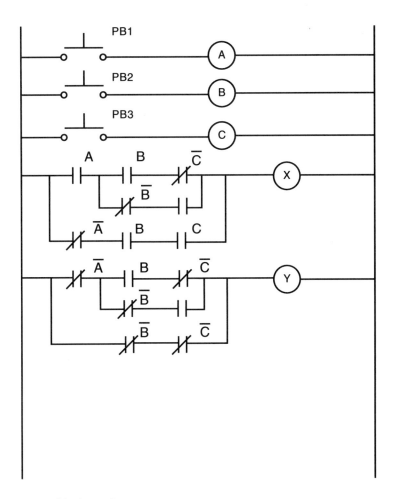

Figure 8.3.9 Ladder logic diagram

A ladder diagram can be easily made according to the Boolean equation. Figure 8.3.9 shows the ladder diagram for this example in which relays A, B, and C are used to be directly controlled by PB1, PB2, and PB3.

Step 7: Verify the results. Review the state diagram by tracing the circuit throughout each of the possible input combinations and verify that the result is the same as indicated in the state table.

Step 8: Add the Memory Element to the circuit. In many applications, "instant" switches are used as the input device. In this case, if the actuation of an output must be maintained even after the input element has been released, a

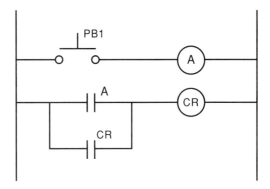

Figure 8.3.10 Power-holding circuit

memory element is normally added to provide the maintaining function. The memory is usually a normally open contact controlled by the output relay, as shown in Figure 8.3.10. When the output relay CR is energized by pressing PB, the normally open contact CR is closed to form a holding circuit to continuously energize the output relay.

The memory circuit shown in Figure 8.3.10 needs to be improved because there is no way to de-energize the relay coil CR. A normally closed PB needs to be added to the circuit to build a complete memory circuit. Figure 8.3.11 shows the complete circuit. This circuit is often referred to as the power-holding circuit.

If the circuit needs to be transferred in order to act as the power-holding circuit, the above method can be used to make the transfer. If the X output of the

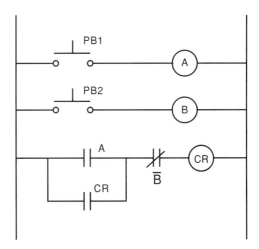

Figure 8.3.11 Completed power-holding circuit

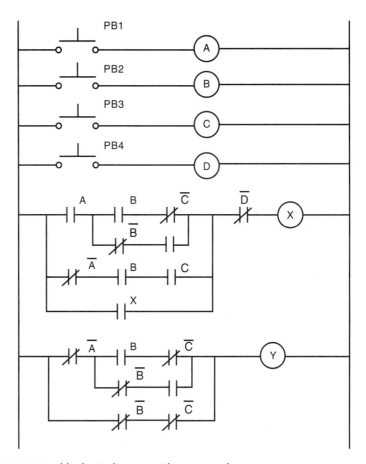

Figure 8.3.12 Ladder logic diagram with memory element

previously mentioned circuit needs to be transferred to act as the power-holding circuit, then a normally closed relay and a repetition memory unit must be connected to the ladder diagram. Figure 8.3.12 illustrates this principle.

In summary, the logical relationship of a specified control task of a PLC can be represented directly by a state table. The state table is straightforward so that any mistakes of the control logic can be easily avoided and any complications can also be readily expressed. Based on the completed state table, a Boolean equation can be obtained easily and correctly. By using the Karnaugh map method, the Boolean equation can be simplified directly from the state table. Using simplified Boolean expressions to draw a ladder diagram is very easy and can make logical sense of representing the state table. In essence, this procedure makes the construction of a ladder logic diagram much easier and more reliable for industrial applications.

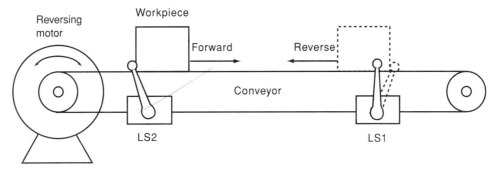

Figure 8.3.13 Conveyor apparatus for Example 8.2

Example 8.2 Figure 8.3.13 shows that a workpiece is loaded onto a conveyor belt and operates between two limits of travel. When limit switch LS2 is activated, the conveyor moves forward. When limit switch LS1 is activated, the conveyor changes direction. Pressing the Start button causes the motor to run in the forward direction, and pressing the Stop button stops the motor.
      a. Draw the ladder logic diagram.
      b. Write the sequence of operations.

Solution

      a. Ladder logic diagram as shown in Figure 8.3.14.
      b. The sequence of operations is as follows:

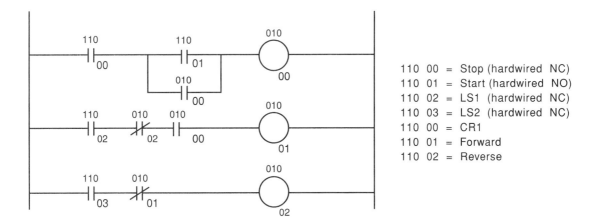

Figure 8.3.14 Ladder logic diagram for Example 8.2

1. When the Start button is pressed, control relay CR1 is energized and the conveyor moves in a forward direction.
2. When the workpiece makes contact with limit switch LS1, the motor is reversed and the conveyor moves in the opposite direction.
3. When the workpiece makes contact with limit switch LS2, the motor returns to the forward direction.
4. The Stop button stops the motor regardless of in which direction it is turning.

Flowchart programming language is also used with some PLC's. A **flowchart** is a pictorial language that shows the interconnections of variables within a process. It is a pictorial description of an algorithm, showing the point of decision, relevant operations, and the sequence in which they should take place to solve the problem correctly. Flowcharts typically use four basic symbols: the oval, parallelogram, diamond, and rectangle. These symbols are linked in a sequential manner using arrows to indicate direction. Table 8.1 shows flowchart symbols, definitions, and number of arrows flowing in or out of each symbol.

Table 8.1 Flowchart Symbols and Arrow Properties.

Symbol	Meaning	Possible number of IN arrows	Possible number of OUT arrows
Start	Indicates the beginning of the flowchart	None	One only
(parallelogram)	Input/output takes place	A minimum of one	One only
(diamond)	Decision point usually of magnitudes of numbers	A minimum of one	Two (Yes and No)
(rectangle)	Computation or other type of operation	A minimum of one	One only
Stop	Indicates the end of the flowchart	A minimum of one	None

Figure 8.3.15 shows an example of a process control flowchart for measuring temperature using an analog input. If the temperature is greater than 300°F, a heater coil is energized. The first step in the chart after START is to input the preset values for the decision-making process. A push button (PB) is pressed each time a temperature reading is required. When the PB is pressed, the temperature is sampled by the analog input device and the data is stored in a memory location.

8.4 PLC PROGRAMMING TERMINALS

A programming terminal provides the user with a method of entering and monitoring a program in the CPU's memory. Programming terminals vary widely in complexity, ease of use, and cost. There are two basic methods of entering a program into a PLC. It can be done with either a dedicated programming terminal or with a computer using PLC **software.** In both cases, the programming terminal is connected to the CPU via the connector on the processor. This connector is typically a 25-pin RS-232C serial communications port, which is compatible with most IBM computers.

A dedicated programming terminal is still the most common method of entering a program into a CPU. Although the computer is popular with engineers for PLC programming on the floor, a dedicated programming terminal is capable of withstanding more-rigorous use. For example, a portable laptop computer is in itself very rugged and durable, but the hard disk required to store programs is sensitive to shock and corrosion. If a hard disk is dropped, it can easily be destroyed, whereas a dedicated terminal is capable of withstanding tremendous abuse.

Although most PLCs will perform the same basic functions, the most noticeable difference between products is the keyboard. For example, Texas Instruments and many other software-based programs use a set of function keys marked F1 to F8 to access a wide variety of operations, including access to timers, counters, sequences, math functions, and others. AEG, Modicon, and Allen-Bradley utilize shift key functions to increase the number of tasks on the keyboard, as well as providing keys, such as search and supervisory, that are multifunction keys.

Dedicated Programming Terminals

A dedicated programming terminal is a device that is used exclusively for program entry and PLC monitoring. It is dedicated to operate with only one brand of PLC and is generally limited to the range of functions it can control. A typical dedicated terminal consists of a Cathode Ray Tube (CRT), keyboard, and the electronic circuitry required for developing, modifying, and loading a program into CPU memory. A CRT provides the user with a visual representation of programs and ranges in screen sizes from 5 inches to 12 inches (diagonally). Figure 8.4.1 shows a photograph of a dedicated programming terminal.

There are two basic types of dedicated programming terminals: dumb and intelligent. Dumb terminals have been extensively used over the past twenty years and are relatively inexpensive programming devices. This type of terminal relies

240 Control Systems

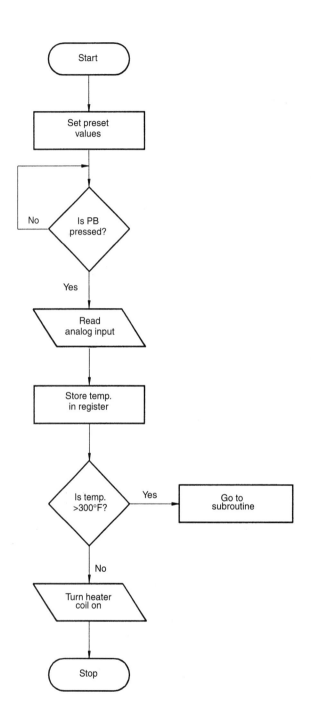

Figure 8.3.15 Flowchart for monitoring temperature

8.4 PLC Programming Terminals

Figure 8.4.1 Dedicated programming terminal *(Source: Allen-Bradley)*

entirely on the CPU for software documentation, and it must be connected to a processor to make any changes in the control program.

An intelligent dedicated programming terminal is a microprocessor-based device that has its own memory and software. This type of device allows programs to be developed and altered without the programming terminal being connected to a CPU. Some intelligent terminals also have the feature of network interfacing, which allows the terminal to be connected to a manufacturer's Local Area Network (LAN). This configuration allows the programming terminal to access any PLC in the network, and perform editing and monitoring functions. In most cases, an intelligent dedicated programming terminal will also have a floppy-disk port on the front of the terminal. This port allows programs to be loaded and retrieved from the PLC by the terminal itself. Consequently, programs can be stored and written anywhere, and then loaded into the CPU via the floppy-disk port. Figure 8.4.2 shows an intelligent dedicated programming terminal.

The keyboards of all dedicated terminals share a certain amount of common functions, such as coil, contacts, timers, counters, force, print, cursor movement, insert, delete, get, put, and so on. The keyboard shown in Figure 8.4.3 is for the Allen-Bradley T-3 dedicated programmer. This keyboard allows access to a wide variety of coil functions directly from the keyboard itself, including the MCR function, latching relays, timer and counter-coils, get and put instructions, as well as math functions. Because Allen-Bradley PLCs have special requirements for

242 Control Systems

Figure 8.4.2 Intelligent dedicated programming terminal *(Source: Texas Instruments)*

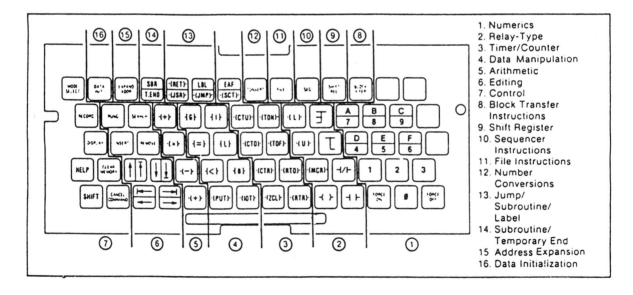

Figure 8.4.3 Keyboard for Allen-Bradley T-3 terminal *(Source: Allen-Bradley)*

Table 8.2 SEARCH Function Capabilities

Function	Mode	Keystrokes	Description
Locate first run of program	Any	[SEARCH][ ]	Moves cursor to first line of program.
Locate last rung of program	Any	[SEARCH][ ]	Moves cursor to end of program.
Locate specific address	Any	[SEARCH][8][address]	Locates address in program.
Print	Any	[SEARCH][43]	Prints single line of program.
Print	Any	[SEARCH][44]	Prints entire program.
Display first 20 lines of data table configuration	Any	[SEARCH][5][0]	Displays first 20 lines of data table on CRT.
Change CPU status to RUN/PROGRAM mode	R/P	[SEARCH][590]	Allows on-line programming change.

entering rungs or ladders, the dedicated keyboard also has branch end and branch start instructions.

The T-3 also contains program instructions for writing subroutines to allow the CPU to jump to different points in a program and branch back and forth. The SEARCH instruction on the T-3 programmer is used for a wide variety of tasks. Table 8.2 illustrates some SEARCH function capabilities. The Mode column indicates what operation the CPU should be in: Program, Run/Program, or Run. If the word *any* appears below the Mode column, it indicates that the CPU may be in any of the three possible states. The Allen-Bradley T-3 terminal also contains a series of HELP directories that are designed to assist in programming a PLC.

Computer-Based Programming Terminals

In recent years, a definite trend has been developing toward the use of personal computer-based program-development systems. Some manufacturers offer products that use an industry standard personal computer (i.e., IBM) as the base of their product offerings. Some offer the personal computer as part of the product, whereas many offer just the software to be used with the user's choice of personal computer hardware. Personal computers made by PLC manufacturers are generally much more rugged and durable than those found in typical office environments. For example, computer keyboards are often sealed to prevent spills and other dirt from entering and damaging the electronics of the system. There are also some types of filter systems that protect the disk drive of the computer, as well as extended temperature and humidity tolerances similar to those that the PLC would be exposed to in normal operating conditions.

Programming software is the essential software required for the creation of contacts and coils using a personal computer.

Documentation software is used in conjunction with the programming software to allow for the customizing of each coil and contact in a program.

Data collection and analysis software is becoming increasingly popular in personal computers for industrial control. This software is based on a spreadsheet format. That is, it has the ability to collect data from single or multiple PLC CPUs. By using this type of software, data can be displayed in more traditional computer-related formats, such as graphs, charts, and similar formats.

Real-time operator interface software, typically referred to as a man-machine interface (MMI), is another newer type of personal computer software for PLC applications. This type of software allows supervisory control to be established, using images on a CRT screen to advise an operator of process conditions and to provide a warning system in the event of process failure. Using real-time software, an operator can enter input information to control process by a keyboard. Some software packages are also combined with touch screens to allow an operator to make process revisions by touching a CRT screen at the appropriate point.

Simulation software is also available to allow the computer to be used to simulate the operation of a process-control program. The advantage of this type of software is that it allows an existing system to be effectively measured and analyzed, and it permits process designers to simulate various process conditions prior to the actual construction of a particular PLC system.

A host computer is defined as a multiuser system that is centrally located in a facility and is used for a variety of tasks. These tasks can range from the supervision of numerous PLC systems to shop-floor scheduling, maintenance and tool management, and broader tasks including production cost analysis, and so on. In very large PLC-based control environments, it is quite common for a host computer to be used with associated software and communications hardware for monitoring and editing programs from a centralized location. Figure 8.4.4 shows a typical host computer–based programming system.

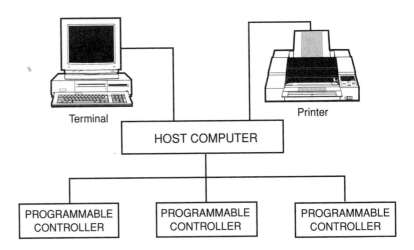

Figure 8.4.4 Host computer–based programming terminal

Hand-Held Programming Terminals

A hand-held programming terminal is an inexpensive portable means of programming small PLCs. These devices are often the size of hand-held calculators and usually have LED or dot matrix 2-line by 16-character LCD displays. The keyboard consists of numeric keys, programming instruction keys, and special functions keys.

Hand-held programming terminals can be either intelligent or dumb terminals depending on their make and model. A dumb terminal is very similar to a large CRT terminal in that it has basic features for entering and editing programs. An intelligent hand-held programming terminal is microprocessor-based and provides the user with features such as system diagnostic routing, message displays, and the like. Figure 8.4.5 shows a typical hand-held programming terminal.

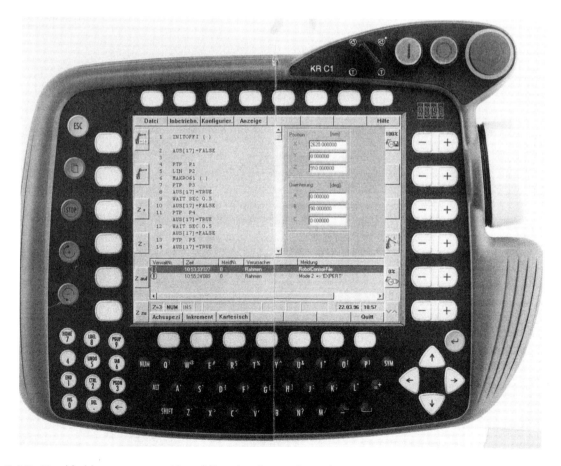

Figure 8.4.5 Hand-held programmer with multifunctional control panel *(Source: KUKA Welding Systems and Robot Corp.)*

8.5 PROPORTIONAL-INTEGRAL-DERIVATIVE (PID)

All the processes discussed previously for the PLC are not the continuous type. They are either ON or OFF, or travel linearly between two points. By continuous process we mean one in which the output is a continuous flow, like the pneumatic or hydraulic system of a robot manipulator. Process control for continuous process cannot be accomplished fast enough by a PLC On-Off control. Furthermore, analog PLC control is also not effective or fast response.

The control system most often used in continuous processes is the proportional-integral-derivative control. The PID control can be performed by mechanical, pneumatic, hydraulic, or electronic control systems as well as by PLCs. Many medium-size PLCs and all large PLCs have PID control functions, which are able to execute process control effectively. In this section, we will discuss the basic principles of the PID control and some of its typical functions.

PID is an effective control system for continuous processes that performs two main tasks to provide feedback to a PLC:

1. PID control keeps the output at a set level even though varying process parameters may tend to cause the output to vary from the desired set point.
2. PID promptly and accurately changes the process level from one set point to another set point level.

The characteristics of each of the PID control components and types of feedback used in PLC controlled systems are briefly discussed next.

On-Off feedback is used to send a full-ON or full-OFF signal to the controller to indicate whether a discrete output has been energized. Controlling activity is achieved by the period of On-Off cycling action.

Proportional control, also known as ratio control, is a control system that corrects the deviation of a process from the set level back toward the set point. The correction is proportional to the amount of error. For example, suppose we have a set point of 575 CFM in an airflow system. If the flow rises to 580 CFM, a corrective signal is applied to the controlling air vent damper to reduce the flow back to 575 CFM. If the flow somehow rises to 585 CFM, twice the deviation from set point, a corrective signal of four times the magnitude would be applied for correction. This larger corrective signal theoretically gives a faster return to 575 CFM. Proportional control does not usually work effectively by itself, resulting in an offset error. Therefore, a feedback signal is sent back from the output to the input of the PID module to correct the error.

Integral control, also known as reset control, is a control system that returns the flow to its original setting, because with proportional control there is only output error from the original set point. Integral control senses the process error and the time the error has persisted. Then a control signal is used in conjunction with the proportional corrective signal to reduce the error that caused the output deviation. Integral feedback is used when offset signals must be taken into account. An offset can be an error that is adversely affecting process control.

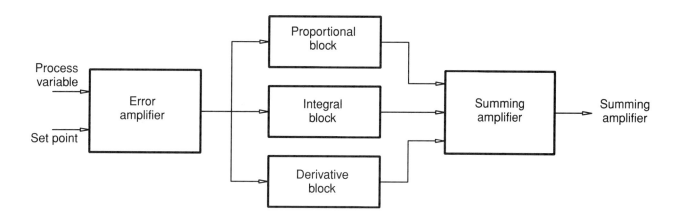

Figure 8.5.1 Block diagram of a typical PID controller

Derivative control, also known as rate control, produces a corrective signal based on the rate of change of the signal. The faster the change from the set point, the larger the corrective signal. The derivative signal is added to the proportional-integral system. This gives a faster action than the proportional-integral system signal alone. Derivative feedback is used when sudden changes in signal occur that can drastically affect system operation. This type of feedback is used when a fast response is required to reduce the possibility of overshoot. A typical PID control system is shown in block diagram form in Figure 8.5.1.

Essentially, a PID control system has a feedback loop that is proportional, plus derivative, plus integral. In this type of system, the proportional mode is not necessary because of the integral mode, which automatically accommodates the offset and nominal setting. The integral action of a PID control system is generally used when the system is trying to maintain the process variable at its nominal operating value and where changes in the process variable will only take place as a result of changes in the load. The derivative mode changes the control signal according to how fast the control error changes. If the error is not changing, the derivative mode does nothing. A slow-changing error signal produces an output signal that changes slowly. If the controlled variable is changing quickly, the control signal is quickly increased to compensate. The derivative action is used to increase the speed of response, whereas the integral action prevents steady-state errors from taking place in the process flow rate or actuator position. PID control is generally used on processes that exhibit rapid and large disturbances in error signals. Figure 8.5.2 shows a typical block diagram of process control system with PID module.

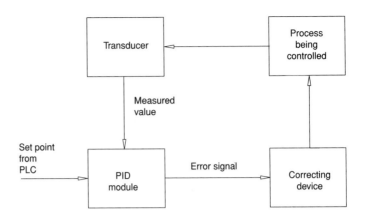

Figure 8.5.2 Block diagram of process control system with PID module

8.6 COMPUTER NUMERICAL CONTROL (CNC)

The first Numerical Control (NC) was built in 1952 at the Massachusetts Institute of Technology (MIT).

Numerical control is a method of controlling the movements of machine components by directly inserting coded instructions in the form of numerical data (numbers and letters) into the system. The system automatically interprets this data and converts it to output signals similar to robots with sensory perception capabilities. These signals, in turn, control various machine components along specific paths comparable to robot manipulator movements.

Computer numerical control (CNC) is a system in which a microcomputer or microprocessor is an integral part of the control panel of a machine or equipment. The part program may be prepared at a remote site by the programmer, similar to robots' off-line programming, and may incorporate information obtained from drafting software packages and machining simulations to ensure that the part program is bug free. The operator can modify the programs directly, prepare programs for different parts, and store the programs. Because of the availability of small computers with large memory, microprocessors, and program editing capabilities, CNC systems are widely used today.

The two types of control circuits used in CNC and robot systems are open loop and closed loop. In the open-loop system, as shown in Figure 8.6.1, the signals are given to the servomotor by the processor, but the movements and final destinations are not checked for accuracy. The closed-loop system, as shown in Figure 8.6.2, is equipped with various transducers, sensors, and counters that measure the position of the robot manipulator (or table at CNC) accurately. Through feedback control, the position is compared against the signal. Movements are terminated when the proper coordinates are reached.

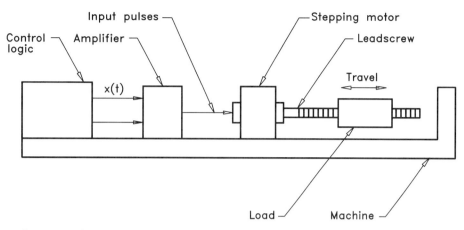

Figure 8.6.1 Open-loop control system for NC

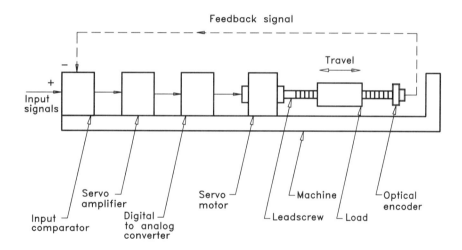

Figure 8.6.2 Closed-loop control system for CNC

8.7 MICROPROCESSOR UNIT (MPU)

A **microprocessor** is a computer central processing unit that is manufactured on a single integrated-circuit chip utilizing large-scale integration technology. The microprocessing unit is the CPU of a microcomputer with its external memory, input/output interface devices, buffer, clock, and driver circuits. A microprocessor is the arithmetic/logic unit and control section of a computer scaled down so that it fits on a single IC chip.

The microprocessor plays a dominant role in computer technology and has contributed uniquely in the development of many new concepts and design tech-

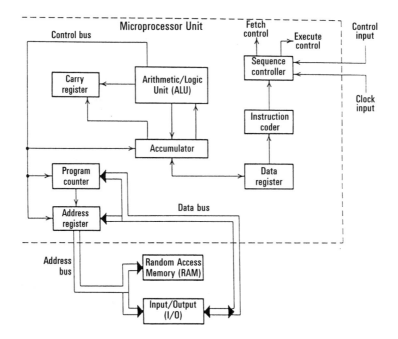

Figure 8.7.1 Microprocessor block diagram

niques for modern industrial systems. The development of the microprocessor has allowed the robot industry to develop to its present state.

A microprocessor is a digital device that is designed to receive data in the form of 1's and 0's. It may then store this data for future processing, perform arithmetic and logic operations in accordance with previously stored instructions, and deliver the results to an output device. In a sense, a microprocessor is a computer on a chip.

A block diagram of a typical microprocessor, as shown in Figure 8.7.1, contains a number of basic components connected in a rather unusual manner. Included in its construction are the arithmetic/logic unit, an accumulator, data register, address registers, program counter, instruction decoder, and sequence controller.

Arithmetic-Logic Unit

The **arithmetic-logic unit (ALU)** (also known as arithmetical element, arithmetic section, and logic section) is the section of the computer that carries out all arithmetic and logic operations on the data words supplied to it. The ALU is a calculator chip made to work automatically by control signals developed in the instruction decoder.

Accumulator

The **accumulator** of a microprocessor is a specific register in which the results of an ALU operation is formed; here numbers are added or subtracted, and certain operations such as sensing, shifting, and complementing are performed. The accumulator is designed to store operands that are to be processed by the ALU. Before the ALU can perform, it must first receive data from an accumulator. After the data register input and accumulator input are combined, the logical answer or output of the ALU appears in the accumulator.

Data Register

The **data register** of a microprocessor serves as a temporary storage location for information applied to the data bus. Typically, this register will accommodate an 8-bit data word. An example of a function of this register is to produce effective storage for the ALU input. In addition, it may be called on to hold an instruction while the instruction is being decoded or it may temporarily hold data prior to the data being placed in memory.

Address Register

The **address register** is the register in which the address part of an instruction is stored by a computer. In some units this register may be programmable. This means that it permits instructions to alter its contents. The program can also be used to build an address in the register prior to executing a memory reference instruction.

Program Counter

The **program counter** (also known as location counter, instruction counter, program counter, or sequence counter) is the counter that indicates the location of the next computer instruction to be interpreted. This unit simply counts the instructions of a program in sequential order.

Instruction Decoder

A **decoder** is a device that translates coded characters into a more understandable form. Each specific operation that the MPU can perform is identified by an exclusive binary number (0 and 1) known as an instruction code. Eight-bit words are commonly used for this code in which digits are interpreted as coefficients of successive powers of the base 2. Exactly 2^8 or 256_{10} separate or alternative operations can be represented by this code. After a typical instruction code is pulled from memory and placed in the data register, it must be decoded. The **instruction decoder** simply examines the coded word and selectively decides which operation is to be performed by the ALU. The output of the decoder is first applied to the sequence controller.

252 Control Systems

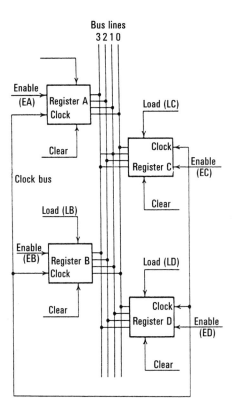

Figure 8.7.2 Registers connected to a common bus line

Sequence Controller

The **sequence controller** performs a number of very vital functions in the operation of a microprocessor, using clock inputs to maintain the proper sequence of events required to perform processing tasks. After instructions are received and decoded, the sequence controller issues a control sign that initiates the proper processing action. In most units, the controller has the capability of responding to external control signals.

Buses

The registers and components of most microprocessors are connected by a bus-organized type of network. The term **bus**, in this case, is defined as a group of conductor paths that are used to connect data words to various registers. A simplification of registers connected by a common bus line is shown in Figure 8.7.2.

8.7 Microprocessor Unit (MPU) **253**

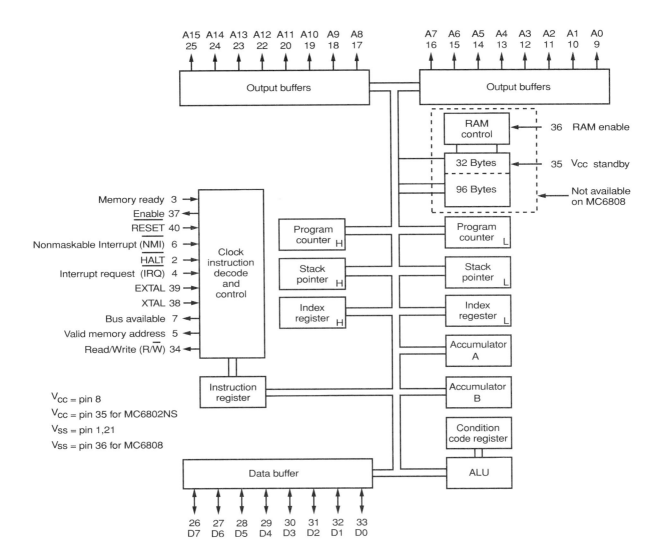

Figure 8.7.3 Block diagram of the Motorola 6802/6808 microprocessor

Microprocessor Architecture

The physical layout or **architecture** of a microprocessor is generally much more complex than the present version described. In practice, it is not really necessary to know what goes on inside a microprocessor other than to have some idea of its primary functions. An expanded block diagram of the Motorola 6802/6808 microprocessor is shown in Figure 8.7.3 to compare its architecture with the simplified unit presented in Figure 8.7.1.

8.8 UNIVERSAL ROBOT CONTROLLER (URC)

The robot needs programs for its controller to execute its motions. Programs for robots can be divided into two areas: operating systems and application programs. Operating systems, which are sometimes called control programs, are furnished by the manufacturer and are fixed for each robot model. They cannot be changed or exchanged with other manufacturers' robots.

Operating systems tell a robot how to execute some motion or move the manipulator. In other words, operating systems supply the intelligence that enables the robot to receive, understand, and carry out tasks that are given to it. In contrast, application programs tell the robot when to do these motions in order to accomplish some specific task. Application programs are also known as user's programs. These programs contain positional data, axis velocity, geometric axis moves, gripper instruction, and interface instructions to other devices and equipment. Operating systems are usually encoded in some type of read-only memory (**ROM**), whose contents are built-in by the manufacturer and cannot be changed.

The complexity and extensiveness of the operating systems program's treatment of the robot's motion today depend mostly on the type of manipulator and controller used. The URC is a PC-based (Windows NT), open architecture, multiple-machine controller that just has emerged into the market. It has been developed by a U.S. company named Robot Workspace Technologies, Inc. The universal robot controller, which has advanced processing capabilities, will be able to fulfill current and future needs by eliminating the robotic islands of automation and substituting all factory controllers with this new device for all robots. The URC provides real-time, dual-Pentium processing power; application accessibility; built-in functionality; intuitive operator **interface**; and integration capabilities. Existing robots can easily, and cost effectively, be retrofitted with the new URC. The manufacturer's specifications for the URC are listed in Table 8.3.

8.9 INTERFACING

Interfacing is the interaction between two or more devices or systems that handle data in different ways, such as in different codes or formats, to communicate with each other.

Interfacing, therefore, is the communication between various components of the robot workcell that establishes a link between the peripheral devices, the controller, and the manipulator's end effector.

Two types of signals are involved in the interfacing of the robot controller: input signals and output signals. The input signals are developed through a receiver circuit within the controller, and the output signals are connected through transistors or dry contact relays.

The input and output signals are called digital signals because the signals are either high or low. The high digital signal operates from a 5 V or 12 V DC source, or even 120 V AC. The low digital signal operates from a 0 V DC reference point. The times at which these signals are either ON or OFF are also controlled by the robot's controller.

The interfacing operation requires the programmer of the robot to assign various lines as either input lines or output lines. If the programmer wants special

8.7 Microprocessor Unit (MPU)

Table 8.3 Universal Robot Controller Specifications

CPU Specifications	Two (2) Intel Pentium 133 MHZ CPUs, slot segmented backplane, MB RAM, 1 2 GB hard disk, 1 44 MB floppy disk drive, two (2) serial ports, parallel port, VGA card.
Communications	1 RS232, 1 Configurable (RS422, RS485), Ethernet. RS232 or backplane Integrate vision, parallel printer, RS232/485 PLC interface. IBM compatible.
Operating System	Windows NT or MS-DOS
Programming Language	Windows NT: The ARMotion Library provides RWR's full range of motion control without restricting the user to a proprietary robot programming language. The ARMotion library is a WIN322 dynamic link library (DLL) that allows the user to program the robot in Visual Basic, Visual C/C++, Excel, Delphi, or any third-party Windows-compatible programming language to develop multimedia, database driven, CAD interface, or sophisticated communication applications. MS-DOS: The ARBasic programming language provides an ANSI-compatible BASIC language with extensions for robot control. ARBasic provides point, tool, and frame manipulation, path control, path coordinated I/O, and other sophisticated robotic control features.
RWT Motion Control Software	Includes robot language interpreter, URL (Universal Robot Language) run-time license, path planner, trajectory generator including kinematics, servo driver, object I/O system. Supports frames of reference motion types, dynamic and static frame/tool offsets, motion termination types, and joint and Cartesian coordinated system conversions. Motion commands; speeds infinitely variable, Cartesian coordinate system conversion. Motion commands: speed configurable accel/decel, digital filters, torque/speed, optimization, relative and ablolute path coordinated I/O, frame & tool shifts unaimed TCP's mirror image, continuous path, collision detection axis and Cartesian moves, velocity modulation analog output.
Operator Panel	10.4″ active matrix TFT 640 × 480 color flat panel display, resistive touch panel, 83 key membrane keyboard trackball.
Teach Pendant	12″ × 8″, D display, deadman switch, 25ft. cable. Weighs 3 lb. Functions include step, teach.
Digital I/O	32 bits, individually configurable. Connector compatible with industry standard relay mounting racks PBB, PB16, PB24, or PBPB32. Pamux optional, up to 512 I/O points.
Analog I/O	A/D - (B) channels, 8 bit, 0–5 volt, differential or single ended. D/A–(6) canailles bit, +/–10 volt optional Analog speed reference output available For modulation of press parameters (flow rate, etc.).
Number of Axes	Multiple machine capability. Six (6) axes per robot, up to (f) robots per controller, up to 24 axes maximum. Fully coordinated motion, tracking.
Motion Types	Joint, linear, circular, tool. Teach motion coordinates include world, part, and tool.
Motor Types	AC servo, DC servo, stepper. Plus or minus 10 volt analog output or step and and direction outputs available for drive interface. Optical encoder or resolver feedback. AC servo drive commutation available with resolver feedback option. Velocity or torque mode control.
Interlocks	9 pin 0 connector suitable for interface to I/O modules. Motor power enable output, brake release output, motor power input, shutdown input. Watchdog timer and servo process monitors assure compliance with the latest in worldwide safety standards.
Loop Closure Time	Less than 10 microseconds per axis plus 50 microseconds overhead.
Servo Fainters	Configurable for each axis, millihertz resolution, automatic optimization, scale factor, pole, zero, notch fitted, position gain, velocity loop bandwith, velocity loop integrator.
Cabinet Dimensions	51.25 in. high × 23.75 in. wide x 19.75 in. deep (1300mm high x 600mm wide × 500mm deep)
Controller Weight	Approximately 300 lb.
Utility Requirements	Manipulator dependent
Operating Temperature Range	20–60 degrees C. Relative humidity range 10–90 percent.

lines, then these lines must be connected so that the interfacing operation can be controlled through programming methods.

The interfacing operation may be developed through a remote connection. This process allows the operator of the workstation to control different work cells from a central location. Status signals from the robot can be sent to the remote workstation, and signals from the remote workstation can be sent to the robot controller. These signals are wired through cables that connect two locations.

In many high-technology automated systems, computers are interfaced with the controller. The programs are written on the computer and then downloaded to the robot controller. The information is sent by using an RS-232C or an RS-422 format for the data. External memory is another type of interfacing for the robot controller. External memory may be a magnetic tape that is connected to the controller or a bubble memory cassette. These external memory devices allow the user to download and upload data for the operation of the robot.

8.10 WORKCELL CONTROL

A robotic workcell may be defined as a cluster of one or more robots and several machine tools or transfer lines that are interconnected in such a way that they work together in unison. All the necessary accessory equipment is embraced within the workcell and, together, establishes a particular work environment. Figure 8.10.1 illustrates a typical robot workcell.

Criteria for establishing and utilizing a robotic workcell include:

1. It must be capable of performing required machining functions on a limited number of parts of sizes within predetermined limits.
2. One worker must be capable of operating the workcell with minimal skill requirements.
3. Selection of operator personnel from the shop floor must be possible.
4. Manual parts fabrication in the workcell must be feasible in case of system failure.
5. Compliant end effectors must be used because of inherent inaccuracies of available commercial robots.
6. Part programming must be done on line.
7. Safety sensors must be installed for protection of personnel, equipment, and work in process.
8. Consistent quality must be maintained on parts.
9. Productivity must be increased to make the system economical.
10. Early implementation must be made for quickest payoff.

The workcell controller performs several important functions in the robot installation. These functions can be divided into three main categories:

1. Sequence control
2. Operator interface
3. Safety monitoring

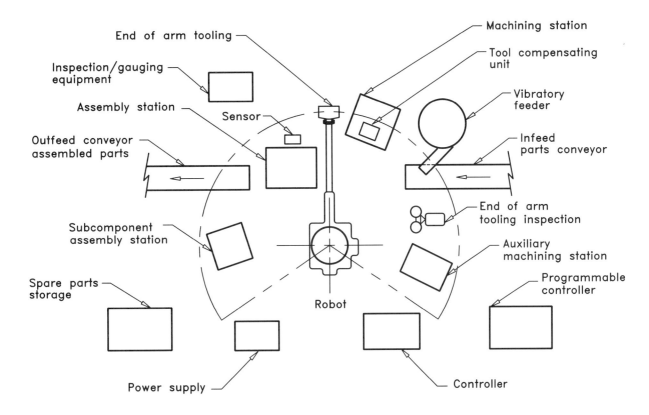

Figure 8.10.1 Typical workcell: A robot is positioned to service a number of machine clusters.

Sequence Control

Sequence control is the basic function of the workcell controller. It is concerned with regulating the sequence of activities in the cell. The sequence is determined not only by controlling the activities as a function of time; it is also determined by using interlocks to ensure that certain elements of the work cycle are completed before other elements are started. In a loading and unloading application input/output, interlocks in sequence control are used for purposes such as the following:

1. Making sure that the part is at the pickup location before the robot attempts to grasp it.
2. Ensuring that the part is properly loaded into the machine before the processing cycle begins.
3. Indicating to the robot that the machine cycle is completed and the part is ready for unloading.

Additional functions within the scope of sequence control include logical decision making, computations, and control of irregular elements that must occur with a certain frequency.

Decisions associated with the work cycle are typically determined according to the value of an input interlock. Computations are sometimes required to support a robot work cycle. Programs for palletizing operations are most efficiently written by calculating the positions, pickup or drop-off points.

Operator Interface

Means for the operator to interact with the robot cell must be provided. Reasons for establishing such interface include the following:

1. Programming the robot.
2. Participation in the work cycle. An operator and the robot each perform a portion of the work in the cell. The operator typically accomplishes tasks that require judgment or sensory capabilities that the robot does not possess.
3. Data entry, such as part identification or part dimensions.
4. Emergency stopping of the cell activities.

Safety Monitoring

Emergency stopping of the robot cycle requires that an alert operator be present to notice the emergency and take positive action to interrupt the cycle. Safety emergencies are not always so convenient as to occur when an alert operator is present. A more automatic and reliable means of protecting the cell equipment and people who might wander into the work zone is called safety monitoring.

Safety monitoring is a workcell control function in which sensors are used to monitor the status and activities of the cell and detect the unsafe or potential unsafe conditions.

Various sensors can be used to implement a safety monitoring system in a robot cell. These sensors include simple limit switches that detect whether the movement of a particular component has occurred correctly, temperature sensors, pressure-sensitive floor mats, light beams combined with photosensitive sensors, and machine vision systems.

The safety monitoring system is programmed to respond to various hazard conditions in different ways. These responses might include one or more of the following:

1. Complete stoppage of a cell
2. Slowing down the robot speed to a safe level
3. Warning buzzers to alert maintenance personnel of a safety hazard in the cell

4. Special program subroutines to permit the robot to recover from a particular unsafe event

8.11 SUMMARY

A control system is an internal part of a workcell of the entire enterprise network. The major goal of a control system in automated manufacturing is to integrate various operations to improve productivity, increase product quality and uniformity, minimize cycle time and effort, and reduce labor cost. The four classifications of control systems in robotics are:

1. Limited sequence
2. Playback with point-to-point path
3. Playback with continuous path
4. Intelligent

The advancements in control system technology brought improvements in manufacturing operations by integrating the process control and machine control as one distributed system for the entire manufacturing plant. This was accomplished by the evolution of the Microprocessor Unit (MPU), the communication network or Manufacturers Automation Protocol (MAP), the Local Control Units (LCUs), and the Local Area Networks (LAN).

A modern device used widely for controlling systems is the Programmable Logic Controller (PLC). A PLC is a microcomputer-based device that uses stored instructions in programmable memory to implement logic, sequencing, timing, and arithmetic control functions through digital or analog input/output modules for controlling various machines and processes.

Ladder logic is a term used to describe the format of a schematic diagram to be entered into a PLC.

A programming terminal is a device used for entering and monitoring a program in the CPU's memory of a computer system. There are two basic methods of entering a program into a PLC. It can be done by a dedicated programming terminal or computer-based programming terminal.

The major components of a PLC are:

1. Input module (I)
2. Output module (O)
3. Central processing unit (CPU)
4. Power supply (typically 115 V DC)
5. PLC memory (for logic and sequencing instructions)
6. Programming detachable device (for entering the program)

Proportional-Integral-Derivative control, or PID, is used for any process condition to provide feedback to a PLC. PID is a mechanical equation that provides closed-loop control of a process system. It is a sophisticated analog control network for inputs and outputs that is constantly changing.

Computer numerical control (CNC) is a form of programmable automation in which the mechanical actions of a piece of equipment are controlled by a program containing coded data supplied by a microcomputer or microprocessor.

The operating principle of CNC is to control the motion of the workhead relative to the work part and to direct the sequence in which the motions are carried out. The basic robot system is an example of a machine that fits the definition of CNC because the robot's manipulator movements are controlled by a program supplied by a microcomputer similar to CNC.

A microprocessor is the central processing unit of a microcomputer. It uses a memory-mapped technique to address I/O devices. It supports interface devices such as codes, serial A/D converters, and others. It also provides buffer, clock, and driver circuit assistance, and in general its numerical applications are very suitable in robotic computations, based on algorithms for kinematics and dynamic movements.

Interfacing is the interaction between two or more devices or systems with different data to communicate with each other. In interfacing external components, devices and systems communicate with a robot's workcell, peripheral equipment, controllers, and manipulator's end-effector sensors.

There are two types of signals involved in interfacing of the robot controller: input signals and output signals. The input signals are developed through a receiver circuit within the controller; the output signals are connected through transistors.

In industrial applications, robots involve and operate with other pieces of equipment, material-handling devices, and even human beings. A means of coordinating the activities of the different equipment must be established. This is done by using workcell control to provide the sequencing and coordination of the workcell equipment and devices.

8.12 REVIEW QUESTIONS

8.1 What is a control system in a workcell, and how can its primary purpose be defined?
8.2 What are the four types of control systems used in robotics?
8.3 What technological advancements brought improvements in manufacturing operations?
8.4 What is a programmable logic controller?
8.5 List the components of a PLC and describe the function of each.
8.6 Compare the function and operation of an open-loop positioning system and a closed-loop positioning system.
8.7 Under what circumstances is a closed-loop positioning system preferable to an open-loop system?
8.8 List the six basic points regarding ladder logic diagrams.
8.9 What is the main difference between ladder logic and relay logic?
8.10 What do the horizontal and vertical lines represent on a ladder diagram? Explain.

8.11 What is a computer numerical control and how can its operating principle be defined?
8.12 How is an industrial robot similar to numerical control?
8.13 Describe the memory systems used in the robot controllers.
8.14 What is a microprocessor? Explain its function.
8.15 What does an ALU in a microprocessor unit do?
8.16 What is an accumulator and what function does it perform?
8.17 Describe the purpose of a data register and address register.
8.18 Explain the architecture of a microprocessor.
8.19 Describe how communication lines are established between the controller and the peripheral devices.
8.20 Name two types of input signals for the interfacing operation.
8.21 What is the logic voltage level for an input signal?
8.22 What type of computer port on the controller is able to transmit data as well as receive data?
8.23 Define the difference between an input interface and an output interface.
8.24 What is the function of the programming device or programming terminal?
8.25 Describe the functions that are performed by the workcell controller, their purposes and interactions.
8.26 What are the two basic methods of entering a program into a PLC?
8.27 What does an intelligent dedicated programming terminal contain?
8.28 What are the two advantages of using computer-based programming terminals?
8.29 What are the four types of software packages available with computer-based programming terminals?
8.30 What is a host computer?
8.31 What is the difference between open-loop and closed-loop control?
8.32 Why is feedback used in process control systems?
8.33 Describe the four main types of feedback used in process control systems.
8.34 Describe the basic operation of a PID control loop
8.35 What principle do PID modules employ to anticipate changes in control variables?

8.13 PROBLEMS

8.1 Draw a ladder logic diagram showing two stop push-buttons and two start push-buttons controlling one motor. Include the overload contacts in your diagram and use the following I/O addresses: $Stop_1$ = A, $Stop_2$ = B, $Start_1$ = C, $Start_2$ = D, Overload = E, and Motor = F.
8.2 Figure 8.13.1 shows how one limit switch is used to control two output devices. Redraw the circuit in ladder logic form without using an NC contact.
8.3 Redraw the circuit in Figure 8.13.2 in ladder logic form, eliminating the vertical contact.
8.4 Convert the program of Figure 8.13.3 to a circuit that will function on any PLC system.

262 Control Systems

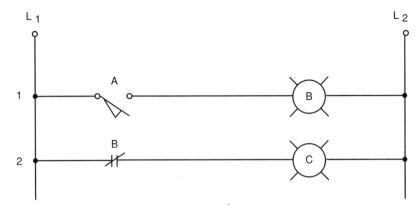

Figure 8.13.1 Relay logic diagram for Problem 8.2

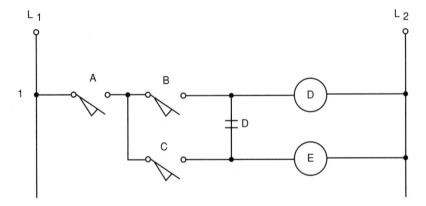

Figure 8.13.2 Relay logic diagram for Problem 8.3

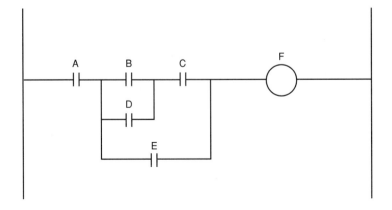

Figure 8.13.3 Ladder logic diagram for Problem 8.4

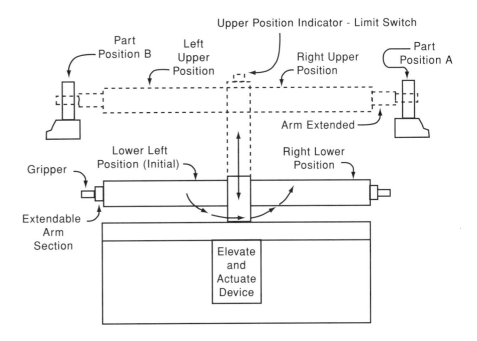

Figure 8.13.4 Basic pick-and-place robot for Problem 8.5.

8.5 Refer to the robot shown in the Figure 8.13.4. Develop a pattern for the sequence of operation. You must move a part from position A to position B. The robot starts at the lower-left, initialized position.

8.6 The workcell in Figure 8.13.5 is similar to Problem 8.5. It differs only in that it has six active positions instead of four. It has a positioning solenoid for each of its placements and also an arm-extend and gripper-close actuators. Develop a ladder logic PLC program to move a part from LM to UR and also a program to move a part from UR to LL.

8.7 For the robot shown in Figure 8.13.6 develop a timed sequence program (time of your choice) to accomplish the following order:

1. Starting position as shown.
2. Pick up part at A5.
3. Move part to B4.
4. Return to initial starting position.

264 Control Systems

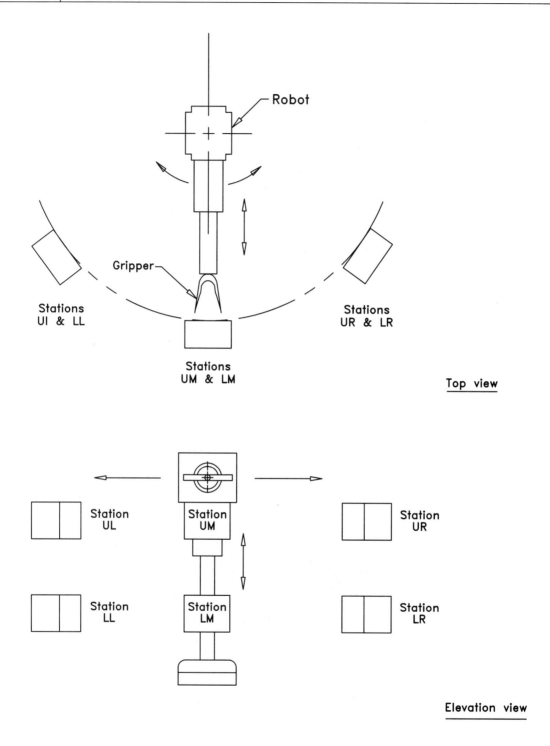

Figure 8.13.5 Workstation/robot for Problem 8.6

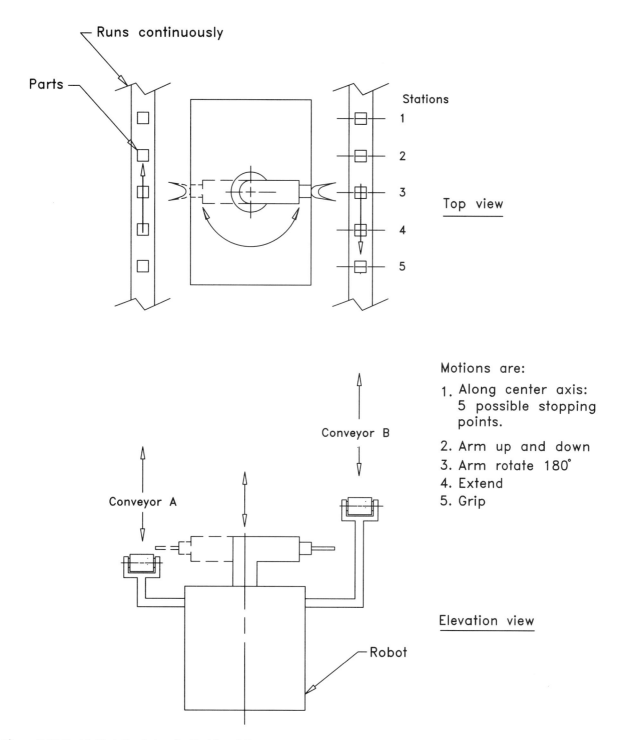

Figure 8.13.6 Multi-station/robot for Problem 8.7

8.14 REFERENCES

Amic, P. J. *Computer Numerical Control Programming.* Englewood Cliffs, NJ: Prentice Hall, 1997.

Asfahli, C. R. *Robots and Manufacturing Automation.* 2d ed. New York: John Wiley & Sons, Inc., 1992.

Babb, M. IEC 1131-3, "A Standard Programming Resource for PLCs." *Control Engineering* (February 1996): 67–68.

Chang, C. H., and M. A. Melkanoff. *Machine Programming and Software Design.* Englewood Cliffs, NJ: Prentice Hall, 1989.

Fievet, J. P. "The Use of PLCs in Paper Mill Load Shed Systems." *Tappi-Journal* 80, no. 3 (1997): 105–109.

Friedman, S. B. *Logical Design of Automation Systems.* Englewood Cliffs, NJ: Prentice Hall, 1990.

Gray, J. O. *Advanced Robotics and Intelligent Machines.* London, U.K.: IEEE, 1996.

"How PLC Opens New Automation Opportunities." *Electrical World* (February 1997): 46–47.

Martin, V. D. "Programmable Logic Controllers," *Electronics Now* (April 1996): 37-38.

Morris, H. M. "Operator Interfaces Let Users and PLCs 'Talk' to Each Other." *Control Engineering* (April 1996): 69–70.

Noaker, P. M. "Down the Road with DNC." *Manufacturing Engineering* (November 1992): 35–38.

Stevenson, J. *Fundamentals of Programmable Logic Controllers, Sensors, and Communications.* Englewood Cliffs, NJ: Prentice Hall, 1993.

Tsafestas, S. A. *Microprocessors in Robotics and Manufacturing Systems.* Boston, MA: Kluwer Academic Publishers, 1991.

Webb, J. W., and R. A. Reis. *Programming Logic Controllers—Principles and Applications.* 3d ed. Englewood Cliffs, NJ: Prentice Hall, 1995.

CHAPTER 9

Programming

9.0 OBJECTIVES

After studying this chapter, the reader should:

1. Be acquainted with robot programming
2. Understand programming methods
3. Recognize programming languages
4. Be able to define the levels of robot programming
5. Be familiar with space position programming
6. Be informed of motion interpolation
7. Be aware of program statements
8. Comprehend sample programs

9.1 ROBOT PROGRAMMING

A robot must be programmed to do useful work and perform its motion in a workcell. A robot program can be defined as a path of movements of its manipulator, combined with peripheral equipment actions to support its work cycle. The peripheral equipment actions include operation of the end effector, making logical decisions, and communicating with other pieces of equipment in the workcell.

Programming, therefore, represents one of the most advanced areas in robot technology, particularly as compared with earlier programming methods. Earlier programming techniques were largely associated with replacing the skills of human operators with a playback electromechanical device as a source of programming information. Of historic interest, diagrams illustrating earlier versions of the teach-and-playback method design by Unimation are shown in Figure 9.1.1 and Figure 9.1.2. These systems were used in the early 1970s and provided very little programming flexibility. Today, the playback systems in use are the same in principle but utilize powerful digital computers in place of earlier electromechanical parts.

Many automated systems today are controlled by computers with multiple processors to perform logical decision and to permit high-level machine communication with other pieces of equipment. This type of computer furnishes the following functions in the robot environment:

1. Robot manipulation
2. Sensing
3. Logical decision making
4. Data processing

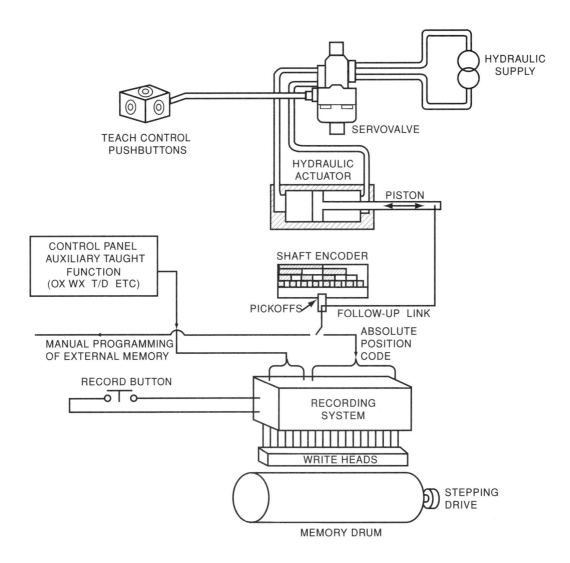

Figure 9.1.1 Programming mode in the early 1970s used by Unimation, Inc.

Robot manipulation is the control motion of all robot joints and work-envelope actuators, including position, velocity, and path control devices and actuators.

Sensing is the collection of information within the physical workcell, including sensory and control of peripheral equipment.

Logical decision making is the ability to use information gathered from the workcell to modify a system operation or to designate various preprogrammed paths and routines.

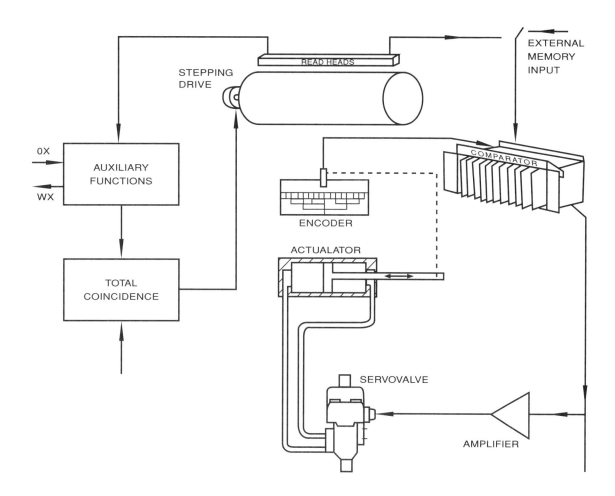

Figure 9.1.2 Playback mode in the early 1970s used by Unimation, Inc.

Data processing is the capability to use databases to communicate with other intelligent machines and keep records, exchange programs, generate reports, and control activity in the work environment.

Many automated manufacturing systems use all four functions to accomplish large production. However, for small-bench production, this is not recommended because of cost incentives.

In this chapter, we discuss the method by which robot programming is accomplished, some common programming languages, the levels of robot programming used, and finally, we examine motion interpolation with program statements and sample programs.

9.2 PROGRAMMING METHODS

A robot is programmed by entering the programming commands into its controller memory. There are five methods of entering the commands:

1. Manual
2. Teach pendant
3. Walk-through
4. Computer terminal
5. Off-line

Manual programming is used for robots with point-to-point open-loop controllers. This method is associated with limited-sequence pick-and-place robots. The sequence in which the motions occur is regulated by a sequencing device (stepping switch, cam, and the like). This device determines the order in which each joint is actuated to form the complete motion cycle. Setting up and adjusting the necessary end-stops, switches, and cams and wiring the sequence are more of a manual type of machine setup than a programming method. This type of programming is typical of the first-generation approach. However, these robots appear to be simple in nature and are capable of performing many manufacturing tasks.

Today, and in the foreseeable future, nearly all industrial robots will have digital computers as their controllers together with comparable storage devices as their memory units.

Manual programming is usually simple. It does not require a skilled operator, and the capital investment and maintenance costs are low. These robots are capable of high operating speeds and have good accuracy and repeatability. However, their flexibility is limited. They have only two or three degrees of freedom. Control of intermediate points along the path is generally not available, and only two positions are programmed for each axis.

Teach pendant programming and robot language programming are the two methods most commonly used today for entering commands into computer memory. **Teach pendant programming** involves the use of a teach pendant (hand-held control box) that has toggle switches or contact buttons for controlling the movements of the manipulator joints. Figure 9.2.1 illustrates a teach pendant.

By using toggle switches or contact buttons, the programmer power drives the robot arm to the desired positions in sequence, and records the motions into memory. During the playback testing, the robot moves through the sequence of positions under its own power. This method of programming is commonly used for playback robots with point-to-point control. The teach pendant is simple to learn and suitable for programming many tasks found in industry. It does not require a skilled operator. However, complex motions and applications requiring close tolerances may require a lengthy programming time. The robot must be operating during programming. The program cannot be entered into the pendant while the robot is off-line (away from the robot cell).

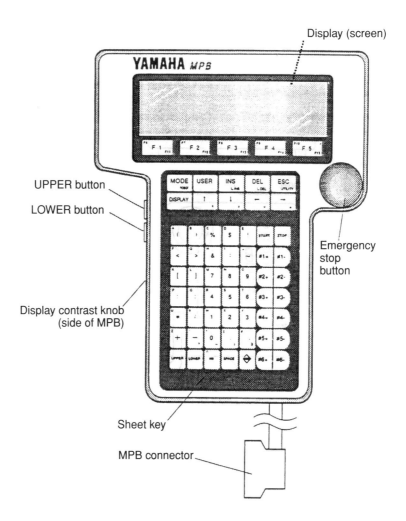

Figure 9.2.1 Teach pendant Yamaha model MPB features an easy-to-see 40-character by 8-line large LCD display and built-in error-history display and I/O monitoring. *(Courtesy of Yamaha Robotics Corp.)*

Walk-through programming is used for playback robots with continuous path control, irregular motion patterns, and for roughly locating the tool center point for some robots. Spray painting, arc welding, grinding, deburring, and polishing are the most common industrial applications. This programming method requires the operator to physically grasp the end effector and manually move it through the motion sequence, recording the path into memory. If the robot arm itself is difficult to move, due to its weight, a special device often replaces the actual robot for the teach procedure. The device has similar joint configurations to the robot, and it is equipped with a trigger handle or control switch that is activated by the operator, recording motions into memory. The motions are recorded as a series of closely spaced points. During playback testing, the actual robot arm goes through the same sequence of points. In walk-through programming, a highly skilled operator is required to perform the precise motion tasks if the robot cannot be programmed off-line. Older programming methods recorded locations at a defined sampling frequency, say 3 points/sec, but the main problem was that the programming points could be filled or exceeded.

Computer terminal programming can be done off-line or on-line (at the robot's cell). Final testing or playback of the program is done at the job site.

The difference between teach pendant and computer terminal programming is that the first uses manual movements to record the robot's actions and the second uses textual language commands that are written in English-like statements to perform the programming. The program, therefore, is written and debugged in the actual computer. Many systems provide a combination of both methods.

Computer programming provides greater flexibility and high efficiency. The robot does not have to be taken out of operation while the program is being written and debugged. Therefore, productivity is not affected.

Digital computers took over the control function in robots. High-level computer languages allow programming of more-complex operations and tasks to be performed by the robots.

However, as mentioned previously, this requires the operator to be experienced in the use of computers, high-level languages, and to have a good knowledge in programming logic.

Off-line programming permits the robot program to be prepared at a remote computer terminal and downloaded to the robot controller for execution. The advantages of off-line programming is that simulation techniques and new programs may be used to refine and modify operations without taking up valuable robot time. In the past, off-line programming packages had to make adjustments for geometric differences between three-dimensional computer models and the actual physical cell.

Today, vision and sensor systems have been used to update the three-dimensional computer models and avoid the necessity for the calibration step in current off-line programming methods. This approach is called world modeling and is associated with high-level language statements in which the programmer would specify a task to be performed without giving details of the procedure used to perform the task.

Off-line programming applications are increasing today and will grow more in the future because of the graphical computer simulations used to validate program development.

Off-line programming systems, generally providing a computer graphic interface, allow robots to be programmed without access to the robot itself during programming. Off-line programming also serves as a natural vehicle to tie **com-**

puter-aided design (CAD) databases used in the design phase of a product to the actual manufacturing of the product. In some cases, this direct use of CAD data can dramatically reduce the programming time required for the manufacturing process.

Off-line programming indicates that the programming is done in advance by the use of a robot programming language rather than being done in a teaching mode. The advantages of off-line programming are:

1. It can prepare programs without using the robot so that the robot is available for other uses.
2. Layout and cycle time of the operations can be optimized in advance.
3. Previously worked-out procedures and routines can be incorporated in the program. It is not necessary to redesign each operation every time it is used.
4. Sensors can be used to detect external stimuli, and appropriate action can be taken in response. This action would increase the complexity of programming, and the robot would be considered to be in the adaptive mode of operation.
5. Existing computer-aided design (CAD) and **computer-aided manufacturing (CAM)** information can be incorporated into the control functions.
6. Programs can be run in advance to simulate the movements actually programmed without incurring the risk of damage. There are computer display techniques, for example, that simulate the movements of the robot on screen to assist in program development.
7. Robots can be used to manufacture individual units by calling on previously developed routines. It would be impractical to do this by guiding and attempting to match different paths.
8. Engineering changes can be incorporated quickly by substituting only that part of the program that controls a particular design characteristic.

Controlling a single robot arm to carry out a task allows the program to be greatly simplified. In a nonadaptive system, there is no sensory feedback from the environment, so the only inputs are the position sensors in the joints. Therefore, a simple program design can be used. The VAL II program, as well as several others, has been found to be very useful in this mode of operation.

Off-line programming systems are important, both as aids in programming present-day industrial automation and as platforms for robotics research.

An off-line programming system includes a spatial representation of solids and their graphical interpretation, automatic collision detection, incorporation of kinematics, path planning and dynamic simulation, simulation of sensors, concurrent programming, translation by postprocessors to various target languages, and workcell calibration.

Because the scope of this text does not allow for advanced languages, the motivated reader should refer to the user's guide from the list of manufacturers at the back of this text for a more complete description with examples of existing languages.

9.3 PROGRAMMING LANGUAGES

Robot programming language is a set of words and rules governing their use, employed in constructing a program. The program provides the control instruction(s) required for the robot to perform its intended task.

There are many programming languages available, but most are limited to use on one or only a few robots. Most of these languages are explicit, meaning that each of their instructions gives the robot a specific command to do a small step (e.g., open gripper) as a part of a larger task. However, the advanced high-level languages can give instructions at the task level (e.g., assemble 100 of product xyz).

High-level language is a programming language, such as BASIC, C, or C++, that is not dependent upon the machine language of a computer. High-level language has been designed to allow the use of words similar to those in the English language.

Low-level language is a programming language, using symbolic code, that is based on the language of a particular computer and requires an assembler to translate it into actual machine language.

Machine language is a programming language comprised of a set of unique machine codes that can be directly executed by a given computer. Each computer has its own machine language. Programmers rarely use machine languages today because instructions and data must be in binary notation. High-level languages are preferred because they can be easily translated into machine language by the compiler. Loosely, machine language is called assembly language.

Most robot programming implemented today uses a combination of language programming and teach-pendant control programming. The high-level language programming is used to define the logic and sequence of the program for the robot's actions; the specific point locations or positioning in the workplace are defined using teach pendant control.

The first robot programming language was WAVE, developed in 1973 as an experimental language for research purposes by the Stanford Artificial Intelligence Laboratory.

The second development of a subsequent language began in 1974 at Stanford again. The language was called AL, and it could control multiple arm coordination. The concepts of WAVE and AL, which was known as Stanford arm, went into development of the first commercially available robot textual language. VAL (for Victor's Assembly Language—after Victor Scheinman, the developer of Stanford arm) was introduced in 1979 by Unimation Inc. for its PUMA robot series, which was a combination of WAVE and AL languages. VAL II was an upgraded language of VAL and released in 1984.

Many other languages have been developed since then for various applications and robot control enhancements. There are as many robot programming languages as there are robot manufacturers. No set standards have been established yet in the industry for interchangeability of programs among manufacturers.

Some common languages and their sources are:

- AML—IBM
- HELP—General Electric
- VAL & VAL II—Unimation, Adept
- RAIL—Automatrix
- MCL—McDonnell Douglas
- RPL—SRI
- AL—Stanford University
- AR-BASIC—American Cimflex
- Robot-BASIC—Intelledex
- Karel—GM, Fanuc Robotics
- JARS—NASA's Jet Propulsion Laboratory
- T^3—Cincinnati Milacron

Even though robot programming languages have been developed by different manufacturers, many are similar in structure. According to J. J. Craig (1989), the main difference found is the choice of key words and commands. Following are two example programs written in AL and KAREL for palletizing applications.

Palletizing Application Written in AL

BEGIN "Palletizing sample program"
    FRAME in_pallet, out_pallet, part;
    COMMENT
        The (1 , 1) positions of the pallets and
        grasping position of the parts;
    VECTOR del_r1, del_c1;
    VECTOR 'del_r2, del_c2
    COMMENT Relative displacements along the rows and columns;
    SCALAR r1, c1, ir1, ic1;
    SCALAR r2, c2, ir2, ic2;
    COMMENT counters;
    EVENT in_pallet_empty, in_pallet_replaced;
    EVENT out_pallet_full, out_pallet_replaced;
    COMMENT
        Here insert the frame definitions for IN_PALLET
        and OUT_PALLET and the vector value for displacements
        along the rows and columns. These would be taught and
        recorded using the robot. FRAME definitions are typically
        unreadable by humans;
    COMMENT
        Now define the procedure PICK and PLACE called in the

```
                    main program later on;
            PROCEDURE PICK;
            BEGIN
            FRAME pick_frame
            ir1 : = ir1 + 1 ;
            IF ir1 GT r1
            THEN
            BEGIN
                ir1 : = 1;
                ic1 : = ic1 + 1;
                If ic1 GT C.
            THEN
            BEGIN
                SIGNAL in_pallet_empty;
                WAIT in_pallet_replaced;
                ic1 : = 1;
            END;
       END;
            pick_frame : = in_pallet+(ir1-1)*del_r1+(ic1-1)*del_c1;
            MOVE BHAND TO pick_frame
            CENTER BARM;
            AFFIX part TO BARM;
       END;
       PROCEDURE PLACE;
       BEGIN
            FRAME place_frame
            ir2 : = ir2 + 1;
            IF ir2 GT r2
            THEN
            BEGIN
                ir2 : = 1;
                ic2 : = ic2 + 1;
                IF ic2 GT c2
                THEN
                BEGIN
                    SIGNAL out_pallet_empty;
                WAIT out_pallet_replaced;
                    ic2 : = 1;
                END;
            END;
            place_frame : = out_pallet+ (ir2-1) *del_r2+ (ic2-1) *del_c2;
            MOVE part TO place_frame;
            OPEN BHAND TO 3.0*IN;
            UNFIX part FROM BARM;
       END;
```

```
        COMMENT The main program;
        OPEN BHAND TO 3.0*IN;
        WHILE TRUE DO
        BEGIN
            PICK;
            PLACE;
        END;
END;
```

Palletizing Application Written in KAREL

```
program PALLET
--   Transfers workpieces from one pallet to another.
var
--   Variables for the input pallet:
     BASE1             : position      -- (1 , 1) position on pallet
     IR1 , IC1         : integer       -- counters for rows & cols
     NR1 , NC1         : integer       -- limits for rows & cols
     DR1 , DC1         : vector        -- delta between rows & cols
     ISIG1 , OSIG1     : integer       -- signals for pallet changing
--   Variables for the output pallet:
     BASE2             : position      -- (1 , 1) position on pallet
     IR2 , IC2         : integer       -- counters for rows & cols
     NR2 , NC2         : integer       -- limits for rows & cols
     DR2 , DC2         : vector        -- delta between rows & cols
     ISIG2 , OSIG2     : integer       -- signals for pallet changing
```

```
routine PICK
--   Pick a workpiece from the input pallet.
var
     TARGET            : position      -- target pose
begin
     IR1 = IR1 + 1
     if IR1>NR1
     then
         IR1 = 1
         IC1 = 1 + 1
         if IC1>NC1
         then
             IC1 = 1
                       -- get a new pallet
             , dout [OSIG1] = true
                       -- notify pallet_changer
             wait for din [ISIG1] +
```

```
                                        - - wait for input line to go high,
                                        - - meaning pallet has been changed
                            dout [OSIG1] = false
                                        - - turn off our output signal
                                        - - compute target pose
                endif
        endif
        TARGET = BASE1
                                        - - start with (1 , 1) pose
        shift (TARGET , (IR1-1) *DR1 + (IC1-1) *DC1)
                                        - - shift for row and col offset
                                        - - get the part
        move near TARGET by 50          - - move to 50 mm away from TARGET
        move to TARGET
        close hand 1
        move away 50                    - - back away from TARGET by 50 mm
end PICK
```

```
routine PLACE
- - Place a workpiece on the output pallet.
var
        TARGET  : position              - - target pose
begin
        IR2 = IR2 + 1
        if IR2>NR2
        then
                IR2 = 1
                IC2 = IC2 + 1
                if IC2>NC2
                then
                        IC2 = 1
                                        - - get a new pallet
                        dout [OSIG2] = true
                                        - - notify pallet-changer
                        wait for din [ISIG2] +
                                        - - wait for input line to go high,
                                        - - meaning pallet has been changed
                        dout [OSIG2] = false
                                        - - turn off our output signal
                                        - - compute target pose
                endif
        endif
        TARGET = BASE2
                                        - - start with (1 , 1) pose
```

```
        shift (TARGET , (IR2-1) *DR2+ (IC2-1) *DC2)
                        - - shift for row and col offset
    move near TARGET by 50        - - move to 50 mm away from TARGET
    move to TARGET
    open hand 1
    move away 50       - - back away from TARGET by 50 mm
end PLACE
```

MAIN PROGRAM

```
begin
            IR1 = 0 ; IC1 = 0      - -initialize counters
            IR2 = 0 ; IC2 = 0
    - -   initialize other variables:
    - -
    - -   BASE1 , NR1 , NC1 , DR1 , DC1 , ISIG1 , OSIG1
    - -
    - -   BASE2 , NR2 , NC2 , DR2 , DC2 , ISIG2 , OSIG2
    - -
    - -   numerical pose definitions omitted here
            open hand 1
            while true do         - - loop
                 pick
                 place
            endwhile
end PALLET
```

These two palletizing examples from Gruver and Soroka (1988) show the actual code of how to accomplish the task in different robot programming languages.

Each of these programs solves the same scenario: Pick a part from a pallet with r1 rows and c1 columns and put it into a pallet of r2 rows and c2 columns; signal or wait for presentation and removal of full or empty pallets. These programs are documented or self-documenting so that they should be able to be followed with a careful reading.

Basic Robot Programming

Robot programs can fall into three categories:

1. Operating system programs
2. Application (or user's) programs
3. Special programs

Operating system programs control the basic operation of the robot system. They are developed by the manufacturer, and they give the controller the necessary routines for the robot's general operation. Such programs cannot be altered by the user without damage to the robot operating system.

Application programs are the ones that the users write to meet their applications. They contain information about axis velocity data, location geometry, and the application of specific information for the manipulator when it is at the programmed location.

Special programs are often used in fully automated workcells to program the flow of information through the cell controller and to establish network communication between intelligent devices and standard programs in a production environment. Because such programs are written either for a specific application or are developed around an existing software, they are written by the user or by a third party.

The development of a program for a robot system involves four steps:

1. Defining the type of robot used
2. Defining the task of the robot
3. Identifying the sequence of events by which the program must be structured
4. Identifying the conditions of the program

For the construction of an in-house program, the same procedures are applied as in a computer program. The program is developed through the four steps just as outlined here. Then, a flowchart or **pseudo code** (which enlists statements about the program) is written to correspond to the program and to identify the programming process. Next, the flowchart is converted into machine codes, and finally, the program is entered into the controller through a device for teaching the information to the robot. After the program has been entered, it can be recalled from memory to replay the stored steps.

9.4 LEVELS OF ROBOT PROGRAMMING

Today, the majority of robots in industry use hierarchical control programming. However, the trend is toward task-oriented programming, which simplifies the programming task. In Chapter 4, we discussed hierarchical control positions into a number of different levels. Each level accepts simpler commands to the next lower level. Figure 4.2.1 illustrates the control structure and how the hierarchy is divided into simple tasks.

Because some authorities recognize the four basic levels of robot programming and controlling systems as shown in Figure 9.4.1, we are going to limit our discussion only to a brief description.

Level 1 programming concentrates on the physical control of robot motion in terms of joint or axis, leading-by-hand teaching (manual) and front-panel programming. Input/output is not incorporated in the programs. This level of programming was used in the early 1970s. Applications include some spot welding and pick-and-place tasks.

9.4 Levels of Robot Programming

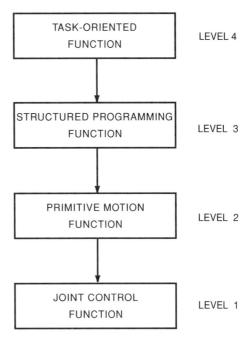

Figure 9.4.1 Levels of robot programming

Level 2 programs are written in simple robot-programming languages, and their techniques assist programmers in entering motion, branching, coordinate-transformation, and signal instructions. Such systems may also provide a number of the lead-by-hand and control-panel operations that level 1 systems provide. These programming capabilities permit the running of user programs that are more complex than level 1 systems. Representative applications are found in some palletizing and arc welding uses.

Level 3 programs are the most modern and growing methods of robot programming. The programming languages incorporate extended capabilities, including structured constructs, full arithmetic functions, external robot-path modifications, and supervisory computer-communications support. Because these systems support the functions and features also found on the two lower levels, level 3 systems can handle level 2 applications as well as modify the robot arm's path, based upon data transmitted from external sensing devices, including machine vision.

The primary function of level 4 programming is to distinguish from the user commands and program structure what usually must be written by the programmer.

In today's manufacturing environment, where low-cost operations are critical for success, the user needs higher levels of control to solve his or her problems. However, each higher level takes more-complex task commands and higher-level sensory information to perform satisfactorily. Programming at level 4 should have the following characteristics:

1. Higher levels of control for positioning commands.
2. If path control or robot motion is not specified, one must be determined.
3. Programming is to be permitted in natural language.
4. A world modeling system should assist the robot to keep track of objects.
5. Programming should permit accident-free motion.
6. Robot teaching should be simplified.

However, today some limited programming at this level is available. A great deal of research is underway at universities and industrial laboratories, but much more is needed before such a system can be implemented. Industrial applications indicate that level 3 robot programming is the most frequently used today. Figure 9.4.2 shows a robot system coordinated by means of a structured programming language.

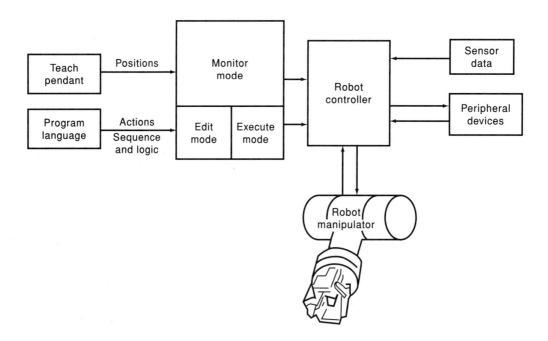

Figure 9.4.2 Diagram of robot system coordinated by means of a structured programming language

9.4 Levels of Robot Programming 283

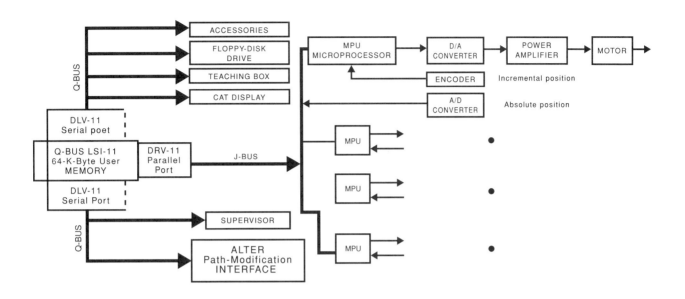

Figure 9.4.3 VAL-II robot controlling system. A Q-bus was used by Unimation, Inc. to connect the LSI-11 to the components of a robot system. A J-bus connects the 64K byte user memory to the processor-control system.

A good example of an advanced level 2 programming system is VAL-II. The VAL-II programming system, developed by Unimation, provides the functionality of some level 3 systems. As shown by Figure 9.4.3, the basic hardware comprises a standard 19-inch Q-bus backplane, a servo control subsystem, a trajectory-human-interface subsystem, and a power-distribution system. VAL-II, which supplements an earlier VAL-I system, has 64K bytes of C-MOS memory with battery backup (about twice that of the earlier VAL-I system). Another 64K bytes of C-MOS memory contain the operating system. The system also includes a double-sized/double-density disk unit that stores up to 10,000 program steps per diskette, or about 1 megabyte of memory. VAL-II also makes eight serial ports available to users.

In addition, the VAL-II operating system uses a time-sliced control scheme. During operation, about half of each major clock cycle is devoted to arm-trajectory planning and computation. The remaining portion is divided into time slots for the CRT interface, the supervisory computer interface, external path modification, monitor-command execution, main-program step execution, and process-control program step execution. Each task is scheduled to run for a given amount of time during each major clock cycle to ensure that the system can respond to time-critical events as quickly as possible. Any event will be received and acted upon within 28 milliseconds of occurrence.

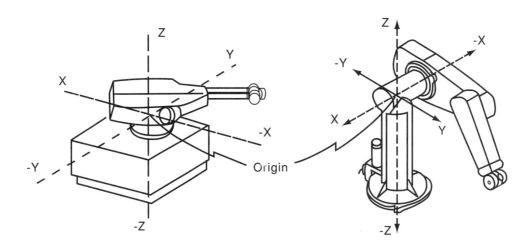

Figure 9.4.4 World coordinates identified on a spherical and articulator style manipulator as Cartesian coordinates from the origin

Path Modification is accomplished through the use of an interface, named *Alter*. An **RS-232C** standard interface is used for this purpose. An RS-232C is a communication standard computer interface for connecting peripheral equipment to CPU devices that are specifically designed to meet the requirements of serial transmission.

Robot paths can be modified in either of two modes, world or tool. The world-coordinate mode refers to the fixed x, y, z directions, which are predetermined by a default procedure for each robot. Figure 9.4.4 shows the world-coordinate system of robots using Cartesian coordinates. The tool coordinate mode refers to the direction in which a robot is pointing at any given instant. (Figures 2.3.9 and 2.3.10 indicate the location of these two modes.)

Robot motion can be viewed in real time by simply using the "WHERE" command, which transmits data consisting of rectangular positional offsets in the x, y, z directions and rotational changes about the three axes. Figure 9.4.5 illustrates such flow of commands and status information in a real-time control system. A vision system can correct the position of a robot by using data received from the machine vision system's fields of view.

VAL-II also can be interfaced to a supervisory computer system through Digital Data Communications Message Protocol (DDCMP), used by Digital Equipment Corporation in its network communications. The physical connection is a standard RS-232C serial link running a 9,600 bits per second.

Process Control Program, which is a background task that contains no arm-motion instructions, also can be run on the VAL-II system concurrently with a motion-control program. The process-control task also can be used to modify the

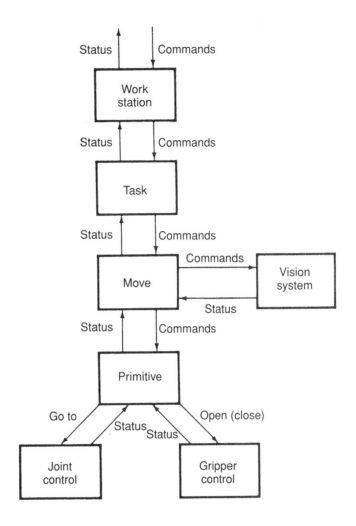

Figure 9.4.5 The flow of commands and status information in a real-time control system

robot path in real time by using the "WHERE" command. This path modification can be calculated within the VAL-II system or based on various external inputs. See Table 9.1 for such a program.

Program Statements are a number of instruction sequences that can be used with VAL-II language to logically organize sections of user programs and to control the order in which they are executed. Such program statements include: IF...THEN...ELSE...END, WHILE...DO...END, DO...UNTIL, and others. The format of the IF...THEN...ELSE...END is conventional except that the key

Table 9.1 VAL-II Program for Drawing a Circle in the X-Y Plane and Modifying It in the Z Plane.

- PROGRAM circle
1. PROMPT "input radius in mm," rad
2. TYPE "Move to center of circle"
3. TYPE "Press COMP button when done."
4. DETACH; Allow user to use Teach Pendant
5. ; Wait for COMP button on Teach Pendant
6. DO
7. UNTIL PENDANT (2) BAND 20
8. ATTACH ; Regain control of arm
9. DECOMPOSE c[ ] = HERE ; get XYZDAT data for center
10. TYPE /B, "Moving in 5 seconds" ; Beep terminal
11. DELAY 5
12. PCEXECUTE pc.alt, -1.0 ; Start ALTERing program
13. Set internal ALTER, WORLD mode, cumulative
14. ALTER (-1, 19)
15. WHITE SIG (1032) DO ; Continuously make circles
16. ; until signal stops it
17. FOR ang = 0 TO 360 STEP 5
18. x = rad*COS(ang) + c [1]
19. y = rad*SIN(ang) + c [2]
20. MOVES TRANS(x,y,c [3], c [4], c [5], c [6])
21. END
22. END
23. PCEND ; Finish up ALTERing program
- END

- PROGRAM PC . ALT
1. ; This program will ALTER the z component of the
2. ; circle, dependent upon inputs from the external
3. ; binary signal lines 16-21
4. ;
5. ; E.G. If the binary signals are set up like:
6. ; BITS 1021 1020 1019 1018 1017 1016
7. ;   0   1   0   1   1   0
8. ;
9. ; Then this corresponds to a 26(8) = 16(H) = 22.
10. ; correction in the z direction. Because
11. ; this value will be divided by two (see ALTOUT).
12. ; the correction would be 11 mm.
13. ;
14. ALTOUT 0,0,0,BITS(1016,6)/2,0,0,0

Note: VAL-I and VAL-II programs were developed by Unimation (Westinghouse).

Table 9.2 Structured VAL-II program.

- PROGRAM robot

```
1    ;  Program to execute one piece of code for odd-numbered
2    ;     rows, and a different section of code for
3    ;     even numbered rows until all rows are done
4            10 row = 1 ; Start at row #1
5              DO
6                IF row MOD 2==0 THEN ; Check for an even row
7    ; ODD row
8                  FOR part = 1 TO col, odd, end
9                    MOVE odd [row*col . odd.end + part]
10                 END
11               ELSE
12   ; EVEN row
13                 FOR part = 1 TO col.even.end
14                   MOVE even [row*col.even.end + part]
15                 END
16               END
17             UNTIL row == row.end + 1
```

- END

Note: VAL-I and VAL-II programs were developed by Unimation (Westinghouse).

word END is used to signify the end of the sequence of statements. The same key word is used to indicate the termination of the WHILE...DO sequence. The DO...UNTIL operates in the reverse sequence with respect to the WHILE...DO statement because the logical condition is tested after the statements are executed. Table 9.2 illustrates an example of such a program statement.

VAL-II also has procedural motion capability. This allows users to define robot motion with mathematical formulas.

Future expansions of VAL-II include the ability to pass parameters to subroutines, allowing users to work with languages more familiar to them. Also, the ability to interface and move robots in tandem with welding positioners and gantry

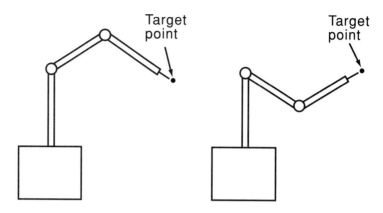

Figure 9.5.1 Two alternative axis configurations to achieve the target point

systems will be enhanced. Welding positioners, which hold the piece the robot is welding, are used when the robot's range does not extend sufficiently far to reach the whole area to be welded.

9.5 SPACE POSITION PROGRAMMING

As discussed in Chapter 3, there are four basic robot anatomies: polar, cylindrical, Cartesian, and jointed arm. Each one has three axes associated with the arm and body configuration and two or three additional joints associated with the wrist. The arm and body joints determine the general position in space of the end effector, and the wrist determines its orientation. If we think of a point in space in the robot program as a position and orientation of the end effector, there is usually more than one possible set of joint coordinate values that can be used for the robot to reach that point. For example, there are two alternative axis configurations that can be used by the jointed-arm robot shown in Figure 9.5.1 to achieve the target point by both of the alternative axis configurations; there is a difference in the orientation of the wrist with respect to the point. This concludes that the specification of a point in space does not uniquely define the joint coordinates of the robot.

Example 9.1

Assume that we are programming a point-to-point Cartesian robot with two axes, and two addressable points for each axis. Determine the robot's workspace and the sequence of the program.

Solution

Figure 9.5.2 shows the four possible points in the robot's rectangular work space. A program for this robot to start in the lower-left-hand corner and traverse the perimeter of the rectangle could be written as shown in Table A.

9.5 Space Position Programming

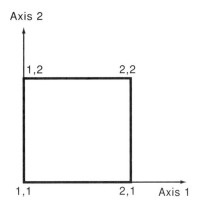

Figure 9.5.2 Robot work space for Example 9.1

Table A

Step	Move	Comments
1	1,1	Move to lower-left corner
2	2,1	Move to lower-right corner
3	2,2	Move to upper-right corner
4	1,2	Move to upper-left corner
5	1,1	Move back to start position

The point designations correspond to the x, y coordinate positions in the Cartesian axis system, as illustrated in Figure 9.5.2. Therefore, in this example the definition of points in space corresponds exactly with the joint coordinate values.

Example 9.2 Using the same robot as in Example 9.1, perform the program shown in Table B.

Solution

This program is the same as in Example 9.1 except that the point in the upper-right corner (2, 2) has not been listed. In this program, the move from point 2,1 to point 1,2 requires both joints to be moved.

The question arises: What path will the robot follow in getting from the first point to the second? One possibility is that both axes move at the same time and the robot will therefore trace a path along the diagonal line between the two points. Therefore, the path that is followed involves a slow motion (as described in Chapter 3), which is along the diagonal, as illustrated in Figure 9.5.3.

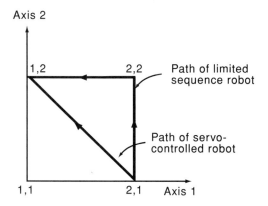

Figure 9.5.3 Robot path for Example 9.2

Table B		
Step	Move	Comments
1	1,1	Move to lower-left corner
2	2,1	Move to lower-right corner
3	1,2	Move to upper-left corner
4	1,1	Move back to start position

As illustrated by the preceding discussion of Example 9.2, it is possible for the programmer to make certain types of robots pass through points without actually including the points in the program. The key phrase is *pass through*. These are not addressable points in the program, and the robot will not actually stop at them in the sense of an addressable point.

Methods Defining Positions in Space

Irrespective of robot configuration, there are several methods that can be used by the programmer during the teach mode to actuate the robot arm and wrist. The three methods of defining positions in space are:

1. By joint movements
2. By x, y, z coordinate motions (also called world coordinates)
3. By tool coordinate motions

The joint movement method is the most basic and involves the movement of each joint, usually by means of a teach pendant. This method of teaching points is often referred to as the joint mode. Successive positioning of the robot arm in this way to define a sequence of points can be a very tedious and time-consuming way of programming the robot.

The x, y, z coordinate motions or world mode allows the wrist location to be defined using the conventional Cartesian coordinate system with origin at some location in the body of the robot. In the case of the Cartesian coordinate robot, this method is virtually equivalent to the joint mode of programming. For polar, cylindrical, and jointed-arm robots, the controller must solve a set of mathematical equations to convert the rotational joint motions of the robot into the Cartesian coordinate system. These conversions are carried out in such a way that the programmer does not have to be concerned with the substantial computations that are being performed by the controller. The x, y, z method of defining points in space was illustrated in Figure 9.4.4.

Tool coordinate motions can be defined as a Cartesian coordinate system in which the origin is located at some point on the wrist and the x, y plane is oriented parallel to the faceplate of the wrist. The z axis is perpendicular to the faceplate and pointing in the same direction as the tool, as was shown in Figures 2.3.9 and 2.3.10.

The physical limitations and characteristics of changing movement from joint to world to tool coordinates are as follows.

In the joint movements, due to the large work envelope defined by the sequence of points, the motion is tedious and time-consuming for programming the robot. To overcome this disadvantage, many robots can be controlled during the teach mode to move in x, y, z coordinate motions by applying the world coordinate system, which transforms a smaller rectangular work envelope, and the programmer does not have to be concerned with the substantial computations that are being performed by the controller. The wrist is usually being maintained by the controller in a constant orientation.

In the tool movements, again due to the small work envelope defined by the sequence of points, a significant amount of computational effort must be accomplished by the controller in order to permit the programmer to use the tool motions for defining the orientation. To overcome this disadvantage, the same procedures as previously mentioned can be applied by the programmer for defining the appropriate motions and points.

Therefore, from the preceding we can see clearly that the programming motion is not related to the execution motion. It is possible for the programmer to make certain types of robots pass through points without actually including those points in the program.

Reasons for Defining Points

Two main reasons exist for defining points in a program:

1. To define a working position for the end effector
2. To avoid obstacles

The first category is the most straightforward. This is the case where the robot is programmed to pick up a part at a given location or to perform a spot-welding operation at a specified location. Each location is a defined point in the program. This category also includes safe positions that are required in the work cycle.

The second category is used to define one or more points in space for the robot to follow, which ensures that it will not collide with other objects located in the workcell.

Speed Control

Most robots allow for their motion speed to be regulated during the program execution. A dial or group of dials on the teach pendant are used to set the speed for different portions of the program. It is considered good practice to operate the robot at a relatively slow speed when the end effector is operating close to obstacles in the workcell, and at higher speeds when moving over large distances where there are no obstacles.

The speed is not typically given as a linear velocity at the tip of the end effector for robots programmed by lead-through methods. There are several reasons for this. First, the robot's linear speed at the end effector depends on how many axes are moving at one time and which axes they are. Second, the top speed of a polar coordinate robot will be much greater with its arm fully extended than with the arm in the fully retracted position. Finally, the speed of the robot will be affected by the load it is carrying due to the force of acceleration and deceleration. All these reasons lead to considerable computational complexities when the control computer is programmed to determine wrist end velocity.

9.6 MOTION INTERPOLATION

Interpolation is a process used to estimate an intermediate value of one (dependent) variable that is a function of a second (independent) variable when values of the dependent variable corresponding to several discrete values of the independent variable are known.

Example 9.3 Suppose that we were programming a two-axis servo-controlled Cartesian robot with eight addressable points for each axis. Accordingly, there would be a total of sixty-four addressable points that we could use in any program that might be written. The work volume is illustrated in Figure 9.6.1.

Solution

Assuming the axis sizes to be the same as our previous limited sequence robot, a program for the robot to perform the same work cycle as Example 9.1 would be as follows:

9.6 Motion Interpolation **293**

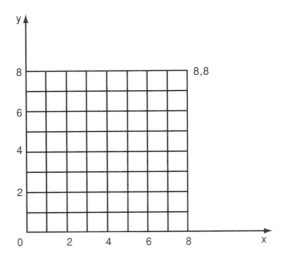

Figure 9.6.1 Robot work area with 8 x 8 addressable points

Step	Move
1	1,1
2	8,1
3	8,8
4	1,8
5	1,1

If we were to remove step 3 in this program (similar to Example 9.2), our servo-controlled robot would execute step 4 by tracing a path along the diagonal line from point 8, 1 to 1, 8. This process is referred to as interpolation.

As indicated in Chapter 3, there are different interpolation schemes that can be specified for the robot to get from one point to another. On many robots, the programmer can specify which type of interpolation scheme to use. The possibilities include:

1. Joint-interpolation
2. Straight-line interpolation
3. Circular interpolation
4. Irregular smooth motions (manual teach-pendant programming)

For many commercially available robots, joint-interpolation is the default procedure that is used by the controller. That is, the controller will follow a joint-interpolated motion between two points unless the programmer specifies straight-line or some other type of interpolation.

Circular interpolation requires the programmer to define a circle in the robot's workspace. This is most conveniently done by specifying three or four points that lie along the circle. The controller then constructs an approximation of the circle by selecting a series of addressable points that lie closest to the defined circle. The movements that are made by the robot actually consist of short-straight-line segments. Circular interpolation, therefore, produces a linear approximation of the circle. If the gridwork of addressable points is dense enough, the linear approximation looks very much like a real circle. Circular interpolation is more readily programmed using a based-on programming language than with teach-pendant techniques.

In manual teach-pendant programming, when the programmer moves the manipulator wrist to teach spray painting or arc welding, the movements typically consist of combinations of smooth motion segments. These segments are sometimes approximately straight, sometimes curved, and sometimes back-and-forth motions. We are referring to these movements as irregular smooth motions, and an interpolation process is involved in order to achieve them. To approximate the irregular smooth pattern being taught by the programmer, the motion path is divided into a sequence of closely spaced points that are recorded into the controller memory. These positions constitute the nearest addressable points to the path followed during programming. The interpolated path may consist of thousands of individual points that the robot must play back during subsequent program execution.

9.7 PROGRAM STATEMENTS

Robots usually work with something in their work space. In the simplest case, it may be a part that the robot will pick up, move, and drop off during execution of its work cycle. In more-complex cases, the robot will work with other pieces of equipment in the workcell, and the activities of the various equipment must be coordinated.

Nearly all industrial robots can be instructed to send signals or wait for signals during execution of the program. These signals are sometimes called interlocks or program statements. The most common form of interlock signal is to actuate the robot's end effector. In the case of a gripper, the signal is to open or close the gripper.

To accomplish this coordination, we introduce two commands that can be used during the program. The first command is SIGNAL M, which instructs the robot controller to output a signal through line M (where M is one of several output lines available to the controller). The second command is WAIT N, which indicates that the robot should wait at its current location until it receives a signal on line N (where N is one of several input lines available to the robot controller).

Example 9.4 The program shown in Table C is to accomplish a press unloading task with the gripper beginning in the open position, as shown in Figure 9.7.1.

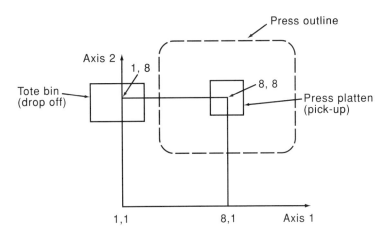

Figure 9.7.1 Press unloading operation for Example 9.4.

Table C

Step	Move or signal	Comments
0	1,1	Start at home position
1	8,1	Move to wait position
2	WAIT 11	Wait for press to open
3	8,8	Move to pickup point
4	SIGNAL 5	Signal gripper to close
5	8,1	Move to safe position
6	SIGNAL 4	Signal press to actuate
7	1,1	Move around press column
8	1,8	Move to tote pan
9	SIGNAL 6	Signal gripper to open
10	1,1	Move to safe position

Solution

Each step in the program is executed in sequence, which means that the SIGNAL and WAIT commands are not executed until the robot has moved to the point indicated in the previous step. The operation of the gripper is assumed to take place instantaneously so that its actuation would be completed before the next step in the program was started.

An alternative way to address this problem is to cause the robot to delay before proceeding to the next step. In this case, the robot would be programmed to wait for a specified amount of time to ensure that the operation had taken place.

The form of the command for this second case has a length of time as its argument rather than an input line.

$$\text{DELAY X SEC}$$

This command indicates that the robot should wait X seconds before proceeding to the next step in the program.

Example 9.5

Shown in Table D is a modified version of Example 9.4, using delay time as the means for assuring that the gripper is either opened or closed. This command is indicated after the signals close or open.

Table D

Step	Move or signal	Comments
0	1,1	Start at home position
1	8,1	Move to wait position
2	WAIT 11	Wait for press to open
3	8,8	Move to pickup point
4	SIGNAL 5	Signal gripper to close
5	DELAY 1 SEC	Wait for gripper to close
6	8,1	Move to safe position
7	SIGNAL 4	Signal press that hand is clear
8	1,1	Move around press column
9	1,8	Move to tote pan
10	SIGNAL 6	Signal hand to open
11	DELAY 1 SEC	Wait for gripper to open
12	1,1	Move to home position

9.8 SAMPLE PROGRAMS

Consider the robot illustrated in Figure 9.8.1 This is an electric-drive, overhead gantry robot that incorporates tactile sensor technology. As designed, the machine is a completely self-contained workstation that allows either stand-alone or integrated assembly-line configuration. It has a high level of dexterity of the kind required for the assembly of disk drives, printed circuit boards, wire harnesses, and telecommunication and electromechanical devices. Automatic laser inspection will be added to the system at a later date.

As shown in Figure 9.8.1, the intelligent gripper ("Adaptive Touch"™) incorporates both optical sensing and a strain gauge design that allows force-sensing thresholds to be programmed in three rectangular axes. Parts that are out of tolerance or that are not properly oriented are detected quickly, enabling the robot to respond as required. It will take intelligent action based on the specific "force

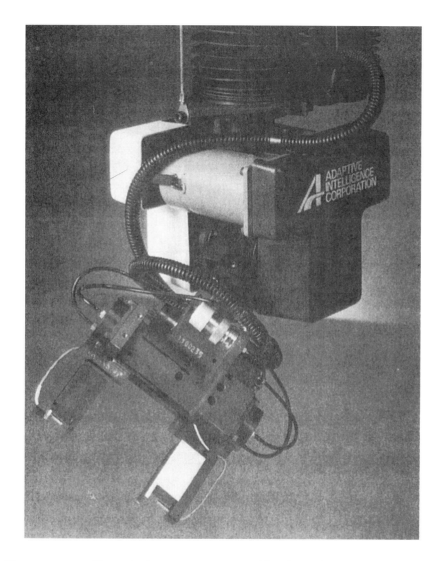

Figure 9.8.1 Intelligent, high-precision parts-assembly robot, with grippers, that incorporates both optical sensing and strain gauge sensors *(Source: Adaptive Intelligence Corp.)*

signatures" programmed into the gripper for recognizing and accommodating specific assembly tasks. The design of the machine allows expandable modular configurations of three, four, and six axes. These axes handle payloads of 17, 15, and 9 pounds, respectively, at a maximum speed of 40 inches/second. The six-axes configuration consists of the three linear axes (x, y, z) and three rotary axes (yaw, pitch, and roll). All are driven by DC servomotors. Accuracy and repeatability are ±0.002 and ±0.001 inch, respectively. Resolution is specified as 0.0005 inch.

298 Programming

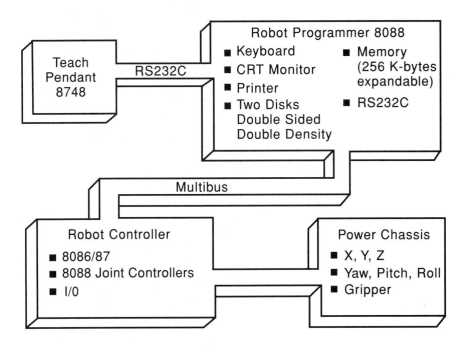

Figure 9.8.2 Architecture of AAMPL system consists of four separate subsystems (Source: Adaptive Intelligence Corp.)

A teach pendant is provided for on-line programming and reprogramming to accommodate production line changes. A hand-held tool allows production personnel to control all robot movements, time delays, high-level commands for self-calibration, and to call up complex subroutines.

Supporting the teach pendant, including on-line and off-line programming through an IBM compatible computer, is the firm's Adaptive Assembly Machine Programming Language (AAMPL^{TM}). Programs can be written and tested interactively on-line or written off-line and downloaded from disk for implementation. The system's architecture is shown in Figure 9.8.2 and consists of (1) a teach pendant, (2) the robot programmer, (3) the robot controller, and (4) the robot power chassis.

The AAMPL program allows for three possible situations as delineated in Table 9.3. A sample robot program is shown in Table 9.4, and a summary of robot control with I/O control statements is given in Tables 9.5 and 9.6.

Table 9.3 Adaptation of AAMPL Language to Three Major Installations

1. Within Its Own Work Space

The robot's multi-tasking software enables the control of all activities within the work envelope, including:

    (a) Gripper arm movement to a given site with a specified speed, acceleration, clamp force, and wait on arrival.

    (b) Use of force feedback and/or optical sensing to test for particular conditions, or to determine precise location of an object whose position may vary.

    (c) Exception recognition by monitoring how the arm travels and what patterns of forces it encounters and automatically comparing actual values with anticipated values.

    (d) Exception handling by means of responses prescribed by the user.

    (e) Logging of production statistics and exception data.

    (f) Switching from one batch-oriented procedure to another simply by changing the tooling plate and the assembly program and executing a self-calibration routine.

2. Within Its Vicinity

Like the robot controller, the integrated I/O controller has multi-tasking software and a dedicated command set. With input from 64 sources and output to 32 devices, it can:

    (a) Program the AARM's interactions with conveyors, auto-guided vehicles, and other ancillary parts-feeding equipment.

    (b) Schedule multiple events concurrently and time I/O to be synchronous or asynchronous with the motions of the gripper.

    (c) Operate two or more robots as a cluster, passing information, parts, and tools between them.

3. Within a Factory-Wide System

The full upward compatibility of the system permits:

    (a) Integration of the robot into an existing manufacturing structure. Software is available to interface the AARM to a master factory computer or CAD/CAM system. Instructions can be downloaded to the robot, and data compiled by the robot can be uploaded.

    (b) The AARM can be incorporated into a true manufacturing network. Full compatibility (hardware and software) with standards set by MAP (Manufacturing Automation Protocol).

Table 9.4 Sample Robot Program with AAMPL Language

COMMENT: FIRST PICK UP THE SPECIAL RING ASSEMBLY TOOL
    MOVE TO OVERTOOL
    MOVE TO TOUCHING, ACC=10, SPD=10
    MOVE TO LOCKON, ACC=5, SPD=5
    MOVE TO WITHDRAW, ACC=25, SPD=25

COMMENT: NEXT GO TO GET THE RING
    MOVE TO OVERRING
GETRING
    MOVE TO TOPRING

COMMENT: THE GRIPPER WILL SOON BE HOLDING THE RING...PROGRAM WILL BRANCH WHEN GRIPPING FORCE IS 0.5 POUNDS INDICATING RING IS PRESENT
    MONITOR ENABLE, USING PINCH 0.5 TO HAVRING
    MOVE TO GRABRING, SPD=25

COMMENT: WE GET HERE IF THERE IS NO RING
    MOVE TO OVERRING
    DISPLAY WE HAVE GONE TO PICK UP THE RING BUT IT IS NOT THERE
    GO TO GETRING
HAVRING
    MONITOR DISABLE USING PINCH
    MOVE TO LIFTRING

COMMENT: NOW MOVE TO POSITION OVER THE POST
    MOVE TO OVERPOST
    MOVE TO POSTBASE, SPD=15

COMMENT: GRIPPER OPENS TO DROP THE RING, THEN LIFTS THE TOOL TO CLEAR THE POST
    MOVE TO RELEASE
    MOVE TO OVERPOST

COMMENT: NOW RETURN THE TOOL TO ITS PARK
    MOVE TO WITHDRAW
    MOVE TO LOCKON, ACC=10, SPD=10
    MOVE TO TOUCHING, ACC=5, SPD=5
    MOVE TO OVERTOOL, ACC=25, SPD=25

Table 9.5 Summary of Robot Control with AAMPL Programming Language

DATA TYPES:		
	Site	A point in space defined by coordinates for X, Y, Z, Yaw, Pitch, Roll and width of gripper opening.
	Counter	An integer used to count parts, errors, or repetitions of a procedure.
COMMANDS:		
Motion Control	Move	Moves the arm to a predefined site. You can specify acceleration, speed.
	Stop	Stops the current motion of the gripper.
	Home	Places the gripper at a predefined "home" position in the work envelope.
Site Modification	Origin	Used to execute a set of moves relative to a chose location.
	Inc./Dec	Changes the location of a site along an axis, e.g., decrementing a Z value to get to the next lower part in a vertical pallet. Also used to increment or decrement counters.
	Store	Moves site data from one location to another, or from the current AARM position to another site. Also used to set a counter value.
Robot Program Control	Repeat	Repeats a series of AAMPL statements. A counter specifies the desired number of repetitions.
	If	Compares a counter value to a constant and redirects program flow depending on the result.
	Label	Marks a spot in the program to which control can be transferred from commands like Go To.
	Go To	Transfers program control to the specified label.
	Perform	Executes a sequence of program steps, then returns program control to the statement following the Perform command.
	Monitor	Looks for input from the gripper sensors (a light beam between the fingers, or Pinch, Tip or Side sensing with the fingers), or from up to 64 external intelligences. When input occurs, the motion of the gripper stops and program control resumes at the specified label.
	Call	Executes an AAMPL sub-program stored on disk.
Operator Interface	Display	Displays a message for the operator, who must press CLEAR key to resume program execution.
	Comment	Annotates a program listing with information about the intent of the instructions.
	Log	Records date and time along with a label in a disk file. Reports exception and production data.
SUB-ROUTINES:		
	Findpost	With the gripper fingers positioned on either side of a post, locates the exact center of the post. This position can be stored as a site.
	Findtop	With the gripper fingers positioned over an object, locates the top of the object.
	Grasp	With the gripper fingers on either side of the object but not necessarily centered on it, closes the gripper and moves it so as to grasp the object.

Table 9-6 Input/Output Control Statements with AAMPL Programming Language

DATA TYPE:		
	Channel	Collection of bits comprising input from a sensor or output control for an external device.
COMMANDS:		
I/O Activity	Input	Reads the information from an input sensor and stores the value into a computer.
	Ouput	Sends a value to an output channel.
Task Control	Task	Defines an independent set of I/O statements which can execute concurrently with AARM motion and with other I/O tasks.
	Start	Starts and I/O task.
	Stop	Stops and I/O task.
Time Control	Wait	Suspends task execution until a specified input channel reaches some value.
	Delay	Suspends task execution for a specified time period.
I/O Program Control	Loop	Restarts a task a specified number of times.
	Compare	Starts one of three other tasks depending on whether a given input channel is less than, equal to, or greater than some value.
Operator Interface	Notify	Sends an event notification to the robot programmer. A Monitor statement can watch for the event and redirect program flow when it occurs.

9.9 SUMMARY

In this chapter, we have examined the topic of robot programming. Programming represents one of the most advanced areas in robotic technology.

Programming is the process of preparing a detailed sequence of operating instructions to solve a particular problem, testing it to ensure its accuracy, and preparing documentation to be run on a digital computer.

Robot programming can be defined as a path in space through which the manipulator is directed to move. A program is a logically arranged set of programming instructions.

Programs for robots can be subdivided into two categories: the operating system programs that tell the robot how to do something and the application programs that tell the robot when to do something. The intelligence of a robot comes from the operating system. Robot operating systems vary from mechanical to very sophisticated multiple processor digital computers. The capabilities of a robot operating system depend on the type of controller, the arm configuration, and the method used to train the robot.

There are five methods to enter the programming commands into a controller memory: manual, teach pendant, walk-through, computer terminal, and off-line.

Robot programming language is a set of words and rules governing their use, employed in the construction of a program.

Many special programming languages can be used to write robot-operating-system programs, such as mechanical language, machine language, assembly language, and high-level programming languages such as BASIC, COBOL, and C.

Programming through mechanical setup requires the least work by the robot's controller; task programming requires the most.

There are four levels of robot programming and controlling systems:

1. Joint control
2. Primitive motion
3. Structured constructs
4. Task-oriented

Level 1 programming, which is through mechanical setup, requires the least work by the robot's controller; task-oriented programming, level 4, requires the most.

At present, the most frequently used method is the structured programming category, level 3. Receiving and interpreting information from sensors increases the system's intelligence needs further.

9.10 REVIEW QUESTIONS

9.1 What are the four functions furnished by the computer to high-technology robots?
9.2 What are the five methods used to enter the programming command into the controller memory?
9.3 Describe teach-pendant programming and its applications.
9.4 Describe computer terminal programming and its applications.
9.5 Describe off-line programming and its future potential.
9.6 What are robot programming languages?
9.7 What are the two main robot programs and what are their functions?
9.8 What are the four levels of robot programming?
9.9 What are the six advantages of task-oriented programming?
9.10 Which level of programming is most frequently used? Why?
9.11 How can you increase the system's intelligence?
9.12 What is the purpose of using workcell control?
9.13 What are the four reasons for using operator interface?
9.14 What are the three responses of safety monitoring?
9.15 Describe the operation of an intelligent high-precision parts-assembly robot.
9.16 Describe the three major installation situations of a robot.
9.17 Describe robot control programming language by a case study example.
9.18 Describe I/O control programming language by a case study example.

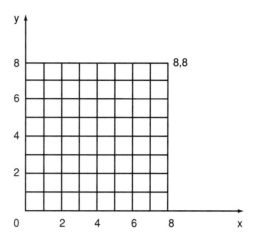

Figure 9.11.1 Robot work area for Problem 9.1.

9.11 PROBLEMS

9.1 Using the 8 × 8 square grid in Figure 9.11.1, show the path taken by a Cartesian coordinate robot if it is directed to move between the following sets of points in the grid using linear interpolation:
   a. point (1, 1) and point (6, 6)
   b. point (2, 1) and point (8, 2)
   c. point (2, 2) and point (7, 5)

9.2 Using the gridwork in Figure 9.11.2 for a robot with one rotational axis and one linear axis, show the path taken by the robot if it is directed to move between the following sets of points in the grid using joint interpolation:
   a. point (1, 1) and point (6, 6)
   b. point (2, 1) and point (8, 2)
   c. point (2, 2) and point (7, 5)

9.3 Using the gridwork illustrated in Problem 9.2 for a robot with one rotational axis and one linear axis, show the path taken by the robot if it is directed to move between the same sets of points in the grid using linear interpolation.

9.4 Rewrite the program shown in Table E so that it includes the use of WAIT instructions to make sure that the gripper has opened and closed properly before the next step in the program has been executed. Use the encoding format we have adopted in this chapter to write the program.

9.5 The robot is to be programmed to pick up a part from point A and move it to point B, followed by a move to a neutral position. Points A and B are to be defined by the programmer within the robot's work volume. Then the robot should pick up the part at point B and move it back to point A, followed

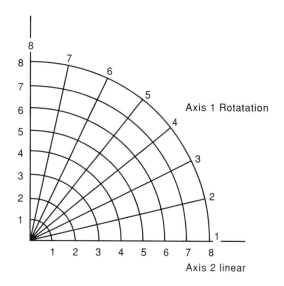

Figure 9.11.2 Robot gridwork area for Problem 9.2.

Table E

Step	Move or signal	Comments
0	1,1	Start at home position
1	8,1	Move to wait position
2	WAIT 11	Wait for press to open
3	8,8	Move to pickup point
4	SIGNAL 5	Signal gripper to close
5	8,1	Move to safe position
6	SIGNAL 4	Signal press to actuate
7	1,1	Move around press column
8	1,8	Move to tote pan
9	SIGNAL 6	Signal gripper to open
10	1,1	Move to safe position

by a move to the previous neutral position. The robot can be operated continually in the "run" mode to repeat the motion pattern over and over.

9.6 As an enhancement to Problem 9.5, if the capability exists on the robot used (e.g., Microbot TeachMover), program the robot to check each time it attempts to pick up the part to determine whether or not it has closed on the part. If the part is in the gripper, then continue the program. If the part is not in the gripper, the robot should move to the neutral position, provide a signal of some kind (e.g., light or buzzer), and wait five seconds. It should then attempt the pickup again.

9.7 Program the robot to pick up two blocks (the blocks are of different sizes) from fixed positions on either side of a center position, and to stack the blocks in the center position. The larger block will always be on one side of the center, and the smaller block will always be on the other side of the center position. The smaller block is to be placed on top of the larger block.

9.8 This exercise is similar to Problem 9.7 except that the positions of the two blocks can be exchanged at random. It is not known whether the larger block is on one side of the center or the other. The robot must be programmed to always pick up the larger block first, place it at the center position, and then pick up the smaller block and place it on top of the larger block.

9.9 The robot is to be programmed to pick up a part from a known fixed position on a conveyor and to place it at an upstream location on the conveyor so that the conveyor will deliver it back to the pickup point. The fixed pickup position is established by means of a mechanical stop along the conveyor so that the part is always in the same orientation and location for the robot.

9.12 REFERENCES

Ahmad, S., and A. Luo, "Coordinated Motion Control of Multiple Robotic Devices for Welding and Redundancy Coordination through Constrained Optimization in Cartesian Space." Proceedings of the IEEE Conference on Robotics and Automation, Philadelphia, February 1988.

Amic, P. J. *Computer Numerical Control Programming*. Upper Saddle River, NJ: Prentice Hall, 1997.

Chang, C. W., and M. A. Melkanoff. *NC Machine Programming and Software Design*. Englewood Cliffs, NJ: Prentice-Hall, 1989.

Craig, J. J. *Introduction to Robotics*. 2d ed. Reading, MA: Addison-Wesley Publishing Company, 1989.

Gruver, W., and B. Soroka. "Programming, High Level Languages." In *The International Encyclopedia of Robotics*, R. Dorf and S. Nof, Editors. New York: Wiley Interscience, 1988.

Lee, M. H. *Intelligent Robotics*. New York: Halsted Press, 1989.

Luggen, W. W. *Flexible Manufacturing Cells and Systems*. Englewood Cliffs, NJ: Prentice Hall, 1991.

Rosheim, M. E. *Robot Wrist Actuators*. New York: Wiley, 1989.

Seames, W. *Computer Numerical Control: Concepts and Programming*. 2d ed. Albany, NY: Delmar Publishers Inc., 1990.

Sharon, D., J. Harstein, and G. Yantian. *Robotics and Automated Manufacturing*. Aulander, NC: Pittman, 1989.

SILMA Inc. "Programming in Sil." Available from SILMA Inc., 1601 Saratoga-Sunnyvale Rd., Cupertino, CA 95014, 1988.

Webb, John W., and R. A. Reis. *Programmable Logic Controllers: Principles and Applications*, 3d ed. Englewood Cliffs, NJ: Prentice, 1995.

Wells, L. K., and J. Travis. *Graphical Programming Made Even Easier*. Upper Saddle River, NJ: Prentice Hall, 1997.

CHAPTER 10

Artificial Intelligence

10.0 OBJECTIVES

After studying this chapter, the reader should:

1. Be acquainted with intelligent systems
2. Understand the elements of artificial intelligence
3. Recognize the system architecture
4. Be familiar with applications of advanced robots
5. Be aware of fuzzy logic controls
6. Be able to define advanced concepts and procedures
7. Perceive future developments
8. Realize the impact on employment

10.1 INTELLIGENT SYSTEMS

Artificial intelligence (AI) is the development of computers that think and respond like a human. This is one of the fastest growing branches of computer science. The term was coined in the mid1950s by John McArthy at the Massachusetts Institute of Technology. AI languages include list processor (LISP). The system has been designed to imitate intelligence functions or methods to supplement human intellectual abilities. Therefore, AI is that part of science concerned with systems that exhibit the characteristics usually associated with intelligence in human behavior such as learning, reasoning, problem-solving, understanding language, and so on. The goal of AI is to simulate human behavior to a system. One of the greatest results of artificial intelligence thus far is the ability for a computer to understand human speech. So far, however, programming a computer to comprehend anything but the most rudimentary commands has been difficult. The art of bringing relevant principles and tools of AI to difficult problems is known as knowledge engineering. AI is likely to profoundly affect design, automation, and overall economics of manufacturing operations, due in large part to the advances in memory expansion, specifically with the Very Large Scale Integration **(VLSI)** chip design and its decreasing cost. Figure 10.1.1 illustrates the effect of relative cost through the years in electronic and computer components. Scientists believe that artificial intelligence will have a multitude of other applications, including what is presented in this chapter.

10.2 ELEMENTS OF ARTIFICIAL INTELLIGENCE

In general, artificial intelligence applications in manufacturing encompass the following activities:

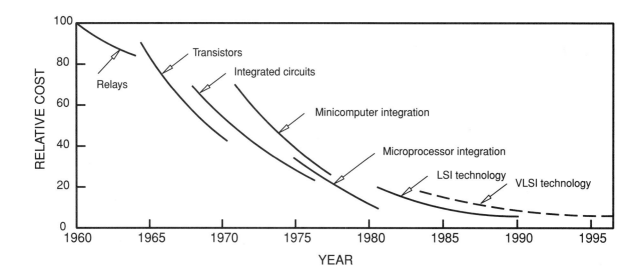

Figure 10.1.1 Relative cost of electronic and computer components through the past thirty-eight years.

1. Expert systems
2. Natural language processing
3. Machine vision
4. Neural networks
5. Machine learning
6. Advanced robots

Expert systems (ES) are intelligent computer programs capable of solving complex problems that normally require a human who has years of education and experience. Artificial intelligence can be used to decide real-life situations, such as the diagnosis of disease based on symptoms or the scheduling of a project that has hundreds of variables. Expert systems can replace certain tasks or add to the knowledge of the user.

An expert system, also called a knowledge-based system, is capable of solving difficult problems using knowledge-based and inference procedures. Figure 10.2.1 shows the basic structure of an expert system.

The goal of an expert system is the capability to perform an intellectually demanding task as well as human experts would. The knowledge required to perform this task is called the domain of the expert system. Expert systems utilize a knowledge base containing facts, data, definitions, and assumptions. They also have the capacity for a heuristic approach, that is, making good judgments by discovery and revelation, and making good guesses, just as an expert would. The

10.2 Elements of Artificial Intelligence

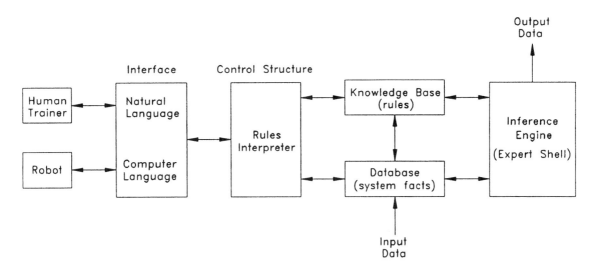

Figure 10.2.1 Basic structure of an expert system

knowledge base is expressed in computer codes, usually in the form of if-then rules, with a series of questions; the mechanism to use these rules to solve problems is called an inference engine. Expert systems can also communicate with other computer software packages.

To construct expert systems for solving complex design problems in manufacturing, one needs both a large amount of knowledge and some mechanisms for manipulating that knowledge to create solutions. Because of the difficulty involved in accurately modeling the many years of experience of an expert or a team of experts, and the complex inductive reasoning and decision-making capabilities of humans, developing knowledge-based systems requires considerable time and effort.

Expert systems work on a real-time basis, and their short reaction times provide rapid responses to problems. The most common programming languages are LISP and PROLOG, although some work is also being done with C++ and other languages. Some more-recent developments are expert system software shells or environments, also called framework systems. These software packages are essentially expert-system outlines that allow a person to write specific applications to suit special needs. Writing these programs requires considerable experience and time. There are many characteristics that separate expert systems from regular computer programs, such as:

1. They can reason like a human and handle uncertainties.
2. They can explain why they have asked specific questions.

3. They can tell how they realized a certain conclusion.
4. They can learn and apply base knowledge from experience.

Several expert systems have been developed and used since the early 1970s, utilizing computers with various capacities, for specialized applications, such as:

- Missile guidance systems
- Problem diagnosis in locomotive engines and various other types of machines and equipment and determination of failures or corrective actions
- Seismic hazard assessment
- Modeling and simulation of production facilities
- Computer-aided design, process planning, and production scheduling
- Management of investment portfolios
- Financial planning
- Management of a company's manufacturing strategy
- Optimizing system designs
- Helping standardized tooling
- Diagnosing diseases

Natural language processing is, traditionally, information that can be obtained from a database in computer memory that has required utilization of a computer program that translates natural language to machine language. Natural language interfaces with database systems are now in various stages of development. These systems allow a user to obtain information by entering English language commands in the form of simple, typed queries. Also, artificial intelligence incorporated into certain types of computers could interpret languages in real time.

Software shells are available and are used in applications such as scheduling material flow in manufacturing and analyzing information in databases. Significant progress is being made on computers that will have speech synthesis and recognition (**voice recognition**) capabilities, thus eliminating the need to type commands on keyboards.

Machine vision basic features were described in Chapter 7. Computers and software for artificial intelligence can be combined with cameras and other optical sensors. These machines can then perform operations such as inspecting, identifying, and sorting parts and guiding intelligent robots that would otherwise require human intervention. Figure 10.2.2 illustrates an expert system as applied to robot guidance by machine vision.

Neural networks are the ultimate goals of AI. Although it is the simulation of human intelligence, a profound difference exists in the fundamentals of the thought processes between computers and the human brain. Human thinking involves very complex interactions of neurons in the brain. Each neuron contacts many other neurons so that very large numbers of neurons coordinate and interact during the thought process. The human brain has about 100 billion linked neurons

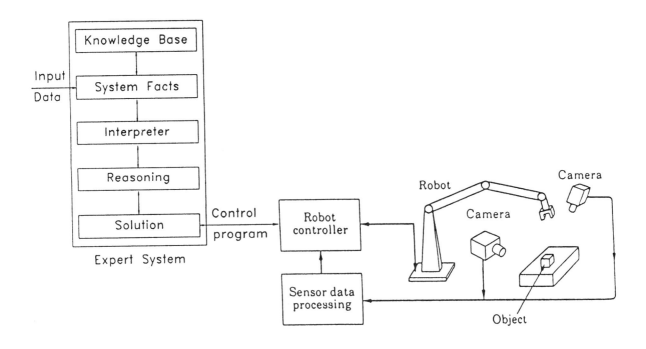

Figure 10.2.2 Expert system is applied to an industrial robot guided by machine vision.

and more than 1000 times that many connections; hence, neural networks are also known as connectionist architecture or connectionism. Computer thinking typically involves the interaction of the processor bit with a single bit of RAM. Many bits can be involved in a decision, but not simultaneously or interactively unless costly nested looping schemes are utilized.

The structure of neural networks makes them nonalgorithmic, massively parallel computing devices in which many inputs are handled simultaneously. However, unlike parallel computers, neural networks do not require complex programming to break up a problem into a form suitable for parallel processing. Thus, because they can classify data and make weighted decisions without firm rules, they are particularly useful for ill-defined problems, or where the data are fuzzy, incomplete, or contradictory, or for problems that require a large number of rules. Other features of neural networks include the capacity (a) to be trained to store, process, and retrieve information, (b) to self-organize and learn to produce a desired goal, and (c) to find good, quick, but approximate solutions to highly complex problems.

Among various applications of neural networks relevant to manufacturing are sensor fusion and signal processing, monitoring and process control, process

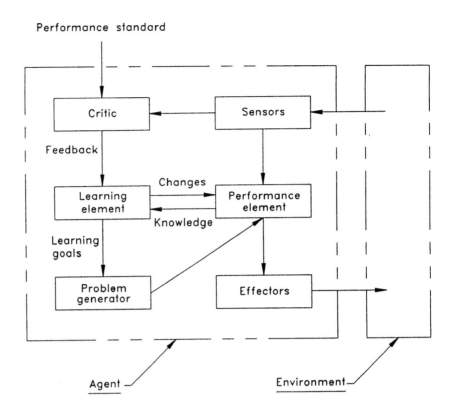

Figure 10.2.3 A general model of learning agents

control simulation and modeling, diagnostics of mechanical equipment, and material handling. There is a great amount of current research taking place concerning this subject and its various applications. Many believe that only through advances in neural network approaches will true artificial intelligence evolve. This field is still very controversial, however, and its ultimate effect on AI will be determined only over a period of time.

Machine learning is the subfield of AI or a subdivision of expert system technology concerned with programs that learn from experience. The idea behind machine learning from observations was accomplished between a hypothetical intelligent agent and the world. The agent receives percepts from the environment and performs actions. An agent may be divided into four conceptual components (as shown in Figure 10.2.3):

1. The learning element
2. The performance element

3. The critic
4. The problem generator

Learning takes many forms, depending on the nature of the performance element, the available feedback, and the knowledge accessible for use. Learning any particular component of the performance element can be cast as a problem of learning an accurate representation of a function. Learning a function from examples of its inputs and outputs is called inductive learning. The performance of inductive learning algorithms is measured by their learning curve, which predicts the accuracy as a function of the number of observed examples. The difficulty, therefore, of learning depends on the chosen representation. Functions can be represented by logical sentences, polynomials, belief networks, neural networks, and others.

Advanced robot is the term that came into general use during the 1980s to describe the emerging developments in sensor-based robotic devices that exploit relatively low-cost computing power to achieve levels of functionality that often appears to mimic intelligent human behavior. Such devices are often semiautonomous in nature with quite sophisticated human-computer interfaces. They clearly represent a significant technical advance on the familiar pick-and-place industrial robot and have a potentially wide range of applications in the manufacturing, nuclear, construction, space, underwater, and health care industries.

Advanced robots can be developed that will see and hear for humans. For example, an unmanned mission to Mars includes a "smart" robot, which is intelligent enough to "know" that it has to stop at the edge of a cliff—without any input from humans. Other such robots can eventually be used in assembly plants to manipulate objects and make certain decisions. Developments in the field utilize results from the domains of cognitive science and artificial intelligence, as well as employing aspects of mechanical and electronic engineering science, real-time computing, control, and sensor techniques. It is thought that the successful integration of the required wide range of enabling technologies to produce viable marketable devices provides one of the most interesting challenges in current engineering.

During the 1980s, a series of initiatives were launched within the European Union to foster the development of studies in and exploitation of advanced robotic activities. These initiatives were broadly in response to technical challenges presented by our Japanese colleagues, but they were also influenced by the increasing availability of low-cost computing power, which would clearly enhance the functionality of existing mechanisms and extend the application domain into new and potentially gainful market sectors.

For the purpose of the initiative, an advanced robot was defined as a machine or system capable of accepting high-level, mission-oriented commands, navigating to a workplace, and performing complex tasks in a semistructured environment with a minimum of human intervention. Such a semiautonomous device would exhibit various attributes of intelligent operations, and its integration would certainly challenge existing capabilities. The application was originally conceived and encompassed the nuclear, space, underwater, construction, and health care industries, general service functions, such as surveillance and cleaning.

In the United Kingdom, a National Center was established and an ambitious core research program was defined. Within a wider European context, various advanced robotic projects were sponsored within the framework of the ESPRIT CIME program, with emphasis being placed on industrial application projects. Parallel work was undertaken in the United States at Carnegie Mellon University and elsewhere, while work in Japan continued at a fast pace. It is perhaps timely to review the progress of the field, compare actual technical achievements with earlier aspirations, and discover what type of market is actually emerging.

Advanced robots are essentially concerned with the development of sensor-based control of mechanisms and automation and the evolution of suitable system architectures to generate appropriate levels of functionality to implement the tasks defined in the original broad specification. This represents a significant advance in existing robotic devices in performing simple tasks, such as path finding and conflict resolution. Modern automation does exhibit to the casual observer aspects of cognitive behavior that may be interpreted as machine intelligence, although it can be argued that few robots developed so far are particularly intelligent. The development of such architectures and the integration of the required techniques of control, actuators, sensors, and artificial intelligence provide some of the most interesting intellectual challenges in current and future engineering research.

10.3 SYSTEM ARCHITECTURE

Modern intelligent robots have specifications that include the ability to accept high-level commands, undertake situation analyses and planned functions, navigate to a required operational point, and perform quite complex mechanical manipulation with the minimum of human interaction. Practical engineering constraints usually require the device to be fault-tolerant and degrade gracefully in the presence of major system failures. Increasingly, there is also a requirement for devices to cooperate with other robots in some organized way so that complex tasks can be performed by coordinating the activities of a group of devices. Such coordination is reasonably straightforward in a well-structured and predictable environment, but less so in open and unstructured terrain.

The engineering specifications usually demand that the device be equipped with a range of sensors, the signals from which must be interpreted, integrated, and utilized for control functions. Such an interpretation cannot generally be direct but must be influenced by other factors, such as an overall situation analysis based on some generated or embedded world model, the fundamental mission directives of the vehicle, and inputs, if any, from the human operator. There is thus a clear need for an overarching structure or schematic in which data flows and control signals can be organized in some logical way and prioritized in an appropriate functional or temporal framework, which can be designed to known standards, verified, commissioned, and maintained as a useful industrial entity.

Various approaches have been adapted to provide a comprehensive structure that will encompass aspects of artificial intelligence. Advanced robot systems com-

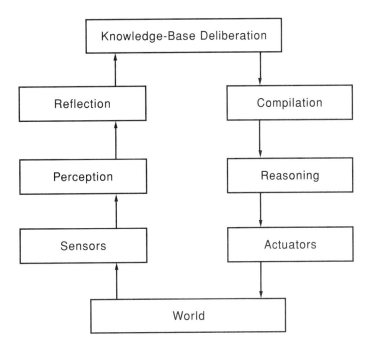

Figure 10.3.1 Architecture for intelligent control

monly incorporate the three major subsystems for machine intelligence: sensors, actuators, and control. Figure 10.3.1 shows the architecture for an intelligent control.

Traditionally, hierarchical deterministic procedures have found acceptance within the industrial environment because of the ready utilization of standard computing procedures and interface. Signal synchronization protocols and the basic transparency of the overall processing system provide the use of finite state machines. A typical implementation of three-layer architecture for real-time intelligent control is shown in Figure 10.3.2.

The lowest or executive level of the figure is concerned with real-time control functions, and it receives **macro** commands, such as course trajectories from the central tactical layer, which generates these commands by interpreting instructions from the highest or strategic layer. In terms of the development of advanced robotic devices, it is this highest layer that is the most interesting. It combines aspects of situation-analysis planning and an interpretation of operator requirements. Such functions generally require a more-or-less complex model of the real world such as task-planning agents, appropriate advisory agents for the human operator, and a communication link with the human-computer interface. A sim-

318 Artificial Intelligence

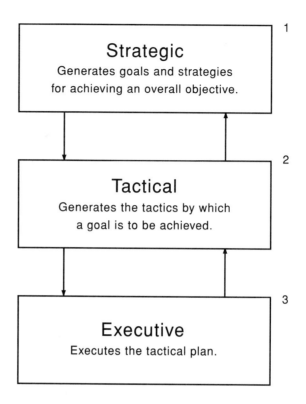

Figure 10.3.2 Three-layer system architecture

plified functional representation of hierarchical architecture is shown in Figure 10.3.3.

System architectures provide a general scheme that defines structural composition and the flow of control and data signals between the various components of the robot, and it can encompass aspects of sensing, modeling planning, and the real-time operation of practical controllers. The system is usually complex and implemented on a distributed set of computing elements linked by networks with suitable bandwidth capabilities.

System design and development are usually nontrivial and, until recently, hindered by a scarcity of suitable software tools capable of integrating AI modules with a real-time control environment. There are, as yet, no performance benchmarks, no universally accepted standards for structures or communication protocols. Commonly, development is undertaken using a number of diverse computing platforms, ranging from powerful workstations for AI system development to sets of distributed microcomputers for data processing and control.

The software support tools used for development have generally been equally diverse and individualistic in nature, and although such a situation is acceptable

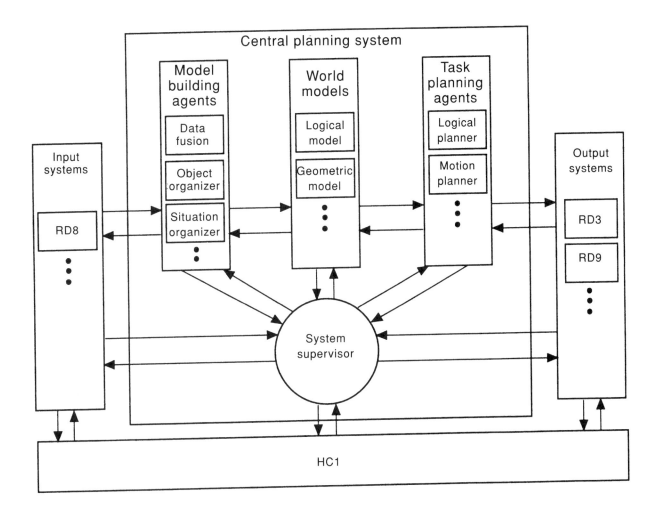

Figure 10.3.3 Generic robot architecture

for research and laboratory prototyping, it has implications for the verification, installation, and maintenance of marketable robotic systems.

In advanced robotics systems, robots are equipped with sensors (vision, tactile, proximity, speech recognition, and voice synthesis), controllers, conveyors, vision-processing equipment, and computers, all networked together through a variety of high-bandwidth and low-bandwidth communication channels. Figure 10.3.4 shows such an ideal domain system.

320 Artificial Intelligence

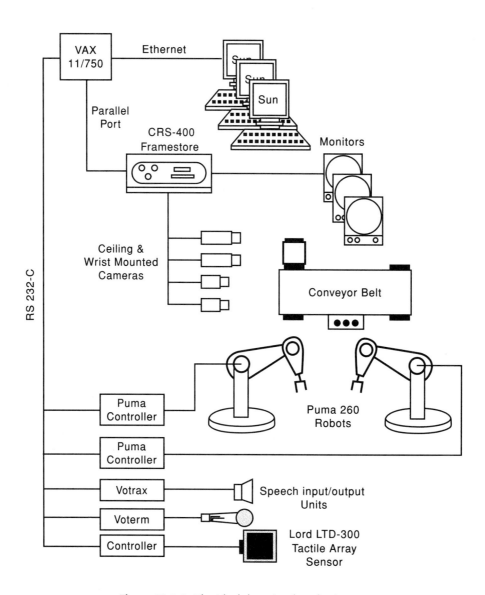

Figure 10.3.4 The ideal domain of a robotic system

Advanced robots should have the ability to operate in unstructured environments where hazardous conditions are encountered. It is also important that their design is inherently robust in every aspect, so that the device can cope with a span of difficult conditions and in very severe performance be degraded in a predictable and safe manner.

In application areas such as health care, transportation, construction, surveillance, agriculture, and the service industries, advanced robotic devices will come into close proximity to humans without the usual safety barriers normally associated with manufacturing robots. Safety and predictability are prime engineering concerns.

Current engineering practice argues for the use of a modular structure that can be readily extended or reconfigured without a major redesign, thus allowing, for example, the use of a range of actuator or sensor modules, as appropriate. The functionality of each module can be precisely specified, and, ideally, standard communication protocols should be utilized for all outputs and inputs. System verification, installation, and maintenance procedures are greatly simplified so that the reliability of operation is enhanced.

The development and burden of standards inevitably constrain transformation, but they also illustrate the growing of the subject domain and the evolution of marketable devices that are fundamentally safe, reliable, and suitable for their purpose.

10.4 APPLICATIONS OF ADVANCED ROBOTS

Over the last five years there has been a gradual expansion in the number of engineering applications of advanced robotic concepts. Although the field is not yet mature, these applications are continuing evidence of a growing confidence in the utility of such devices.

The nuclear industry has always been an obvious application domain for robots with emphasis being placed on remote handling systems. Recent developments include the use of stereo vision and direct force feedback employing six axes-sensitive joysticks based on configurations such as a Bilateral Stewart Platform. Trajectory control, real-time collision avoidance, and the extensive use of geometric modeling procedures for risk assessment and training are technical features now being introduced into material-handling systems.

A number of designs for relatively active mobile robots are now being considered for intervention within unstructured environments. Such a device is shown in Figure 10.4.1 and has mechanical predictable characteristics to operate in a totally stable and safe manner.

There is much scope for the development of innovative computer interfaces for the remote control of such devices, and they will ensure a credible operational situation.

The medical profession tends to be entrepreneurial in its use of technology and has rapidly adapted some of the generic technologies associated with advanced robots. Applications include robot-controlled prostate surgery; the precision robotic machining of bones for hip surgery utilizing geometrical models derived from precisely computed tomography scans; use of autonomous guided vehicles to aid the disabled; and aspects of microsurgery that employ micromanipulators, virtual reality, and force feedback concepts. The safety of such systems and accompanying legal aspects are main issues here and should be addressed in some detail by key people in the field.

322 Artificial Intelligence

Figure 10.4.1 Mobil robot for the nuclear industry

The study of mobile robots has been the focus of much research over the years and now is being utilized in practical applications. The European Union has resulted in the development of practical perception and navigation systems suitable for the control of vehicles—generally, for outdoor use with intelligent autonomous systems for cars, which could have significant implications for modern high-density motorway utilization, both in enhancing car safety and controlling existing traffic systems.

Fully instrumented vehicles have already been tested in realistic situations, and there is commercial interest in the exploitation of the existing technology. In the manufacturing arena, similar generic technologies have resulted in the development of sophisticated AGVs for use in a CIM environment or in the handling of hazardous materials. Automatic inspections are now another common application domain and include practical examples, such as the robot crawler used for monitoring corrosion inside large storage tanks or wall-climbing robots for use in nuclear power plants.

In the United States, free-ranging, fully-instrumented mobile robots are now employed in delivery services within large industrial complexes. In Italy they are employed for delivery of pharmaceutical and other medical products within hospital environments. Autonomous cleaning robots with comprehensive model-based navigation capabilities for use in industrial and commercial areas are now under development in Europe and Japan.

Applications in aerospace combine advanced teleoperated devices and mobility for satellite maintenance and planetary exploration. These encourage the evolution of innovative mobility mechanisms, such as legged robots to cope with difficult terrains. The latter are quite complex in both design construction and operation. But they are generally more active than the conventional multiple articulated tracked vehicle of the type shown in Figure 10.4.1. They have applications in the nuclear industry where there will, inevitably, be buildings with stairs, and in agriculture where a machine with a relative low-footprint area has distinct environmental advantages. Most of the devices produced so far appear to have a relatively limited functionality, but the field is currently being well researched and the work is attracting and increasing the commercial interest.

Economic forecasts produced by the Japanese Robot Society suggest a substantial market for robotic applications in the manufacturing area, particularly with respect to the construction of practical machines for use to perform such tasks as assembly, transportation, and surface-finishing procedures. Such sites are particularly difficult for regular machine operation because they are characterized as very rough unstructured topographies.

Mining is another interesting area for the application of robots because it combines a hazardous environment with relatively structured topography that is certainly suitable for remote teleoperation. In Europe, a project is underway to develop generic system architectures within the ESPRIT ROAD ROBOT project for the control of a set of heavy-duty mobile platforms engaged in such activities as road construction and paving. The objective is to create an outdoor assembly-line operation using conventional CIM principles by adapting components such as large mobile diggers and pavers that are normally operated and coordinated manually.

In a contiguous technical theme the objective of ESPRIT Project ATHENA is to operate a fully instrumented thirty-ton compactor vehicle via a remote, sophisticated computer graphical interface that overlays video and computer-generated geometrical data of landfill sites for waste disposal within which the vehicle moves. The control architecture incorporates both hierarchical and behavioral aspects and uses model-based navigation and sensor fusion techniques. The transfer of advanced robotic concepts into these unconventional domains is supported in each case by strong financial and environmental arguments. There is clearly much scope for future practical implementations of the technology in these and related fields.

Among the manufacturing nations, Japan appears convinced of the economic viability of the totally automated factory. The factory of the future has embedded this concept within the technical spectrum of its recent Intelligent Manufacturing Systems (IMS), which has now been cosponsored by a group of nations as an international study program. The technical spectrum originally defined encompasses almost every aspect of manufacturing from simple metal cutting to advanced anthropomorphic mobile robots. It will operate between individual production cells to ensure the smooth flow of products, even in the event of minor systems failure and uncertainty. Japan is backing this initiative with very substantial research funding, and policymakers there are convinced that a future market exists for such complex automation.

Current international practice tends to focus on the use of sensor-based robots for enhancing conventional production procedures, such as welding and assembly, by providing enhanced operational flexibility and a degree of fault tolerance that can be generated by sensor feedback. The improved technology has, however, allowed the application of automation techniques to production sectors, such as food and agriculture, which previously mainly utilized manual procedures.

The ESPRIT Project ROBOFISH, for example, is concerned with integrating a range of technologies, including computer vision, image processing, and distributed control systems for the automatic processing of freshly caught fish. Other applications include the use of sensor-based robots in agriculture for crop gathering and milk production and the use of CAD/CAM procedures in the clothes industry where some quite innovative gripper techniques have been developed for the precision manipulation of the very thin, light, flexible materials used in manufacturing.

The brief range of applications cited here is not meant to be exhaustive, and some significant application domains, such as underwater surveying and maintenance activities, have not been mentioned. However, the examples given do indicate the extent of early concepts on advanced robot technologies and are now being implemented as practical engineering solutions to significant technical problems. Currently, these engineering implementations are quite diverse in nature, but generic features are emerging that may eventually provide cohesion for the subject area and hasten the evolution of acceptable international engineering standards that, of course, will further promote the acceptance of the technology within a wider range of market sectors.

10.5 FUZZY LOGIC FOR ROBOT ARM CONTROL

Fuzzy logic is the process of solving problems that are filled with ambiguous data, such as whether the room temperature is too hot, too cold, or just right. It uses a multivalued logic to produce an answer that is a best guess and, thus, a more precise and weighted answer. Fuzzy logic is used in computers and other electronic devices for processing imprecise or variable data in place of the traditional binary values. Also, fuzzy logic can employ a larger range of values for greater flexibility. The term was coined in 1965 by L. A. Zadeh, U.S. computer scientist.

Fuzzy logic, in a narrow sense, is a logical system that aims at a formalization of approximate reasoning, and an extension of multivalued logic. In its wide sense, fuzzy logic is the theory of classes with unsharp boundaries.

Fuzzy logic applies to arithmetic, mathematical programming, probability theory, decision analysis, control, neural network theory, topology, and other disciplines.

Fuzzy logic and neural networks have emerged as new methods for decision making and control within the last decade. Nonlinear complex systems can be easily controlled utilizing fuzzy logic principles. The dynamical behavior of robotic systems is complex, especially in the presence of loads. Robotic joints experience and exhibit friction, stickiness, and gear backlash effects. Due to lack of proper linearization of these effects, modern control theory based on state space methods cannot provide adequate control for robotic systems. Furthermore, inversion of Jacobian matrices, particularly when they are near singular, provides another computational problem especially for redundant joint robotic systems.

The Johnson Space Center is investigating the feasibility of applying fuzzy logic–based control for robotic systems. Functions such as tracking, approach, and grapple are typically performed by a robotic arm in a manual mode or semiautomatic mode where a point of resolution is driven to a desired point by kinematic inversion. The center is developing fuzzy logic–based algorithms for semiautomatic mode so that the computational problem of Jacobian inversion can be eliminated. The difference between the desired location and current location is an input vector to the controller that generates joint rate commands. A six-degree-of-freedom robotic arm simulation is used to evaluate the performance of the fuzzy logic–based controller. Figure 10.5.1 shows the simulation flow of the Remote Manipulator System (RMS) forward kinematics and fuzzy controller.

10.6 ADVANCED CONCEPTS AND PROCEDURES

An alternative approach to using a single complex robot to perform a task is to use a group of simpler robots controlled in some distributed manner to achieve the same objective. Each unit in the group could be an essentially autonomous entity, and as interrelationships need not always be precisely coordinated in space and time, the overall behavior is often referred to as cooperative in nature. There are many potential applications for such multiple cooperant robots as, for example,

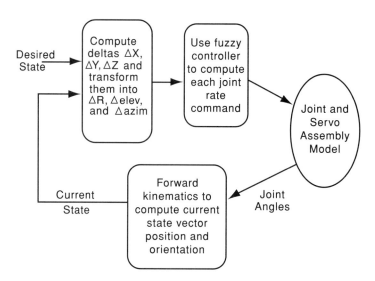

Figure 10.5.1 Simulation flow with Remote Manipulator System (RMS) forward kinematics and fuzzy controller *(Source: R. N. Lee et al., IEEE Technology Update Series, 1994, p. 157,* Fuzzy Logic Technology and Applications)

in the industrial environment where a group of independent devices could be used to move objects that, because of their size, shape, or weight, could not be transported by a single device. Other applications could include the use of multiple cooperating autonomous underwater vehicles for ocean bed surveying, product disassembly (which is quite difficult to achieve with conventional robotic procedures), area cleaning, and surveillance.

A key argument used for the development of such distributed systems is the replacement of a single complex unit by a set of simpler units, thus providing robustness and graceful system degradation as the performance of any one unit is not necessarily affected by the dysfunction of a companion unit, as long as sufficient units are available to provide for redundancy. Each unit in the group need not necessarily be identical, thus allowing a richness in the functionality of any particular combination.

The simplicity of individual robot functionality is not necessarily reflected in the required controlling paradigm, and the field is currently one of particular interest to the research community.

Recent work has focused on the development of architecture, as outlined in Figure 10.6.1, that makes sensible use of a priori knowledge and provides a mechanism for the behavior scripts that control the robots. The concept of a robot to achieve a complex task is of particular interest to the emerging field of microrobotics and will provide many challenges in the future.

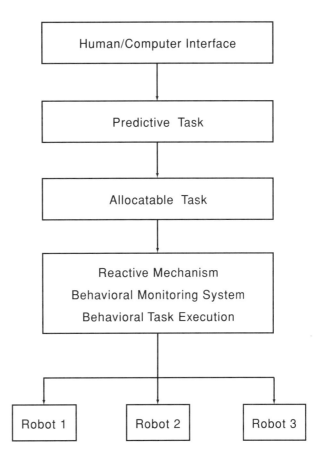

Figure 10.6.1 Hybrid system architecture

10.7 FUTURE DEVELOPMENTS

The widespread demand for and burgeoning availability of inexpensive computing power has rendered feasible the design and implementation of complex robots with a high degree of functionality and possessing architectures. This incorporates the integration of sensor real-time control and innovative concepts based on artificial intelligence. Such devices will be used mainly to complement human skills in applications where it is either undesirable or impossible to use human operators. They will be, for the immediate future, semiautonomous in nature and thus involve human computer interfaces that will be optimized to cope with the complex functionality of the devices. These interfaces will involve aspects of teleoperation, telepresence and virtual reality in the immediate future, and the technology will be embedded to enhance the performance of machine systems.

In order to enhance the capability of the operator, the devices will have an ability to accept and implement high-level commands, execute task and trajectory planning operations, and effect simple maneuvers such as local collision avoidance and beacon-guided automatic docking. Such complexity has implications for traditional engineering concepts of reliability, verification, safety, and maintainability. Many such issues are already well addressed within the aerospace industry, which has a long-established culture in all safety-related matters. This culture and its associated rigor for standards and quality assurance must eventually be adopted by the robot industry as it moves into the market for advanced robots, particularly for safety-critical applications in the nuclear industry and medicine.

Inevitably, there will be a future trend toward enhanced decision-making capabilities within the machine—for example, in the extended use of innovative and flexible architectures and opaque computing procedures based on neural networks or equivalent methodologies. Such a trend will raise interesting engineering, safety, and legal problems that have yet to be addressed, but will be relevant to all types of intelligent systems in whatever way we define the term intelligence.

Clearly, these are significant issues, posing exciting challenges to engineers as they exploit the ever-increasing power of silicon technology in production of devices and processes in the coming decade.

10.8 IMPACT ON EMPLOYMENT

An important consideration is the nature and event of the fully automated factory's impact on employment, with all its social ramifications. The fear of unemployment persists despite strong evidence that advances in technology create more jobs—or at least maintain the same number—rather than eliminate them. Projections indicate that there will be major increases in the number of computer service technicians and maintenance electricians.

Thus, the generally low-skilled, direct labor force engaged in traditional manufacturing will shift to an indirect labor force, with special training or retraining required in areas such as computer programming, information processing, CAD/CAM, and other high-technology tasks that are essential parts of computer-integrated manufacturing. From the operator viewpoint, workers replaced by robots have proved to be retrainable, and little, if any, socioeconomic impact has been experienced where CIM has been adopted.

Opinions about the impact of unmanned factories diverge widely. Consequently, predicting the nature of future manufacturing strategies with any certainty is difficult. Although economic considerations and trade-offs are crucial, companies now widely recognize that in a highly competitive marketplace, rapid adaptability of new technologies is essential to the survival of a manufacturing organization.

10.9 SUMMARY

Artificial intelligence is that part of science concerned with systems that exhibit the characteristics associated with intelligence in human behavior. The elements of artificial intelligence are:

1. Expert systems
2. Natural language processing
3. Machine vision
4. Neural networks
5. Machine learning
6. Advanced robots

Expert systems are intelligent computer programs capable of solving complex problems. Expert systems utilize knowledge-based programs, natural language processing, and robots, and relate to machine vision, movements, and tactile sensing.

Natural language process is, traditionally, information that can be obtained from a database in computer memory that has required utilization of a computer program that translates natural language to machine language.

Machine vision is the acquisition of image data by computers and software combined with cameras and other optical sensors to perform operations such as inspecting, identifying, sorting parts, and guiding robots.

Neural networks are the ultimate goal of AI. However, a profound difference exists in the fundamentals of the thought process between computers and the human brain.

Machine learning is the subfield of AI or a subdivision of expert system technology concerned with programs that learn from experience. The idea behind machine learning from observations was accomplished between a hypothetical intelligent agent and the world. The agent receives percepts from the environment and performs actions. An agent may be divided into four conceptual components.

Advanced robots is the term used to describe intelligent human behavior robots. Applications of advanced robots can be found in many fields, such as the nuclear industry, the medical profession, mobile robots, navigation systems, the aerospace industry, economic forecasts, mining, computer integrated manufacturing, flexible automation, and many other related fields.

System architecture is the general schematic that defines the structural composition and the flow of control and data signals between a robot and controllers.

Advanced concepts and procedures are the alternative approach that replaces complex robots by simpler ones to achieve the same objective.

Fuzzy logic is a logical system with variable processing data and greater flexibility than the traditional binary values. Fuzzy logic applies to arithmetic, mathematical programming, probability theory, decision analysis, control, neural network theory, topology, and other disciplines.

Future development of intelligent systems with flexible architectures is the significant issue in engineering technology for the forthcoming decade. However, the impact on employment is a major issue in the creation of new technologies. Special training and higher education are essential for future adaptability.

10.10 REVIEW QUESTIONS

10.1 What is artificial intelligence? Discuss its use in industry.
10.2 Describe the elements of artificial intelligence.
10.3 Why are expert systems called "expert"? Comment on their capabilities.
10.4 Describe the three elements of expert systems.
10.5 Explain why an expert system is different from a normal computer program.
10.6 List eight of the specialized applications of expert systems.
10.7 What is natural language processing? Explain some of its applications.
10.8 Why is machine vision a part of artificial intelligence?
10.9 What are neural networks and how do they apply to artificial intelligence?
10.10 What are advanced robots? Comment on their capabilities and their history.
10.11 What is a system architecture and how does it apply to intelligent robots?
10.12 List some of the applications of advanced robots.
10.13 What is the alternative of complex robots?
10.14 What progress in artificial intelligence can you foresee in the near future? Why?
10.15 Explain why humans will still need to be in the factory of the future.
10.16 Report on some of the latest technological developments that can enhance AI.
10.17 Will artificial intelligence ever exceed human intelligence? Explain.

10.11 PROBLEMS

10.1 Give examples in manufacturing engineering in which artificial intelligence could be effective.
10.2 Describe your opinion concerning voice recognition capabilities of future machines and controls.
10.3 It has been suggested by some that ultimately artificial intelligence systems will be able to replace the human brain. Do you agree? Explain.
10.4 We stated that neural networks are particularly useful where the problems are ill-defined and the data are fuzzy. Give examples in manufacturing where this is the case.
10.5 With specific examples, describe your own thoughts concerning the state of manufacturing in the United States as compared to the other industrialized nations.
10.6 Consider the problem faced by an infant learning to speak and understand a language. Explain how this process fits into the general learning model, identifying each of the components of the model as appropriate.
10.7 Repeat Problem 10.6 for the case of learning to play tennis (or some other competitive sport with which you are familiar). Is this supervised learning or reinforcement learning?
10.8 Draw a decision tree for the problem of deciding whether or not to move forward at a road intersection given that the light has just turned green.

10.12 REFERENCES

Allen, J. F. *Natural Language Understanding.* Redwood City, CA: Benjamin/Cummings, 1995.

Ayache, N. *Artificial Vision for Mobile Robots.* Cambridge, MA: MIT Press, 1991.

Bloom, P. *Language Acquisition: Core Reading.* Cambridge, MA: MIT Press, 1994.

Boden, M. A. *The Philosophy of Artificial Intelligence* Oxford, England: Oxford University Press, 1990.

Bratro, I. *Prolog Programming for Artificial Intelligence.* Reading, MA: Addison-Wesley, 1990.

Charniak, E. *Statistical Language Learning.* Cambridge, MA: MIT Press, 1993.

Copeland, J. *Artificial Intelligence: A Philosophical Introduction.* Oxford, England: Blackwell, 1993.

Cormen, T. H., C.E. Leiserson, and R. R. Rivest. *Introduction to Algorithms.* Cambridge, MA: MIT Press, 1990.

Dietterich, T. G. "Machine Learning," *Annual Review of Computer Science* 4 (1990).

Dreyfus, H. L. *What Computers Still Can't Do: A Critique of Artificial Reason,* Cambridge, MA: MIT Press, 1992.

Forbus, K. D., et al. *Building Problem Solvers.* Cambridge, MA: MIT Press, 1993.

Ginsberg, M. *Essentials of Artificial Intelligence.* San Mateo, CA: Morgan Kaufmann, 1993.

Heckerman, D. *Probabilistic Similarity Networks.* Cambridge, MA: MIT Press, 1991.

Hirota, K., Ed. *Industrial Applications of Fuzzy Technology.* Tokyo, Japan: Springer-Verlag, 1993.

Jerison, H. J. *Brain Size and the Evolution of Mind.* New York: American Museum of Natural History, 1991.

Judd, J. S. *Neural Network Design and the Complexity of Learning.* Cambridge, MA: MIT Press, 1990.

Kartalopoulos, S. V. *Understanding Neural Networks and Fuzzy Logic.* Piscataway, NJ: IEEE, 1996.

Lowry, M. R., and R. D. McCartney. *Automating Software Design.* Cambridge, MA: MIT Press, 1991.

Luger, G. F., and W. A. Stubblefield. *Artificial Intelligence: Structures and Strategies for Complex Problem Solving.* Redwood, CA: Benjamin/Cummings, 1993.

Marsland, A. T., and J. Schaeffer, Eds. *Computers, Chess, and Cognition.* Berlin, Germany: Springer-Verlag, 1990.

Mason, M. T. "Kicking the Sensing Habit," AI Magazine 14, no. 1 (1993): 58-59.

Partridge, D. *A New Guide to Artificial Intelligence.* Norwood, NJ: Ablex, 1991.

Pomerean, D.A. *Neural Network Perception for Mobile Robot Guidance.* Dordrecht, The Netherlands: Kluwer, 1993.

Rabiner, L. R., and B. H. Juang. *Fundamentals of Speech Recognition.* Englewood Cliffs, NJ: Prentice Hall, 1993.

Rich, E., and K. Knight. *Artificial Intelligence.* New York: McGraw-Hill, 1991.

Russell, S. J., and P. Norving. *Artificial Intelligence: A Modern Approach.* Upper Saddle Pines, NJ: Prentice-Hall, 1995.

Russell, S. J., and E. H. Wefald. *Do the Right Thing: Studies in Limited Rationality.* Cambridge, MA: MIT Press, 1991.

Saraswat, V. A. *Concurrent Constrain Programming.* Cambridge, MA: MIT Press, 1993.

Searle, J. R. *The Rediscovery of the Mind.* Cambridge, MA: MIT Press, 1992.

Shapiro, S. C. *Encyclopedia of Artificial Intelligence.* New York: John Wiley and Sons, 1992.

Winston, P. H. *Artificial Intelligence.* Reading, MA: Addison-Wesley, 1992.

Wos, L., R. Overbeek, E. Lusk, and J. Boyle. *Automated Reasoning: Introduction and Applications.* McGraw-Hill, New York: 1992.

Yoshikawa, T. *Fundamentals of Robotics: Analysis and Control.* Cambridge, MA: MIT Press, 1990.

CHAPTER 11
SAFETY

11.0 OBJECTIVES

After studying this chapter, the reader should:

1. Be acquainted with robot safety
2. Understand safety standards
3. Recognize system reliability
4. Be familiar with human factor issues
5. Be aware of safety sensors and monitoring
6. Realize safeguarding
7. Perceive the important factors of training
8. Apprehend safety guidelines
9. Understand definitions

11.1 ROBOT SAFETY

Safety is the method and technique used for avoiding accidents. The need to safeguard machinery for operator protection has been combined lately with high productivity and automation. Automation is not new and the machinery for automation is unique for each product. However, a new concept has made its way into the workplace in the last decade, challenging our safety considerations by combining automation with programmable versatility. This concept, based on robots, is being termed the workplace of the future, a workplace that is based on computer-integrated manufacturing (CIM).

This workplace of the future is controlled by computers, including the concept of computer-aided design (CAD), production scheduling, computer-aided manufacturing (CAM), inventory controls, maintenance scheduling, and even accounting and purchasing. The integration of all these functions into a flexible manufacturing system (FMS) is what CIM is all about.

The advanced factory, with programmable robots at the heart of the system, is smaller, cleaner, and much quieter than was expected. As workers have been separated from the hazardous tasks, one would expect safety records to be significantly improved. Historically, however, the introduction of automation has increased the number of worker injuries (due in large part to the poor hazard identification with new machinery), and the same has been true with robots. The majority of injuries that have occurred have been caused by contact with the moving parts of a robot. Although there are similarities between robots and conventional machinery, three major differences can be identified that are the concern of safety personnel: (a) speed of movement, (b) predictability of movement, and (c) hazard zones. In conventional machinery, hazard zones may be difficult to recognize, but they are fixed with time. The main difference between conventional machinery and robots is that a robot can be programmed to do different jobs and to react to changes in the process, even making decisions from a limited number of choices. Safety engineering should be applied to robot safety, particularly in the human factors aspects and systems safety approaches.

Robot safety must include the usual considerations of man, machine and **workstations**, environment, and the interface behavior, but it must also consider software. Robot hardware design should be based on sound engineering practices, especially those safeguarding principles that have traditionally been used where humans interact with machines during operation. According to Hammer (1989), accidents are initiated through (a) engineering deficiencies, (b) lack of proper procedures, or (c) inadequate programming. Therefore, emergency stop switches must appear on the control panel and also be added to the pendant used in the teach mode where the operator or programmer may be moving in the robot's work envelope. Of course, the robot's movements are considerably slower in the teach mode, again by design. Comprehensive instruction and operation procedures must also be incorporated through training programs.

Jiang (1986) has summarized systems procedure for robot safety, using the workplace design illustrated in Figure 11.1.1.

In this chapter, we discuss safety standards, system reliability, human factor issues, safety sensors and monitoring, safeguarding, training, safety guidelines, and definitions.

11.2 SAFETY STANDARDS

An important consideration in installing, programming, operating, and maintaining robot systems is safety. Safety may be defined as a judgment of the acceptability of danger, where danger is the combination of hazard and risk. Hazard is defined as an injury producer, and risk is defined as the probability that an injury will occur. The causes of employee injury in a robotic environment vary and include the following:

   a. Parts of the body being caught
   b. Being struck by a part or robot gripper
   c. Falling from equipment or structure
   d. Slipping or tripping on walking or working surfaces
   e. Exposure to dangerous levels of heat or electricity
   f. Excessive physical strain

The Robotic Industries Association (RIA) is the only trade association in North America dedicated to the promotion of robotics safety standards. RIA activities include trade shows, seminars and conferences, standards development, and trade statistics. The American National Standards Institute, Inc. (ANSI) is the clearinghouse for standards developed in the United States. ANSI is also the official organization representing the United States in the interna-

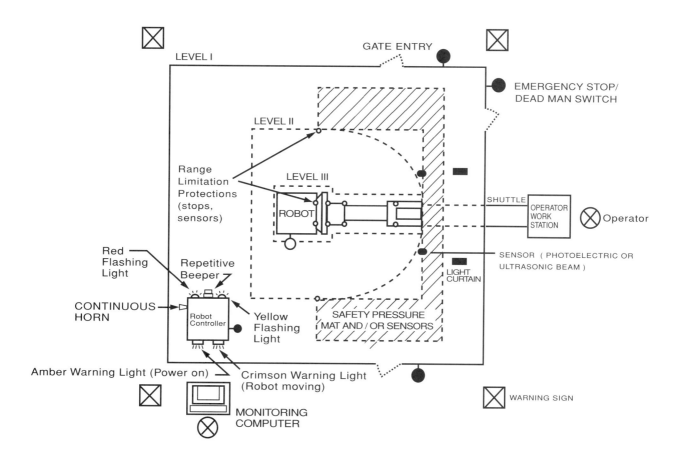

Figure 11.1.1 Robot workplace design *(Source: B. J. Jiang)*

tional standards activities of the International Organization for Standardization (ISO) and the International Electrotechnical Commission (IEC). The Occupational Safety and Health Administration (OSHA) had its safety concerns addressed through the ANSI/RIA R15.06-1992 Robot Safety Standard, which is cited in the OSHA Technical Manual Chapter 18. Additional ANSI/RIA standards help have been created with common specifications for robot end users and suppliers.

The structure of industrial robots safety standards was revised by the RIA Technical Report R15.07/TRI-1993 for the part of simulation/off-line programming. This report includes terms, notations, and data requirements for robot modeling. Also defined are expressions and definitions used to characterize industrial robots and establishes a methodology for the transfer of robot geometric and kinematic information. The contents of the report also provide robot vendors, simulation vendors, and users with a consistent means of information exchange in safety standards.

The update information includes the 7th Annual National Robot Safety Conference Proceedings 1995, which focuses exclusively on robot safety issues with case studies from leading users, updates of ANSI/RIA standards and OSHA guidelines, tips on robot safety training, and more.

In addition to these organizations there is also the National Institute for Occupational Safety and Health (NIOSH), which can offer assistance in preventing the injury of workers by robots. NIOSH is the U.S. federal agency responsible for conducting research to prevent work-related injuries, illnesses, and fatalities. NIOSH's Division of Safety Research (DSR) is the Institute's focal point for research to prevent occupational injuries and fatalities. The mission of the DSR is to reduce the frequency and severity of injuries in industrial workplaces by developing and conducting research projects with real-world intervention potential.

Approximately 85 percent of the accidents involving robots and other machinery are caused by the operator, 5 percent are caused by mechanical or electrical failure, and the remainder are caused by other factors.

Safety standards are very important in the workplace. All the above-mentioned professional organizations have defined the following priorities for eliminating hazards in the workplace:

1. Eliminate the hazard through the machine design stage.
2. Apply safeguarding technology.
3. Use warning signs and labels.
4. Train and instruct the worker, programmer, and maintenance personnel.
5. Prescribe personal protective equipment and devices.

11.3 SYSTEM RELIABILITY

We know that eventually all products fail in some matter or another. Reliability, therefore, is defined as the probability that a product will perform its intended function under stated conditions for a specified period of time. The more critical the application of a particular product, the higher its reliability should be. Industrial robots that fail to perform properly, due to a partial or total functional failure, over a period of time, would not satisfy the economic requirements for implementation in an industrial application. Robot manufacturers, therefore, should strive to make robots or robotic systems as reliable as possible. On the other hand, industrial robots contain many sophisticated components and devices, and the manufacturers cannot be sure that the robot will be used within specified limits. Reliability, of course, also depends on whether a product is properly used and maintained.

The expected reliability of a product depends on the nature of the product and its use. For example, robots and other complex systems are often assumed to have a certain percentage of constant failure during their useful lifetime. This condition is known as series reliability. On the other hand, if bypass equipment is substituted for the failure, the reliability is not critical. This condition is known as parallel reliability. The parallel reliability concept is important in the design of backup systems, which permit a product to continue to function in the event one of its components fails. Electrical or hydraulic systems in an aircraft, for example, are backed by mechanical systems, which are called redundant systems.

Predicting reliability has become an important science and involves complex mathematical relationships. The best insurance for reliable robot operation is preventive maintenance in accordance with the manufacturer's recommendations and backup systems (buffer storage) that will permit the system to continue functioning.

The reliability of an automated complex system is very important because its failure without prevention can result in major economic losses to the manufacturer.

11.4 HUMAN FACTOR ISSUES

With few exceptions, all machines are designed to be used by humans. This is much more true with robots, which need to be programmed by humans. Safety is an important issue and should be incorporated into properly designed robots or any other automated production system.

Human factor issues or engineering is the study of the human-machine interaction and is defined as an applied science that coordinates the design of devices, systems, and physical working conditions with the capacities and requirements of the worker. The machine designer must be aware of the subject and design devices to "fit the worker" rather than expect the worker to adapt to fit the machine. The term *ergonomics* is synonymous with human factors. We often see reference to the good or bad ergonomics of equipment or a household appliance. A machine or robot system designed with poor ergonomics will be uncomfortable and tiring to use, and may even be dangerous. The important social objective of robots and other forms of automation is to remove humans from unsafe and hazardous working conditions.

Besides the size of a robot's work envelope, its speed, its proximity to humans, and interactions with other machinery, many other factors should be considered and investigated, such as:

a. The layout of control panels
b. Teach-pendant accuracy
c. Personnel training
d. Barrier guards
e. Safety devices
f. Interlocks
g. Warnings

In addition, human factor considerations should include evaluation of a robot workstation when an operator enters the workstation for maintenance, programming, and the like.

An important consideration, also, is the speed at which a robot should be allowed to run when a human is in the workstation, particularly during teach-mode programming. The ANSI (1986) has specified the maximum speed of 10 inches per second for teach-mode operations.

Further, in the areas of **teleoperation**, robots have inherent human factor components associated within their scope. In teleoperation, a human attempts to guide a robotic device through some sort of desired configurations, using a high-bandwidth interface. This brings the human in an immediate and integral part of the control loop with potential safety limitations. Likewise, in many laboratory robotics applications, humans and robots are working together in close proximity without the proper safeguards used for industrial installations. This proves that even though all machines are guarded, experience shows that machine designers and builders cannot influence safety to the extent that machine users can.

11.5 SAFETY SENSORS AND MONITORING

In addition to the operator's ability to override the regular work cycle in the event of an observed safety hazard, the workcell controller should also be capable of monitoring its own operation for unsafe or potential unsafe conditions in the cell.

Safety monitoring involves the use of sensors to indicate conditions or events that are unsafe or potentially unsafe. The objectives of safety monitoring include not only the protection of humans who happen to be in the cell, but also the protection of the equipment in the cell. The sensors used in safety monitoring range from simple limit switches to sophisticated vision systems that are able to scan the workplace for intruders and other deviations from normal operating conditions.

Great care must be taken in workcell design to anticipate all possible mishaps that might occur during the operation of the cell, and to design safeguards to prevent or limit the damage resulting from these mishaps. The National Bureau of Standards defines three levels of safety sensor systems in robots:

11.5 Safety Sensors and Monitoring

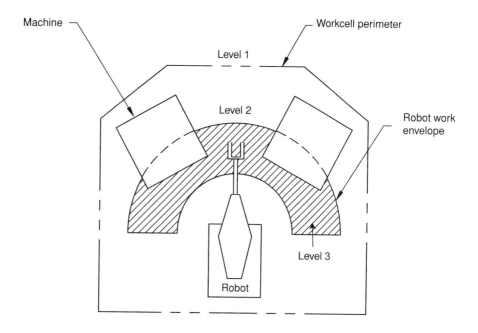

Figure 11.5.1 Three levels of robot safety sensor system.

1. Level 1—Perimeter penetration detection
2. Level 2—Intruder detection inside the workcell
3. Level 3—Intruder detection in the immediate vicinity of the robot

Level 1 systems are intended to detect that an intruder has crossed the perimeter boundary of the workcell without regard to the location of the robot.

Level 2 systems are designed to detect the presence of an intruder in the region between the workcell boundary and the limit of the robot work volume.

Level 3 systems provide intruder detection inside the work volume of the robot.

These sensor systems are intended to protect workers who must be in close proximity to the robot during operation (e.g., during programming). Figure 11.5.1 illustrates the three sensor levels.

The two common means of implementing a robot safety sensing system are pressure sensitive floor mats and **light curtains**. Pressure sensitive mats are area pads placed on the floor around the workcell that sense the weight of someone standing on the mat. Light curtains consist of light beams and photosensitive devices placed around the workcell that sense the presence of an intruder by an interruption of the light beam. Pressure sensitive floor mats can be used for either Level 1 or Level 2 sensing systems. The use of light curtains would be more appropriate as Level 1 systems. Proximity sensors located on the robot arm could be utilized as Level 3 sensors.

The safety monitoring strategies that might be followed by the workcell controller would include the following schemes:

1. Complete shutdown of the robot upon detection of an intruder
2. Activation of warning alarms
3. Reduction in the speed of the robot to a safe level
4. Directing the robot to move its arm away from the intruder to avoid collision
5. Directing the robot to perform tasks away from the intruder

Another system used in safety monitoring is called a "fail-safe hazard detector." The concept of this detector is based on the recognition that some component of the basic hazard sensor system might fail and that this failure might not be found out until some safety emergency occurred. The fail-safe hazard detector is designed to overcome this problem. Figure 11.5.2 shows the operation of a robot with a safety system.

11.6 SAFEGUARDING

It is recognized that most industrial accidents are the results of unsafe acts by the worker. Such acts can occur due to improperly trained operators or careless programmers activating the wrong controls. The remaining accidents result by component failure or other unsafe conditions in the plant. The types of unsafe acts or unsafe conditions are similar to all other machines. However, unlike other machines, robots are not designed for a specific task line fixed automation. Their design includes motion flexibility with many variables in speed, axes, space motion, and many involvements with other equipment. Therefore, the major concern in safety of all personnel involved with industrial robots is very important. According to the National Safety Council (1991) the principal hazards associated with robots are as follows:

1. Being struck by a moving robot while inside the work envelope
2. Being trapped between a moving part of a robot and another machine, object, or surface
3. Being struck by a workpiece, tool, or other object dropped or ejected by a robot

Robot safeguards are further explained by D. L. Goetsch (1996) as follows: The best guard against hazards is to erect a physical barrier around the entire perimeter of a robot's work envelope. This physical barrier should be able to withstand the force of the heaviest object the robot could eject. Various types of shutdown guards can be erected. A guard containing a sensing device that automatically shuts down the robot if any person or object enters its work envelope can be effective. Another approach is to put sensitized doors or gates in the pe-

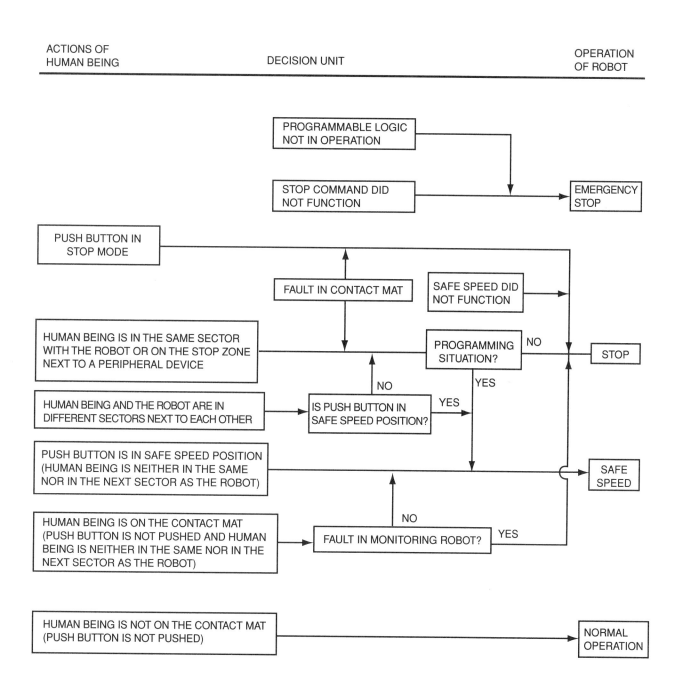

Figure 11.5.2 Robot safety system operation

rimeter barrier that automatically shut down the robot's movement when they are opened. It is difficult to go to a risk-free environment because there is always the possibility of a malfunction or the violation of good safety practices. However, there are ways to minimize the potential of these errors:

1. Safety training
2. Dependable machine design
3. High-reliability controls
4. Proper layout of the work area
5. Safe position and clear visibility for programming
6. Establishing proper maintenance procedures
7. Adequate installation performed in the presence of safety personnel
8. Obeying safety rules and regulations of authoritative organizations

Robots are an inevitable part of our future and the future of our industrial world. Their increased utilization will result in higher productivity, better quality of products, and a safer work environment for all concerned. However, their successful implementation will only happen if their use will only benefit, and not harm, the human. In reality, robots designed today are much safer devices because of our growth in the understanding of safety and equipment. Analyzing the hazards and assessing the risks help reduce or eliminate further accidents in a manner that is still cost-effective.

11.7 TRAINING

Training is an important factor in the successful implementation of any advanced technology in a company. Training should be provided for all personnel, including the management and engineering staff, as well as programming, operating, and maintenance in the plant. According to Hinson (1983), in an article about training programs, robotics training should take place in a company's environment under five categories:

1. Awareness
2. Justification
3. Application
4. Operations and maintenance
5. Safety

Awareness training provides a survey of robotics, including technology, applications, economics, and social implications. It also explores the future trends and research developments taking place in robotics. This category of training is typically presented to managers and engineers to encourage the implementation and the opportunities for applying this technology.

Justification training is intended for managers and engineers who are responsible for implementing robot projects in the company. Justification training deals with economic issues and the unique problems that arise in the justification of robots as compared to other investment projects.

Application training is designed principally for technical people (engineers, production managers, and the like) who must select the applications and plan the installations. Coverage in this category would include basic technology, robot programming, and application issues.

Operations and maintenance are provided for production programming and maintenance personnel. These are designed to give these persons the detailed technical skills and knowledge required to use, program, and service the equipment. Most robot manufacturers provide training programs as part of the sales contract with the customer.

Safety is the goal of the company to provide to all personnel who are involved with robots in the plant an awareness of the potential dangers associated with this technology.

In addition to the implementation issues discussed, the ANSI/RIA Subcommittee (1991) added a new expansion of Section 9 in safety training. The contents of current provisions for training program are as follows:

9.2   Training Program Content. Training as appropriate to each assigned task should include, but not be limited to:

    9.2.1   A review of applicable industry safety procedures and standards.
    9.2.2   A review of applicable robot vendor safety recommendations.
    9.2.3   An explanation of the purpose of the robot system and its operation.
    9.2.4   An explanation of the specific tasks and responsibilities for each person.
    9.2.5   The person or persons to contact when the actions are beyond the training responsibility.
    9.2.6   Identification of the recognized hazards associated with each task.
    9.2.7   Identification of unusual operating conditions.
    9.2.8   An explanation for function testing of safeguarding devices.

11.8 SAFETY GUIDELINES

Researchers have developed many guidelines pertaining to safety issues in robots. Originally, robots have benefited industry by removing employees from dangerous and hazardous work to other industrial safety operations. However, robot applications potentially pose new risks to workers that are unlike those associated with conventional machine tool operations. For instance, although a robot may appear to be idle, a signal from a remote location can cause sudden motion. These kinds of unexpected robot movements are the concern of employees for obtaining further guidelines on robotics safety.

Accordig to K. M. Blache in *Industrial Practices for Robotic Safety* (1991), the following guidelines are for safe use of robots in a production environment:

1. If the robot is not moving, DO NOT assume it is not going to move.
2. If the robot is repeating a pattern, DO NOT assume it will continue.
3. Always be aware of where you are in relationship to the possible positions that the robot may reach.
4. Be aware if there is power to the actuators. Indicator lights will be on when there is power to the actuators.
5. Limit switches or software programming will not be used as the primary safeguard.
6. Teaching, programming, servicing, and maintenance are the only authorized reasons for entry into the work envelope.
7. Safeguards with at least one level of redundancy will be present when employees are required to enter the work envelope.
8. Never climb on, over, or under a barrier for any reason.
9. Before activating power to the robot, employees should be aware of what it is programmed to do, that all safeguards are in place, and that no foreign materials are present within the work envelope.
10. Eliminate any source of stored energy prior to entry into the work envelope.
11. Notify supervision immediately when any unexpected interruption to the normal robot work cycle occurs.
12. Servicing should never occur within the work envelope with power on to the robot.
13. Report any missing or defective safeguard to supervision immediately. Check all safeguards at the beginning of each shift.

Concisely, the social impacts of automation and technology on people and managing change will have a significant impact on robotics safety during the years ahead.

11.9 DEFINITIONS

Robot safety standards present several definitions. The following list of definitions will help to understand better the meaning of safety.

Automatic operation: That time when the robot is performing its programmed tasks through continuous program execution.
Barrier: A physical means of separating persons from the robot.
Control software: The inherent use of control instructions that define the capabilities, actions, and responses of a robot system. This software is fixed by the supplier and usually not modifiable by the user.
Drive power: The energy source(s) for the robot actuators that produce motion.

Emergency stop: A method using hardware-based components that overrides all other robot controls and removes drive power from the robot actuators and brings all moving parts to a stop.

End effector: Part of the robot that holds the tooling at the end of the arm.

Hold: A time period when the robot hardware is motionless, but there is power to the robot and the actuators. The robot may automatically move from the hold.

Industrial robot: A reprogrammable multifunctional manipulator designed to move material, parts, tools, or specialized devices, through variable programmed motions, for the performance of a variety of tasks.

Industrial robot system: An industrial robot system includes the robot(s) (hardware and software) consisting of the manipulator(s), power supply, and controller; the end effector(s); any equipment, devices, and sensors the robot is directly interfacing with; any equipment, devices and sensors required for the robot to perform its task; and any communications interface that is operating and monitoring the robot, equipment, and sensors.

Interruption: A time period when the robot hardware is motionless because the robot system fails or is disabled. There may or may not be power. Usually, an employee must act to eliminate the interruption.

Limited devices: A means for restricting the work envelope that will stop all motion of the robot independent of the control software or the application programs.

Operating work envelope: The volume of space enclosing the movement of robot manipulator, end effector, workpiece, and the robot itself when performing programmed motions.

Pendant: A portable device by which a person can control robot motion.

Presence-sensing device: A device designed, constructed, and installed to create a sensing field or area around a robot(s) that will detect an intrusion into such field or area by a person, robot, or the like. Some examples of presence-sensing devices are light curtains, mats, capacitance systems, proximity detectors, and vision safety systems.

Programming: Providing the instruction required for the robot to perform its intended task.

Proximity detector: A device that senses that an object is only a short distance away, and/or measures how far away it is from the robot.

Repair: To restore the robot system to operating condition after damage or malfunction. Work is usually performed by skilled trades and/or engineering personnel.

Restricted work envelope: The volume of space enclosing the movement of robot manipulator, end effector, workpiece, and the robot itself when the robot is restricted by limiting devices that establish limits that will not be exceeded in the event of any foreseeable failure of the robot or its controls. The maximum distance the machine can travel after the limit device is actuated will be considered the basis for defining the restricted robot operating envelope.

Robot actuator: The cylinder, motor, and so on that directly or indirectly drives the unit; commonly thought of as the arm of the robot minus the end effector.

Safeguard: A guard, device, or procedure designed to protect persons from danger.

Safeguard redundancy level: Combining or arranging of safeguards in an order to provide backup safety features for the robot.

Service: To make fit for use, adjust, repair, maintain. Work is usually performed by the operator or service person.

Software limits: A limit imposed by control software on some motion that inhibits travel beyond a point. This point or limit over a working range can be changed by trained personnel.

Teach: To move a robot to, or through, a series of points to be stored for the robot to perform its intended task.

Vision safety system: A device, such as a camera, that is designed, constructed, and installed to detect intrusion by a person(s) into the robot-restricted work envelope.

Warning device: An audible or visual device used to alert persons to a potential safety hazard.

Work envelope: The volume of space enclosing the maximum designed reach of the robot manipulator, end effector, and the robot itself.

11.10 SUMMARY

Safety is an important component in industrial automation. Understanding the interaction of human beings with machines and the workplace environment is essential for the proper design of machinery and the development of safe and efficient working conditions in manufacturing plants.

Robot safety depends on the size of the robot's work envelope, its speed, and its proximity to humans. It requires a high level of reliability in order to avoid collision and serious damages.

Safety standards have been adopted from nationally recognized organizations and should be obeyed and applied on all occasions.

Reliability of a robotic system is defined as the probability that a system will satisfactorily perform its intended function under given circumstances for a specified period of time.

Human factors engineering is the study of the human-machine interaction that coordinates the design of devices, systems, and physical working conditions with the capacities and requirements of the worker.

Safety sensors and monitoring provide the capability of the workcell controller and its sensors to monitor the operation during unsafe conditions in the cell.

Safeguarding is the prevention of injury or accident in the workplace.

Training is a major factor in the successful implementation of any advanced technology in a company or operation. Training should be provided for all personnel, including management, engineering, programming, operating, and maintenance people in the plant. Training should be conducted in accordance with the ANSI/RIA safety standards.

Safety guidelines have been developed by researchers pertaining to safety issues in robots to reduce or eliminate accidents in a production environment.

The robot safety standards present several definitions to assist your understanding in the importance of safety. Some of these definitions are included in this chapter.

11.11 REVIEW QUESTIONS

11.1 What is robot safety and how do most accidents occur?
11.2 How can a robot workplace design be laid out to eliminate accidents?
11.3 What are robotic safety standards and who are the trade associations dedicated to their promotion?
11.4 Name five causes of employee injury in a robotic environment.
11.5 Name five priorities for eliminating accidents in the workplace.
11.6 What is system reliability and how can it be accomplished in a robotic operation?
11.7 What is human factors engineering and how does it apply to social objectives of robots and other forms of automation?
11.8 What are the eight factors to be considered for safeguarding robots in the workplace? Describe them briefly.
11.9 What is the maximum speed for teach-mode operations?
11.10 What is safety sensors monitoring? Define the three levels of safety sensor systems in robotics.
11.11 What are the five safety monitoring strategies that might be followed by the workcell controller?
11.12 What are the methods or ways that can minimize potential errors in a robotic environment?
11.13 List the five categories of a training program.
11.14 Describe the elements of training for each level of personnel within an industrial organization.
11.15 Name the eight new provisions of training programs added by the ANSI/RIA Subcommittee (1991) for Section 9.
11.16 What are safety guidelines? Name the ones used in a production environment.
11.17 What is an industrial robot system?
11.18 What is a presence-sensing device?
11.19 What is a restricted work envelope?
11.20 What is a vision safety system?

11.12 PROBLEMS

11.1 In a certain robot model, the time between breakdowns has been determined, and the mean time between failures is 360 hours. Repairing the robot has required an average of 9 hours. Calculate the robot's availability.

Hint: Availability $= \dfrac{\text{MTBF} - \text{MTTR}}{\text{MTBF}}$

MTBF = mean time between failures
MTTR = mean time to repair

11.2 Suppose in Problem 11.1 that the mean time between failures is 200 hours, and the mean time to repair when breakdowns occur is 8 hours. If a preventive maintenance program is expected to increase the MTBF to 300 hours and reduce the MTTR to 6 hours, determine the availability of the robot.

11.3 List what you think are two important safety issues concerning robots and how they should be addressed.

11.4 Explain why product liability has become such an important factor in manufacturing.

11.5 Most modern machines and robots are now built with ergonomic considerations in mind, whereas older machines didn't generally have these features. Describe your opinions concerning retrofitting older machines to improve their safe and comfortable use.

11.6 Describe two ways of protecting against unsafe personnel intrusion into a robot's work envelope during robot operation.

11.7 Describe a method of preventing the necessity of personnel intrusion into a robot's work envelope when the worker must share a workstation with a robot.

11.8 How do robots compare with fixed automation in terms of speed?

11.9 What is the potential advantage for supplying buffer storage in the layout of a proposed robotic system workcell?

11.10 What are three principal interfaces between the robot and its environment?

11.13 REFERENCES

ANSI, ANSI/RIA R15.06-1992. *American National Standards for Industrial Robots and Robot Systems—Safety Requirements*. New York: American National Standards Institute, 1992.

Blache, K. M. "Industrial Practices for Robotic Safety." In J. H. Graham, ed. *Safety, Reliability, and Human Factors in Robotic Systems*. New York: Van Nostrand Reinhold, 1991.

Brown, S. *The Product Liability Handbook: Prevention, Risk, Consequence and Forensics of Product Failure*. New York: Van Nostrand Reinhold, 1991.

Goetsch, D. L. *Occupation Safety and Health*. Englewood Cliffs, NJ: Prentice Hall, 1996.

Graham, J. H., ed. *Safety, Reliability, and Human Factors in Robotic Systems*. New York: Van Nostrand Reinhold, 1991.

Hammer, W. *Occupational Safety Management and Engineering.* Englewood Cliffs, NJ: Prentice Hall, 1989.

Hinson, R. "Training Programs Are Essential for Robotics Success." *Industrial Engineering* (September 1983): 26–30.

ISO, ISO/DP 10218. *Manipulating Industrial Robots—Safety.* Geneva, Switzerland: International Organization for Standardization, 1989. (For more information, contact RIA, c/o USA TA9/ISO, TC1184/SC2.)

Jiang, B. J. "A Systematic Procedure for Robot Safety." *Proceedings of the ASSE Annual Professional Development Conference*, New Orleans, LA, 1986. p. 245

Kilmer, R. D. "Safety Sensor Systems for Industrial Robots." *Technical paper MS82-221.* Dearborn, MI: SME, 1982.

National Safety Council. "Robots, Data Sheet 1-717-85." Chicago, 1991, p.1.

OSHA. "Guidelines for Robotic Safety." Occupational Safety and Health Administration, Instruction PUB 8-1.3, Washington, DC, 1987.

RIA, BSR/RIA 15.06. "Proposed American National Standard for Industrial Robots and Robot Systems—Safety Requirements" (Rev. 10). Ann Arbor, MI: Robotic Industries Association, 1997.

CHAPTER 12

Industrial Applications

12.0 OBJECTIVES

After studying this chapter, the reader should:

1. Be acquainted with automation in manufacturing
2. Understand robot applications
3. Recognize material-handling applications
4. Be familiar with processing operations
5. Be informed of assembly and inspection operations
6. Apprehend how to evaluate the potential of a robot application
7. Be aware of future applications
8. Perceive the challenge for the future
9. Be informed of innovations
10. Be acquainted with case studies

12.1 AUTOMATION IN MANUFACTURING

The main goals of automation in manufacturing are to integrate various operations to improve productivity, increase product quality and uniformity, minimize cycle times and effort, and reduce labor cost.

Few developments in the history of manufacturing have had a more significant impact than computers. Computers allow us to integrate virtually all phases of manufacturing operations and maintain a competitive edge in the domestic and international workplace. The result is computer-integrated manufacturing (CIM), which is a broad term describing the computerized integration of all aspects of design, planning, manufacturing, distribution, and management.

True automation began with the numerical control (NC) of machines, which has the capability of flexibility of operations, low cost, and ease of making different parts with lower operator skill. Manufacturing operations are further optimized, both in quality and cost, by **adaptive control (AC)** techniques, which continuously monitor the operation and make necessary adjustments in process parameters.

Great advances have been made in material handling, processing operations, and automated assembly and inspection with the implementation of industrial robots. The role of robots is critical in the implementation of these technologies and other technologies to come.

Flexible manufacturing systems (FMS) integrate manufacturing cells into a large unit, containing industrial robots servicing several machines, all interfaced with a central host computer.

Artificial intelligence (AI) involves the use of machines, computers, and industrial robots to replace the human intelligence.

Expert systems (ES), which are basically intelligent computer programs, are being developed rapidly with capabilities to perform tasks and solve difficult real-life problems.

The factory of the future, in which production takes place with little or no direct human intervention, will use computer-integrated manufacturing (CIM) and industrial robots.

Therefore, the applications of robots in manufacturing are much broader than most people realize. The main objective of this chapter is to list some of the most important robotic applications and inspire the reader to find additional ones.

12.2 ROBOT APPLICATIONS

The following characteristics are important for an industrial environment to promote the use of a robot to replace human labor:

1. Work environment hazardous for human beings
2. Repetitive tasks
3. Boring and unpleasant tasks
4. Multishift operations
5. Infrequent changeovers
6. Performing at a steady pace
7. Operating for long hours without rest
8. Responding in automated operations
9. Minimizing variation

Industrial robot applications that tend to match these characteristics can be divided into four basic categories:

1. Material-handling applications
2. Processing operations
3. Assembly operations
4. Inspection operations

Material-handling applications involve the movement of material or parts from one location to another. Material may be the raw material that goes into production or product ready for shipment. The robot is equipped with a gripper in this case. Material-handling operations include part placement, palletizing and/or depalletizing, machine loading and unloading (such as machine tools, presses, conveyors, and the like), and stacking and insertion operations.

Processing operations require the robot to manipulate a special process tool as the end effector. The applications include spot welding, arc welding, riveting, spray painting, machining, metal cutting, deburring, polishing, and applying adhesives and sealants operations.

Assembly applications involve part-handling manipulation of a special tool and other automatic tasks and operations.

Inspection operations sometimes require the robot to position a workpart to an inspection device or to load a part into an inspection machine for testing; other applications involve the robot to manipulate a device or sensor to perform the inspection.

12.3 MATERIAL-HANDLING APPLICATIONS

A robot, to accomplish the transfer of materials, is equipped with a gripper-type end effector. The gripper must be designed to handle the specific part or parts to be moved in the application. This category includes the following:

a. Part placement
b. Palletizing and/or depalletizing
c. Machine loading and/or unloading
d. Stacking and insertion operations

In modern robot installations, appropriate sensors orientate the gripper of the end effector to deal with random entry of parts into the cell. For material-handling operations, the robot must have the following features:

1. The manipulator must be able to lift the part safely.
2. The robot must have the reach needed.
3. The robot must be a cylindrical coordinate type.
4. The robot's controller must have a large enough memory to store all of the programmed points so that the robot can move from one location to another.
5. The robot must have the speed necessary for meeting the transfer cycle of the operation.

Part Placement

These applications are ones in which the primary purpose of the robot is to pick up parts at one location and place them at a new location. In many cases, a reorientation of the part is desired in the relocation. The basic application in this category is the relatively simple pick-and-place operation, where the robot picks up a part and deposits it at a new location. Transferring parts from one conveyor to another is an example. The requirements of the application are modest; a low-technology robot of the cylindrical coordinate type is usually sufficient. Only two, three, or four joints are required for most of the applications. Pneumatically powered robots are often utilized.

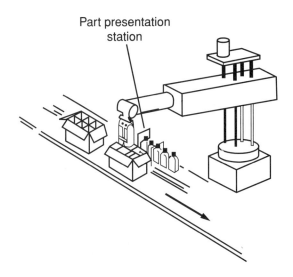

Figure 12.3.1 Robot palletizing bottles into a carton

Palletizing and/or Depalletizing

Many material-handling applications require the robot to stack parts one on top of the other, that is, to palletize them, or to unstack parts by removing from the top one by one, that is, depalletize them. **Palletizing**, for example, could be the process of taking parts from an assembly line and stacking them on a pallet. **Depalletizing** is the opposite. It could be the process of taking the parts off the pallet and placing them on the assembly line. Figure 12.3.1 illustrates a typical palletizing operation that uses a cylindrical coordinate manipulator to place bottles into a carton. The bottles are presented as a pair in front of the manipulator at a parts presentation station. The parts presentation station ensures that the bottles are in the proper position for the manipulator to pick them up and load them into the box.

Another example of palletizing and/or depalletizing is illustrated in Figure 12.3.2. In this application, the robot is operating in conjunction with two conveyor lines and is palletizing or depalletizing two separate areas. Palletizing and/or depalletizing with a robot can also be integrated with the work of a human in critical areas.

Machine Loading and/or Unloading

In machine loading and unloading applications, the robot transfers parts into and/or from a production machine. There are three possible cases:

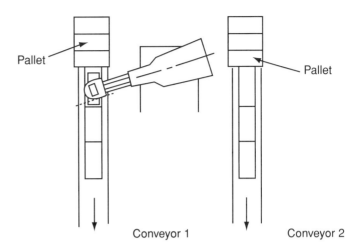

Figure 12.3.2 Robot in a palletizing arrangement operates two lines.

1. Machine loading in which the robot loads parts into a production machine, but the parts are unloaded by some other means. A pressworking operation can be an example where the robot feeds sheet blanks into the press, but the finished parts drop out of the press by gravity.
2. Machine unloading in which the raw materials are fed into the machine without robot assistance. The robot unloads the part from the machine assisted by vision or no vision. Examples in this category include bin picking, die casting, and plastic modeling.
3. Machine loading and unloading that involves both loading and unloading of the workparts by the robot. The robot loads a raw work part into the process and unloads a finished part. A machining operation can be an example in this case.

Industrial robot applications of machine loading and unloading include but are not limited to the following processes:

1. Die casting
2. Plastic molding
3. Metal-machining operations
4. Forging
5. Pressworking
6. Heat treating

One of the difficult problems often encountered in the loading and/or unloading applications is the difference in cycle time between the robot and the

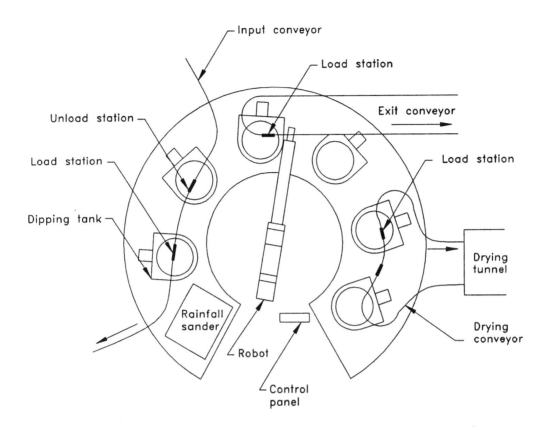

Figure 12.3.3 Robot in an automatic die casting operation.

production machine. The cycle time of the machine may be relatively long compared to the robot's cycle time. Therefore, it is appropriate sometimes for one robot to tend more than a single production machine.

An example of die casting is shown in Figure 12.3.3. In this operation, a wax pattern is dipped into a ceramic slurry and then into a wet slurry of sand. This process may be repeated up to six times so that the wax pattern has the necessary thickness.

For this application, the robot must have certain characteristics. For example, the controller's memory must be large enough to store the various subprograms needed to move the manipulator to the different dip tanks. Also, the velocity of the manipulator's axis must be fast enough to meet the desired production of models during a work shift. The work envelope of the robot must be large enough for the robot to reach several different dipping tanks. Finally, the cell controller must be able to control peripheral devices around the work envelope. The total cycle time for this operation is between three and four minutes from start to finish of the cycle. Each of the steps of the casting process is timed to the exact sequence of the cycle. This robotic operation saves the company many dollars in waste and scrap, and it also removes the worker from a tedious, repetitive task.

Stacking and Insertion Operations

The stacking operation is the process in which the robot places flat parts on top of each other, where the vertical location of the drop-off position is continuously changing with each cycle. The inserting operation is the process where the robot inserts parts into the compartments of a divided carton.

12.4 PROCESSING OPERATIONS

Processing operations are those in which the robot performs a processing procedure on the part. A distinct characteristic of this category is that the robot is equipped with some type of process tooling as its end effector and, in order to perform the process, manipulates its tooling relative to the working part during the cycle. In some processing applications, more than one tool must be used during the work cycle. In those cases, either a gripper or a quick-change mechanism is used to exchange the tools during the operation. Industrial robot applications in the processing operations include, but are not limited to, the following:

1. Spot welding
2. Continuous arc welding
3. Spray painting
4. Metal cutting and deburring operations
5. Various machining operations like drilling, grinding, laser and waterjet cutting, and riveting
6. Rotating and spindle operations
7. Adhesives and sealants dispensing

Spot welding is a metal joining process in which two sheet metal parts are fused together at localized points of contact. Two copper-based electrodes are used to squeeze the metal parts together and apply a large electrical current across the contact point to cause the fusion to occur. The electrodes, together with the mechanism that actuates them, constitute the welding gun in spot welding.

The types of robots used for spot welding are usually large, with sufficient payload capacity to wield the heavy welding gun. Five or six axes are generally needed to achieve the positioning and orientation required. Playback robots with point-to-point control are used, and the programming is accomplished using the powered teach-pendant method. Jointed-arm and polar coordinate robots are the most common anatomies in the automobile spot welding lines.

Continuous arc welding is used to provide continuous welded joints rather than individual welds at specific contact points as in spot welding. The resulting

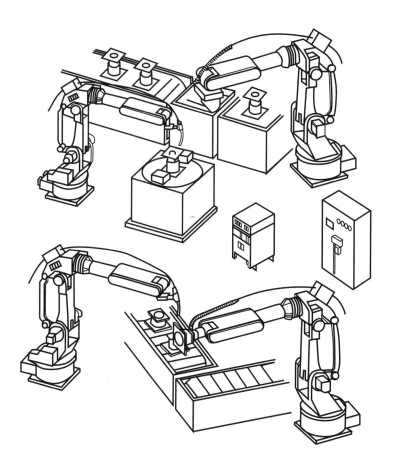

Figure 12.4.1 Welding robot and workpiece-handling robot perform coordinated operation in an arc welding cell.

joint in arc welding is substantially stronger than in spot welding. Because the weld is continuous, it can be used to make airtight pressure vessels and in other fabrication applications where strength and continuity are required. Figure 12.4.1 illustrates an arc welding cell.

The robot used in arc welding jobs must be capable of continuous path control. Programming is typically done by one of the teach-pendant methods. Jointed-arm and Cartesian coordinate robots are frequently used in arc welding applications. The robot should have five or six axes. If it does not, the fixture used to hold the parts during welding should possess several degrees of freedom to establish the required geometric relationship between the robot and the work.

Spray painting is the most common example of a more general class of robot application. The spray painting process makes use of a spray gun directed at the object to be painted. Fluid flows through the nozzle of the spray gun to be dispersed and applied over the surface of the object.

The work environment for human beings who perform this process is filled with health hazards.

The robot applications include spray painting of appliances, automobile car bodies, engines, and other parts, including porcelain coatings and bathroom fixtures. The cell layout is typically an in-line configuration in which the work flows past the robot. The robot must be capable of continuous path control to accomplish the smooth motion sequences required in spray painting. The most convenient programming method is the manual teach-pendant. Jointed-arm robots seem to be the most common anatomy for this application. The robot must possess a long reach in order to access the areas of the work part to be painted in the application.

The use of industrial robots for spray painting applications offers a number of benefits. In addition to protecting workers from a hazardous environment, they also include greater uniformity in the application of paint than people can accomplish manually, reduced use of paint (less waste), lower needs for ventilating the work area because human beings are not present during the process, and greater productivity.

12.5 ASSEMBLY OPERATIONS

Assembly operations are growing application areas for industrial robots. The applications can involve both material handling and the manipulation of a tool. They typically include components to build the product and to perform material handling operations. In some cases, the fastening of the components requires a special tool to be used by the robot.

Assembly operations are traditionally labor-intensive activities in industry. They are also highly repetitive and often boring. For these reasons, they are logical candidates for robotic applications. However, assembly work typically involves diverse and sometimes difficult tasks, often requiring adjustments to be made in parts that do not quite fit together. Sometimes it is necessary to redesign the parts to be assembled.

Assembly operations provide many applications for robots. In general, they are classified as batch assembly or low-volume assembly. In batch assembly, as many as one million products might be assembled. These assembly operations have long production runs and require the same repetitive assembly routine. In low-volume assembly, a sample run of ten thousand or less products might be made.

The assembly robotic cell should be a modular cell. If the production run increases, more modular cells can be used. Figure 12.5.1 illustrates a typical assembly modular cell. Here, the manipulator assembles the components on the assembly line. This manipulator can be reprogrammed for another assembly operation after this job has been finished.

Another concern for the assembly robotic cell is the design of parts that can be assembled by the robot. That is, the robot's end effector must be able to easily

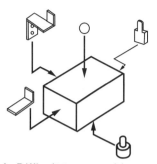 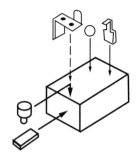

A. Difficult to assemble because of part orientation

B. Preferred assembly because of new orientation

C. Difficult to grip because no lip provided

D. Preferred assembly because lip provided

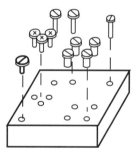 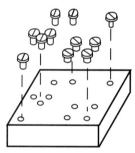

E. Difficult to automate because of too many screw variations

F. Preferred assembly because of fewer screw variations

Figure 12.5.1 Designing parts for robot assembly operation

pick up and assemble the parts. Minimizing the number of parts required in a robotic assembly is a big advantage.

One area that is well suited for robotics assembly is the insertion of odd electronic components. These components could be resistors, diodes, transistors, microprocessors, heat sinks, or a variety of other odd electronic components.

Figure 12.5.2 illustrates a typical overall electronic assembly operation. The cell is identified as flexible because the system can generate a wide variety of manufactured printed circuit boards from one basic system structure.

12.5 Assembly Operations 361

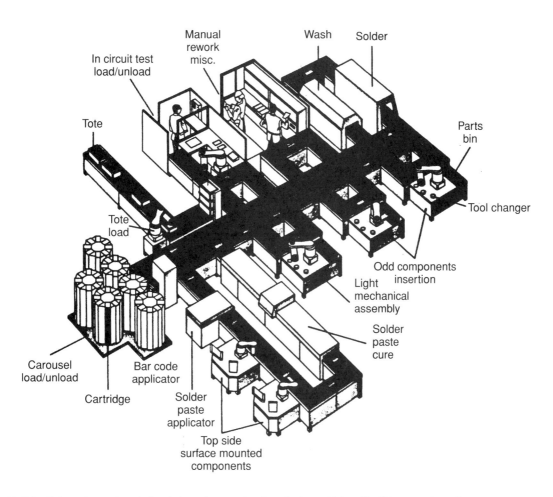

Figure 12.5.2 Robots in a printed circuit board operation interfacing within a flexible manufacturing system.

Printed circuit boards are loaded onto carousels and rotated to be automatically placed on a transport mechanism. A printed circuit board from the carousel is first read by a bar code reader that identifies the type of circuit board and alerts the system that a certain type of board is present in the system. A command will be issued throughout the entire system alerting the system stations to which components must be inserted into the board. The circuit board is then routed to the first station, where solder paste is applied to it. The board is then transferred to the second station, which inserts the electronic components to the top side of the board.

After these two robotic stations complete their task, the board is routed to the next station. Once the operation is completed, a signal is sent from the robot controller to the cell controller. This signal identifies the operation as completed, and the cell controller indexes the printed circuit board to the next operation.

Industrial robots used for the type of assembly operations described here are typically small, with light load capacities. Large proportions of assembly tasks require a robot capable of lifting weights of five pounds or less. The most common configurations are jointed arm, SCARA, and Cartesian coordinate. Programming is often done, using a machine programming language, together with a powered teach pendant to teach locations in the workcell. Accuracy requirements in assembly work are often more demanding than in other robot applications, and some of the more precise robots in this category have repeatabilities as close as ±0.001 in. In addition, the end effector may be required to perform multiple functions at a single workstation in order to reduce the number of robots in the production line.

12.6 INSPECTION OPERATIONS

Some inspection operations require parts to be manipulated, and other applications require that an inspection tool be manipulated. Inspection work requires high precision and patience, and human judgment is often needed to determine whether a product is within quality specifications or not. Because of these complications, the application of robots has not been easy. Nevertheless, the potential rewards are so great that substantial efforts are being made to develop the necessary technologies to achieve success in such applications.

Inspection tasks that are performed by industrial robots can usually be divided into the following three techniques:

1. By using a feeler gauge or a linear displacement transducer known as a linear variable differential transformer (LVDT), the part being measured will come in physical contact with the instrument or by means of air pressure, which will cause it to ride above the surface being measured.
2. By utilizing robotic vision, matrix videocameras are used to obtain an image of the area of interest, which is digitized and compared to a similar image with specified tolerances. This process is complex, although fast, and with the progress made recently can obtain shades of gray versus black-and-white images.
3. By involving the use of optics and light, usually a laser of infrared source is used to illustrate the area of interest. Reflections are captured by receiving optics, which convert the data into digital code. The information can then be compared to acceptable standards. A triangulation technique is used to determine distances.

Furthermore, in this type of inspection, the robot may have either an active or a passive role. In the passive role, the robot feeds a gauging station with a part. While the gauging station is determining whether the part meets the specification, the robot waits for the process to finish. In the active role, the robot is responsible for determining whether the part is good or bad.

12.7 EVALUATING THE POTENTIAL OF A ROBOT APPLICATION

Evaluating the potential of a robot for a selected use requires an extensive and careful examination in the form of a check list as given here:

1. Analysis of the Application
    Long- and short-term objectives
    Manufacturing processes involved
    Space availability
    Budget
    Initial development of automation recommendations
    System objectives
    Functional performance specifications
    Equipment recommendations
2. Feasibility Study
    How a more automated system will affect related operations in the plant
    Material-handling methods
    Potential process changes
    Commercial equipment available
    Updating existing equipment (partial automation) for maximum cost effectiveness
    Preparation of specifications
    CAD layouts of floor space and cell configuration
    CAD cell stimulation
3. System Proposal
    Functional specifications
    System operation
    Robot type
    Tooling
    Fixturing
    Peripheral equipment
    Electronic communications requirements
4. System Design
    Microprocessor controls
    Software
    Multiple levels of control
    Management data
    Preplanned expandability
5. Construction Phase
    It is a good procedure for the system to be set up and thoroughly tested at the supplier's facility. This will minimize the interruption of current production procedures.

6. Installation Phase

   It is a good practice for the supplier to supervise the step-by-step installation of the system.

7. Training and Documentation

   Hands-on robot training should be provided by the supplier for all persons who will interface with the new automated system. Training can take place at the supplier's facility or at the user's plant. The supplier should provide design drawings and documentation for system control, operation, and maintenance.

12.8 FUTURE APPLICATIONS

The ability of the robot to carry out future significant applications will require more major technological breakthroughs in order to be successfully realized. More important, the robot will have to be intelligent in order to make rapid decisions based on current sensory information. It is certain that a set of programmed actions will be totally unsatisfactory for such applications if the significant advances in artificial (or autonomous) intelligence (AI) completely materialize. The areas of vision and tactile sensing are still not completely developed.

The medical applications of robots (e.g., routine examinations, surgery, or prosthetics) are certainly many years away from reality. For example, the "six-million-dollar man" will be possible only with the development of real-time signal-processing techniques that permit the desired signals emanating from the brain and transmitted over nerves to be separated from muscle noise so as to control the prosthesis reliably.

The ability of a robot to carry out surgical procedures or examinations will depend on the development of a variety of external sensors and real-time computer processing techniques.

Underwater applications of robots will involve prospecting for minerals on the floor of the ocean (e.g., nodules of manganese), salvaging of sunken vessels, and the repair of ships either at sea or in dry dock. The military is currently looking at robots for use in a variety of areas. For example, the Air Force and Navy are both interested in mobile firefighters.

The application of robots for surveillance and guard duty is not restricted to the military, because power generating plants, oil refineries, and other large civilian facilities that are potential targets of terrorist groups are being considered also as potential users.

A potential future application of robots would be in the home. Such devices would need to be small, mobile, sensor-based, easy to program, and autonomous.

Another extremely important future use of robots will be as a component in a fully automated factory.

Today, most robots must be programmed by a human being to perform a desired task, and this causes a lot of accidents. However, robot controllers that permit interfacing with database or CAD/CAM systems are not yet completely satisfactory. This development would permit the user to stimulate the workcell, robot, and other machine tools and/or parts feeding mechanisms off-line. The appropriate robot commands will then be easily coordinated with the actions of the other devices operating within the workcell itself. In this way, the manufacturing process can be optimized with respect to cycle time and throughput.

Our discussion of future robotic applications ends by noting that the Committee on Science, Engineering and Public Policy (COSEPUP), composed of experts from industry, academia, and government, gave a series of briefings to a number of federal agency directors in early 1997. The committee predicted that with the development of sophisticated vision systems, tactile sensors, and programming techniques/languages, it would be possible to use robots in industries without fear of accidents.

In summary, some future foreseen applications are listed here by industry and special category:

- Aerospace industry
- Agriculture industry
- Construction industry—road construction, paving, and surface finishing procedures
- Health industry—surgery, rehabilitation, bio-robotics
- Manufacturing industry
- Nuclear industry
- Service industry
- Textile industry
- Transportation industry
- Utility industry
- Lab automation
- Control and guided vehicles
- Underwater surveying and maintenance activities
- Navigation systems
- Surveillance and guard duty
- Firefighting
- The automatic factory of the future
- Household robots

12.9 CHALLENGE FOR THE FUTURE

As we have seen from the previous discussions, much more effort has gone into developing automated systems to improve productivity than has gone into the appropriate matching of people and technology. Now that the speed of technical development is increasing rapidly, the health and safety problems associated with automation, particularly stress-related problems, are also likely to increase. According to a symposium summary report by Kensaburo Tsuchiya (1987) of Japan's University of Occupational Health and Safety, the challenge for the future is "to create jobs which are free from stress and musculoskeletal overloads while at the same time being challenged and interesting for the individual."

According to D. L. Goetsch in *Occupation Safety and Health* (1997), the future holds many problems that we have to deal with in order to meet Tsuchiya's challenge. The most prominent of these problems are the following:

- Increasingly intense international competition may magnify the tendency for companies to neglect health and safety precautions in favor of short-term productivity gains.
- The level of mental stress is likely to increase as the automated manipulation of information forces workers to continually try to handle too much information that is poorly understood.
- Automation and competition are likely to increase the level of anxiety as workers are required to make split-second decisions while knowing that their actions or inactions may have dire consequences.
- New occupational diseases relating to mental, visual, and musculoskeletal problems may arise whose remedies must be sought through a combination of ergonomics, psychology, occupational medicine, and design.
- There is likely to be more unthoughtful introduction of robots into the workplace, which will in turn introduce more unexpected health and safety risks.
- Ignorance may lead to the introduction of automation in an office or factory in forms that do not require workers to think, reason, or make judgments, giving rise to alienation and frustration.

Because they are likely to face these inhibitors, the health and safety professionals of the future must be prepared to deal with them. They need to know what has to happen if the inhibitors are to be overcome. Tsuchiya suggests the following strategies for enhancing the health and safety of tomorrow's automated workplace:

- Technological systems and processes must be designed to take into account the physical, mental, and emotional needs of human workers.
- Workers will need training and continual retraining so that they can effectively and efficiently operate technological systems and interact with them from the perspective of mastery rather than inadequacy.
- Health and safety professionals, management, workers, psychologists, ergonomists, and practitioners of occupational medicine will have to work together as a team in all aspects of the health and safety program.
- The quality of work life and health and safety considerations will have to receive as much attention in the design and implementation of automated systems and processes as do economic and technological concerns.
- Additional research will have to be conducted to determine more clearly the psychological and physiological effects of human interaction with automated technologies.
- Much more comprehensive accident reporting will be needed. Implementation of the "critical incident" reporting system used in com-

mercial aviation might be considered by companies for collecting health and safety data.
- Ergonomists should become involved in accident prevention. They should focus their accident prevention activities on accident/error analysis and simulation of accidents for training purposes.

12.10 INNOVATIONS

Some current innovations of industrial robots with sources of development are listed here:

- How to Make a Robot Smile—developed by Science University of Tokyo under the supervision of mechanical engineering professor Fumio Hara to assist certain handicapped children to overcome the difficulties of manifesting appropriate expressions. Figure 12.10.1

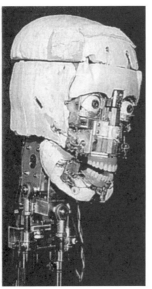
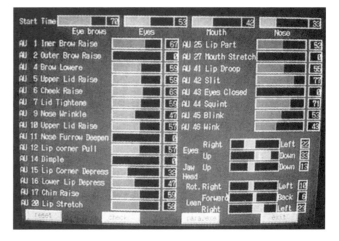

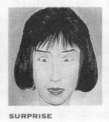

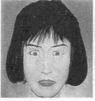

FEAR   HAPPINESS   SURPRISE   SADNESS   ANGER   DISGUST

Figure 12.10.1 Emotive robot with facial expressions (*Source:* MIT's Technology Review, *October 1997*)

illustrates an emotive robot that reads human expressions and responds in kind. It was constructed from an aluminum head fitted with false teeth, eyeballs, rubber mask, wig, and tiny gears that can be controlled by a computer program to contort its face as desired (S. Strauss, in "How to Make a Robot Smile").
- Mouth Robot—under development by Science University of Tokyo. This robot's actuators would realistically mimic lip-movements during speech. Such a robot might help people with speech or language disabilities, because studies show that more than 50 percent of speech understanding stems from facial expression and movements (S. Strauss, in "How to Make a Robot Smile").
- Telepresence Robot—developed by Salford University of England. It's a twin-armed mobile robot system designed for complex tasks in unstructured and hazardous environments. It has high-sensory feedback of visual, audio, and multifunction tactile. The system has been designed for testing and exploration of generic concepts. The principles of the system are shown in Figure 12.10.2 (D. G. Caldwell et al. in "Telepresence Feedback and Input Systems for a Twin-Armed Mobile Robot").
- Haptic Perception Robots—under development by MIT Laboratory for Human and Machine Haptics (Touch Lab)—sensing information that comes from joints and muscles without locating them visually. (See Figure 6.2.2.)
- Medical Robots—funded by the National Research Council of Italy and developed by the Universities of Genoa, Naples, Florence, Bologna, and Polytechnic of Milan. Major applications are (a) surgery, (b) rehabilitation, and (c) biological systems (bio-robotics) that are replicating the function executed by living beings like CCD retina-like sensor, tactile receptors in the fingertip skin, and CCD vision sensor. Figure 12.10.3 illustrates the main applications and prospects of medical robots (P. Dario et al. in "Robotics for Medical Applications").

12.11 CASE STUDIES

The following two case studies have been completed by RMT Engineering for a major pharmaceutical manufacturer and a multinational textile producer.

Case Study 1

RMT Engineering is a designer/integrator of robotic material-handling automation based in Grimsby, Canada. In February 1996, RMT Engineering installed a very compact robotic case palletizer in a constrictive area, adjacent to a "clean room." This study reviews RMT's creative machine design to satisfy this application.

Application requirement. A major pharmaceutical manufacturer produces a medicinal solution packaged in a 4 liter "bag in a box" case, that is produced at up to twelve cases per minute. The case has an offset perforated-slot top for quick

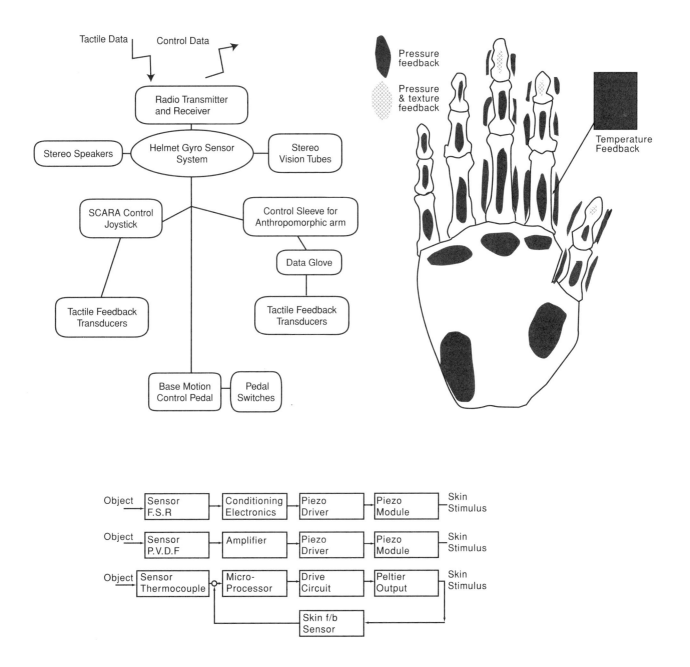

Figure 12.10.2 Telepresence robot design (*Source:* IEEE Robotics & Automation Magazine, *September 1996*)

370 Industrial Applications

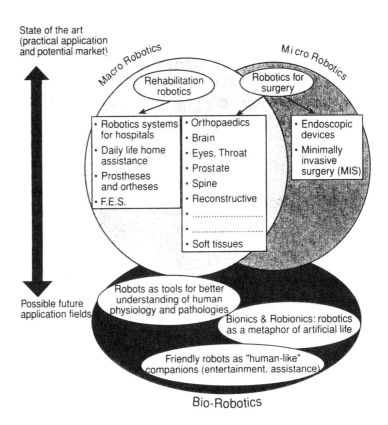

Figure 12.10.3 Robotics for medical applications (*Source:* IEEE Robotics & Automation Magazine, *September 1996*)

removal of the solution's bag for use in blood transfusions. The stacking pattern is a complex interlocked pattern, for stacking stability and center placement of the perforated slot, because the case is tall and "tippy." The primary challenge to automating this process was the extremely tight constraints to locate the case palletizer: The client provided only *60 sq ft between two columns, spanned by an 8' high beam, under which RMT is to palletize up to a 5' high stack.*

    Given data: rate = 8–12 cpm
                    stack pattern = complex interlocked
                    cases per layer = 18
                    layers per pallet = 3 (up to 6)
                    pallet build time = 5–7 minutes
                    infeed—case taper
                    discharge—pallet jack

System solution. Due to the severe space constraints, the motions of the robot would be limited to two axes of motion (x [across] and z [up]), without any y [in/out] or even gripper rotation, because it would protrude into the walking aisle.

The design provided was a customized version of RMT's GP 210 series robot, specifically the GP 212, which is a *low profile mono beam gantry robot*, with two axes of motion.

RMT designed the gantry with a work envelope to contain the 18" wide case infeed conveyor, and the GMA pallet located right beside the conveyor. The machine footprint is $8.5' \times 6.5'$, with overall motions of $x = 5'$, $z = 5.3'$. The robot uses a multifingered clamp gripper to pick and place a row of cases, formed on the infeed conveyor, which has either four or five cases per row. The robot is capable of seven transfers per minute, resulting in a 35 cpm capability.

Once a pallet load is built, the robot moves to a "Park" position over the top of the case infeed conveyor. The operator presses the "work zone entry" button, deactivating robot drive power and the light curtain, and removes the pallet via a pallet jack.

The full pallet is manually exchanged with an empty pallet, the operator presses the "Resume button, and the robot begins the process again, without the need for re-homing the robot. The PC-based servo controls (with NEMA 4 color terminal) are housed in a NEMA 4 stainless steel panel, located 60 feet away in a clean room. An operator push-button station is located on the robot and provides the necessary functions to operate the RMT system.

System supply. RMT Engineering supplied: 2 axis GP 212 gantry robot; side justifying, multifingered clamp gripper; 8 feet of Rapistan lineshaft conveyor; 270 degree indexing, "lift and rotate" case turner; 1 "on floor" pallet guide; NEMA 4 controls with 486 PC interface; integration of 5-foot wide Banner light curtain.

Economic benefits. This drug producer originally began with a one-shift operation at 4 cpm, and the palletizing function was performed by a single person. Within six months after installation, product demand grew to 8–12 cpm on a "round the clock" basis. The robot now lifts an accumulated weight of 38,000 pounds per shift. It is easy to see why the process required automation.

Electrical consumption is about $6 per shift. Operating availability and reliability for the first six months exceeded 98 percent and 99 percent respectively, with little debug time involved.

Summary. This system demonstrates the requirement for accumulation or buffer in any production system for maximum effectiveness. Because there is little buffer either upstream (1 minute) or downstream (0 minute) of the palletizer, an operator must be nearby to perform the pallet exchange or else the production line will shutdown in cascade fashion.

However, even with interrupted production, the RMT Robotic Palletizer has allowed the firm to increase its production rates beyond the capability of what

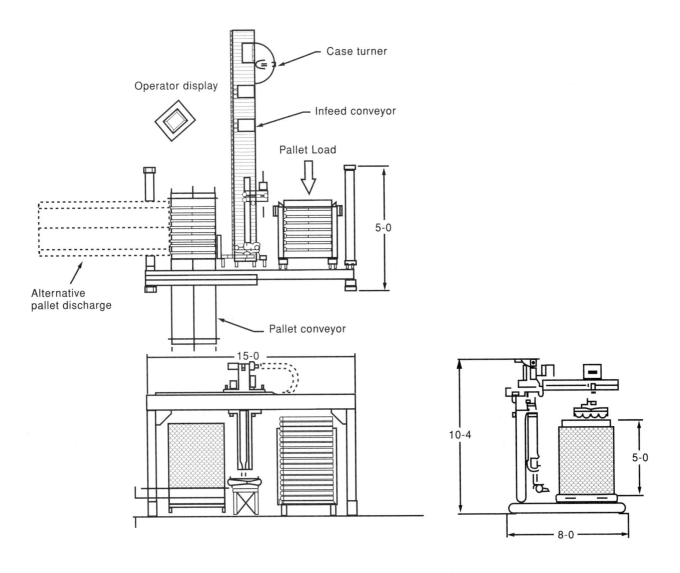

Figure 12.11.1 Pharmaceutical manufacturer case palletizer *(Courtesy of RMT Engineering, Grimsby, Canada)*

even two people would be able to manually palletize. Figure 12.11.1 illustrates the case palletizer.

Case Study 2

RMT Engineering is a designer/integrator of robotic material-handling automation based in Grimsby, Canada. In December 1996, RMT Engineering installed an eleven-station robotic case-palletizing system for simultaneous loading of multiple SKUs onto floor-positioned pallets.

Application requirement. A textile manufacturer produces its patented yarn product at several facilities around the world, one of which is located in Virginia. At its latest "plant of the future," the user installed ten machines to produce the yarn, with plans to add more in the future. In accordance with its ISO 9000 procedures, and its stringent quality standards, RMT's customer required a packaging system that would produce yarn textiles that were completely traceable.

The user contracted RMT Engineering to design/build/commission a system that would palletize up to eleven simultaneously produced, bar-coded cases onto their respective pallets. Cases weigh up to 150 pounds. Once a pallet is built, and removed by forklift, the robot will automatically reload the empty palletizing station with a new user-specific pallet, thus minimizing the incursion into the work envelope. The RMT robotic system is located in a shipping dock environment, where space is at a premium due to traffic. The RMT robotic control system will also operate within Windows NT-Wonderware supervisory system.

Given data: rate = 2 cpm at 150 lb/case
stack pattern = 2×2 column, labels out
cases per layer = 4
layers per pallet = 2
pallet build time
(system) = 8 minutes
infeed = Rapistan case conveyor
discharge = forklift

RMT System solution. The overall design is quite simple: eleven pallets placed side by side, with the case infeed conveyor in the middle, and a static pallet rack at one end.

The ultra large gantry robot provided is a stretched version of RMT's GP 210 series robot, specifically the four-axis GP 214, which is a low profile mono beam gantry robot. The RMT system consists of a robot with a work envelope of $x = 75'$; $y = 5'$; $z = 7'$, yet the robot is less than twelve feet tall.

Because each case is 150 pounds, the stresses on the case bottom are quite high; thus, a bottom supporting gripper tool was provided to minimize case stress, which a normal vacuum gripper or side clamp gripper would not do. However, the tool weighs over 180 pounds, outside the capacity of most robots, especially articulated devices. The total payload is approximately 340 pounds, or only 70 percent of the GP 214's standard capacity.

The pallet-loading operation is straightforward; the tool is designed to also grip the wooden pallets from a static pallet rack. The robot will also accommodate any pallet over/under size disparities by being able to nest the pallets into the floor-mounted pallet guides, and by using the robot's motion control intelligence.

The operation begins by having the robot load all the palletizing stations with a pallet from the rack, which takes approximately forty-five seconds per pallet. Once the pallets are loaded, bar-coded cases are inducted into the robot's work envelope from the Rapistan case conveyor. The user's bar code scanner identifies the case to the robot, which will pick the case from the infeed conveyor and place it onto the respective pallet in a "label out" orientation. Each pallet is coordinated to its production machine. This operation continues, palletizing cases in random order, until a pallet is built, about one every four to eight minutes. The robot broadcasts "pallet complete" messages to several Wonderware operator stations, as well as activating beacons to alert nearby forklift operators.

The robot continues to palletize all other stations (except for the finished SKU), until the forklift operator initiates the "pallet removal" sequence, which causes the robot to go to a "Park" position over the top of the empty pallet rack, and power down, thus letting the forklift operator safely remove the full pallet. Once clear of the work envelope, the operator presses "Resume" and the robot powers up, takes an empty pallet from the rack, and places it into the empty station. The robot then continues to palletize any accumulated cases.

No homing sequence is required. Protection is provided by a light curtain across the seventy-foot-long front face of the palletizer.

System supply. RMT Engineering supplied: 70-feet, six-inch long, four-axis GP 214 gantry robot; hybrid case and pallet gripper-bottom supporting "fork and clamp" gripper, with pallet gripper; eleven "on floor" pallet guides, static pallet rack (ten pallet capacity); Pentium PC–based servo controls; integration of bar code scanner; integration of wide area light curtain; data link into Windows NT platform; systems integration.

Economic benefits. Because this operation is continuous, four operators would be required to do manual palletizing of very heavy cases. As this user is extremely safety conscious, as well as productivity minded, automating the palletizing function is expected, especially with a high return on invested capital. Along with the productivity benefits, the RMT system is fully integrated into the user's data network, thus providing real-time data collection.

Electrical consumption is about $5 per shift. Installation and commissioning were completed in less than three weeks, including integration into the plant's Windows NT/Wonderware supervisory system.

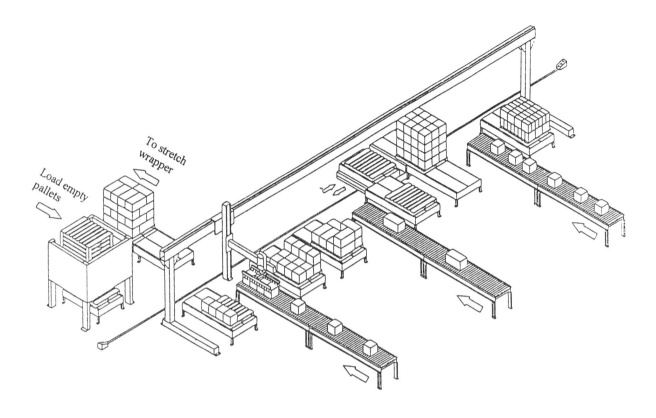

Figure 12.11.2 Textile producer case palletizer. *(Source: RMT Engineering, Grimsby, Canada)*

Summary. This installation is the second of four RMT multi-SKU palletizers for this customer's textile division, and demonstrates the flexibility and machine intelligence available for maximizing manufacturing productivity in multiline production environments. Figure 12.11.2 illustrates the floor plan of this case palletizer.

12.12 SUMMARY

In this chapter, we have taken an extensive in-depth look at the major robotic applications and their special needs. Robots have become commonplace in modern industry. All sectors in automated manufacturing implement the use of industrial robots. Robot applications can be divided into four basic categories:

1. Material-handling operations
2. Processing operations

3. Assembly operations
4. Inspection operations

Material handling operations includes part placement, palletizing, depalletizing, and machine loading and unloading.

Processing operations include spot welding, continuous arc welding, spray painting, metal cutting, and various other machining operations.

Assembly operations include a variety of applications from printed circuit boards assembly to motor vehicles.

Inspection operations are performed by feeler gauge, robotic vision, or the use of optics.

The versatility of the robots used in all the applications is important, as is applying that capability efficiently. Future applications of robots have been explored, and critical areas have been analyzed.

Also, the challenge for the future has been analyzed and evaluated because more effort has gone into developing automated systems to improve productivity than has gone into the appropriate matching of people and technology.

Current innovations and case studies have been included in this chapter to further explore the trends of industrial robot applications.

12.13 REVIEW QUESTIONS

12.1 Describe briefly the impact of automation in manufacturing and the result of this significant progress.
12.2 What are the eight important characteristics in robot applications to replace human labor?
12.3 What are the four basic categories for industrial robot applications?
12.4 What type of applications are included in material handling? Give an example.
12.5 What type of applications are included in processing operations? Give an example.
12.6 What type of applications are included in assembly? Give an example.
12.7 What are the three sensor systems used in inspection tasks? Explain each one in detail.
12.8 What type of robots are used in material-handling applications, processing operations, and assembly operations?
12.9 What are the two main welding areas in which robots are used?
12.10 What type of manipulator is generally used in welding applications and how many axes does it have?
12.11 What is the most efficient arc welding operation?
12.12 For large-bench assembly, which type of automation is best?
12.13 How many axes should a spray-painting manipulator have?
12.14 What are some future applications of robots?
12.15 Why should we challenge the future of technology and what is considered a safe, healthy, and efficient automated workplace?

12.16 Which of the following are characteristics of work situations that tend to promote the substitution of a robot in place of a human worker?
a. Frequent job changeovers
b. Hazardous work environment
c. Repetitive work cycle
d. Multiple work shifts
e. Mobility requirements

12.14 PROBLEMS

12.1 A robot in a batch-processing operation produces 50 units, and each unit requires a sequence of eight operations to complete. The average setup time is 3 hours, and the average operation time per machine in the cell is 6 minutes. Also, the average nonoperation time, due to handling, inspection, and so on, is 7 hours. How many days will it take the robot to produce a batch, assuming that the workcell operates on a 7-hour shift per day?

12.2 A certain part is routed through six machines in a batch production robot cell. The setup and operation times for each machine are given in the following table:

Machine	Setup time (h)	Operation time (min)
1	4	5.0
2	2	3.5
3	8	10.0
4	3	1.9
5	3	4.1
6	4	2.5

The batch size is 100, and the average nonoperation time per machine is 12 hours.
a. Determine the manufacturing lead time.
b. Determine the production rate for operation 3.

2.3 Suppose that the part in Problem 12.2 is made in very large quantities on a flow production line in which an automated work-handling system is used to transfer parts between the machines. The transfer time between stations is 15 seconds. The total time required to set up the entire line is 150 hours. Assume that the operation times at the individual machines remain the same.
a. Determine the manufacturing lead time for a part coming off the line.
b. Determine the production rate for operation 3.
c. What is the theoretical production rate for the entire production line?

12.4 A robot performs a loading and unloading operation for a machine tool. The work cycle consists of the following sequence of activities:

The activities are performed sequentially as listed. Every 30 work parts, the cutting tools in the machine must be changed. This irregular cycle takes 3.0 minutes to accomplish. The uptime efficiency of the robot is 97%; and the uptime efficiency of the machine tool is 98%, not including interruptions for tool changes. These two efficiencies are assumed not to overlap (i.e., if the robot breaks down, the cell will cease to operate, so the machine tool will not have the opportunity to break down, and vice versa). Downtime results from electrical and mechanical malfunctions of the robot, machine tool, and fixture. Determine the hourly production rate, taking into account the lost time due to tool changes and the uptime efficiency.

Activity	Time(s)
1. Robot picks part from incoming conveyor and loads into fixture on machine tool.	5.5
2. Machining cycle (automatic).	33.0
3. Robot reaches in, retrieves part from machine tool, and deposits it onto exit conveyor.	4.8
4. Move back to pickup position.	1.7

12.5 Suppose that a double gripper was used instead of the single gripper indicated in Problem 12.4. The activities in the cycle would be changed as follows:

Steps 1, 4, and 5 are performed simultaneously with the automatic machining cycle. Steps 2 and 3 must be performed sequentially. The same tool change statistics and uptime efficiencies are applicable. Determine the hourly production rate when the double gripper is used, taking into account the lost time due to tool changes and the uptime efficiency.

Activity	Time(s)
1. Robot picks raw part from incoming conveyor in one gripper and awaits completion of machining cycle. This activity is performed simultaneously with machining cycle.	3.3
2. At completion of previous machining cycle, robot reaches in, retrieves finished part from machine, loads raw part into fixture, and moves a safe distance from machine.	5.0
3. Machining cycle (automatic).	4.8
4. Robot moves to exit conveyor and deposits part. This activity is performed simultaneously with machining cycle.	3.0
5. Robot moves back to pickup position. This activity is performed simultaneously with machining cycle.	1.7

12.15 REFERENCES

Asfahl, C. R. *Robots and Manufacturing Automation*, 2d ed. New York: John Wiley and Sons, Inc., 1992.

Caldwell, D. G., A. Wardl, O. Kocak, and M. Goodwin. "Telepresence Feedback and Input Systems for a Twin-Armed Mobile Robot." *IEEE Robotics and Automation Magazine* (September 1996): 29–38.

Dario, P., E. Guglielmelli, B. Allotta, and M. C. Carrozza. "Robotics for Medical Applications." *IEEE Robotics and Automation Magazine* (September 1996): 44–56.

Fuller, J. L. *Robotics: Introduction, Programming and Projects.* Englewood Cliffs, NJ: Prentice Hall, 1991.

Goetsch, D. L. *Occupation Safety and Health.* Englewood Cliffs, NJ: Prentice Hall, 1996.

Groover, M. P. *Fundamentals of Modern Manufacturing.* Upper Saddle River, NJ: Prentice Hall, 1996.

Groover, P., M. Weiss, R. N. Nagel, and N. Odrey. *Industrial Robotics: Technology, Programming, and Applications.* New York: McGraw-Hill, 1986.

Hirota, K., Ed. *Industrial Applications of Fuzzy Technology.* Tokyo, Japan: Springer-Verlag, 1993.

Masterson, J. W., R. L. Towers, and S. W. Fardo. *Robotics Technology.* South Holland, IL: Goodheart Willcox Co., Inc., 1996.

McDonald, A. C. *Robot Technology: Theory, Design and Applications.* Englewood Cliffs, NJ: Prentice Hall, 1986.

Pessen, D. *Industrial Automation.* New York: John Wiley and Sons, 1989.

Rehg, J. A. *Introduction to Robotics in CIM Systems*, 3d ed. Upper Saddle River, NJ: Prentice Hall 1997.

Strauss, S. "How to Make a Robot Smile." *MIT's Technology Review* (October 1997): 14–15.

Tsuchiya, K. "Summary Report on the Fifth University Occupational and Environmental Health International Symposium." In *Occupational Health and Safety in Automation and Robotics*, ed. K. Noro. Chicago: National Safety Council, 1987, p. 422.

GLOSSARY OF SELECTED TERMS

Absolute optical encoder: A device that divides the rotation angle of a shaft into functions of a degree and then reads out the present angle in some binary code.

Acceleration: The change in velocity as a function of time.

Accumulator: A register in the arithmetic-logic unit of a computer used for temporary storage operations and for loading and storing data between the central processing unit and main memory.

Accuracy: A measurement of the ability of a robot end effector that can be programmed to reach a desired point in space.

Acoustical proximity detector: A sensory device with an open-ended cavity that reacts to sound.

Actuator: A device in robots that converts electric, hydraulic, or pneumatic energy into motion.

Adaptability: The ability of a device or robot to alter or modify its program or function in response to changes in its environment without human intervention.

Adaptive control (AC): A method of control in which actions are continuously adjusted in response to feedback.

Adaptive system: A system that modifies its operations in response to sensory information from its environment.

Address: A name, label, or number that is given to a register, memory location, or device.

Addressable point: The designation point of an increment that can be directly accessed by a robot controller through an instruction.

Address register: The part of a microprocessor used to temporarily hold an address of a location in memory.

Air motor: A pneumatic device that converts air pressure to mechanical force.

Algorithm: A given, detailed set of rules, usually in the form of mathematical equations, designed to achieve a specified result in a finite number of steps.

Alphanumeric: A term applied to characters that are either numerals, letters of the alphabet, or special symbols. Such characters are called alphanumeric to distinguish them from characters, special graphics, or machine language symbols.

Alternating current (AC): Electric current that reverses direction periodically. Usually many times per second.

Analog: The representation of measurable quantities by means of continuous physical variables, such as translation, rotation, voltage, or resistance.

Analog information: Information that physically varies continuously.

Analog-to-digital: Often referred to as A/D, this refers to the conversion of continuous qualities to digital quantities.

AND gate: A circuit that has two or more input signal ports and that delivers an output only if and when every input signal port is simultaneously energized.

Angular acceleration: The time rate of change of angular velocity.

Angular actuator: A device whose driving force provides angular rotation.

Anthropomorphic: A word used to describe a robot that has a human shape.

Architecture: The physical and/or logical structure of a computer program or manufacturing process.

Arc welding: A welding process that uses the heat of an electric arc to fuse two metals together along a joint.

Arithmetic-logic unit (ALU): The part of a microprocessor that performs mathematical and logic operations.

Arm: That part of the robot manipulator comprised of an interconnected series of mechanical links and joints that support and move the wrist and end effector through space.

Armature: The part of an electric rotating machine that includes the main current-carrying winding in which the electromotive force is produced by magnetic flux.

Array: A set of storage locations in a computer memory that may be addressed using an index variable.

Articulated: The property of being jointed or connected in such a fashion that the joined parts are movable.

Artificial intelligence (AI): The capability of a computer to perform operations that simulate human intelligence. Examples of such operations are learning, adaptation, recognition, classification, reasoning, self-correction, and improvement.

ASCII: American Standard Code for Information Interchange, an eight-bit code used to represent alphanumeric, punctuation, and special characters for use in control.

Assembler: A computer program that translates mnemonic operators into machine code numbers.

Assembly language: A computer language in which each command can be translated into a machine language command.

Asynchronous: Events, functions, or operations that do not occur simultaneously or in a specifically timed manner.

Automated guided vehicle (AGV): Computer-controlled, battery-operated vehicle that follows an electronic guide path in the workplace.

Automatic tool changer: An adapting device that provides rapid tool-changing capabilities with a minimum amount of time lost.

Automation: The replacement of human or animal labor by machines or the operation of a machine or device actuated by remote control.

Axis: A path of travel in the rotary or translational, prismatic joint in a robot.

Binary: Referring to the property or number system of having only two values.

Binary counter: A device used to count numerical information in binary form.

Binary logic circuit: A computer circuit that makes logical decisions based upon input signals in binary form.

Bistable: A term used to describe any electronic device that can assume one of two stable operational states.

Bit: A single pulse of digital information equal to one binary.

Break point: The point in a computer program at which the program may be interrupted by external intervention.

Buffer storage: A storage device located between input/output channels and main storage control to free the operation in case of any malfunction in the production line.

Bus: The distribution path in a microcomputer.

Byte: A measure of digital computer data, usually shorter than a word (a group of eight bits).

CAD/CAM: Acronym for computer-aided design/computer-aided manufacturing used in projects for designing parts and machinery.

Capacitive proximity sensor: A device used to detect signals based on the air gap between the sensor and nonconductive materials.

Capacitive transducer: A device that measures a change in capacitance and used for sensing pressure and the like.

Cartesian coordinates: Also called rectangular coordinates. A set of three numbers that define the location of a point in a rectilinear coordinate system, referred to as x, y, and z coordinates. A robot manipulator moves in these three directions.

Chip: An integrated circuit on a tiny silicon flake upon which a large number of gates and the paths connecting them are formed by very thin films of metal acting as wires. The chip can be used as main memory or as a CPU.

CIM: Acronym for computer-integrated manufacturing.

Circular interpolation: A control function that defines the points of a circle in a robot's workspace based on a minimum of three specified positions.

Clamp: A function of a pneumatic robot gripper that controls the grasping and releasing of an object.

Clear: A computer command that usually directs the computer to replace information in its storage by zeros or clear the CRT screen.

Closed loop: This is a robot control system that uses feedback to control its operation.

Code: A set of computer instructions to perform a given operation or solve a given problem.

Collect gripper: A gripper used by a robot to pick and place cylindrical parts that are uniform in size.

Collision sensor: A sensing device that provides protection for colliding avoidance. Used for machine tools protection, overload conditions, mobile robots, and safeguarding. It is a contact switch operated with air or electricity.

Compiler: A compiler is a computer language that translates symbolic operation codes into machine operation codes.

Complex sensor: A device that transmits a complex (analog) sensor's signals to a controller after converting them to digital signals or otherwise conditioning them. Such sensors are vision (cameras), sonar, and tactile-sensing array that enable a robot to interact with its enviroment.

Compliance: The ability of a robot to tolerate misalignment of mating parts.

Computer: A device that uses information to perform prescribed operations and supply results of those processes.

Computer-aided design (CAD): The use of a computer to assist in creating or modifying design parameters.

Computer-aided manufacturing (CAM): The use of a computer in the management, control, and operation of manufacturing.

Computer-Integrated Manufacturing (CIM): The process of controlling the entire manufacturing enterprise with the support of a control computing system. The system will collect, generate, store, and transfer data essential to the manufacturing, parts ordering, engineering, and sales process.

Computer Vision: A computer sensing system used to sense spatial relationships.

Contact sensor: A device that detects the presence of an object, or that measures the amount of force or torque applied by the object, through physically touching it.

Continuous path: A method of controlling a robot in which the commands or input specify all points in the desired path of motion.

Contouring: Controlling the path of a robot manipulator between successive three-dimensional positions or points in space.

Control: The process carried out by a control unit to follow the procedure of a production system.

Controlled path: A method of robot motion control in which intermediate points between command points are interpolated by the robot controller. This allows the robot to move in a straight line between programmed points.

Controller: A robot's brain, a computer-instructed system that directs the motion of the robot end effector, such as a microcomputer or some other programmable device.

Control unit: That portion of a computer that directs the automatic operation of the computer.

Convolution: A generally applicable local operation that can be used to perform such operations as noise filtering, edge extraction, and contour following.

Counter: A device that can be set to an initial number and increased by an arbitrary number.

CPS: Abbreviation for (1) characters per second, (2) cycles per second (Hertz).

CPU: Abbreviation for central processing unit.

Cycle: A sequence of operations that is repeated regularly; also, the time it takes for a robot to run through its programmed motions.

Cylindrical coordinates: Spatial coordinates defined by two distances and an angle.

Data: A collection of facts that are processed by a computer.

Data link: Equipment that permits the transmission of information in data format.

Data processing: Any procedure for collecting data and producing a specific result.

Data register: The part of a microprocessor that temporarily stores information applied to the data bus.

Debugging: The process of detecting and eliminating a device's malfunctions of operation.

Deceleration: The rate of decreasing velocity of a motion as a function of time.

Degree of freedom: A motion variable for a robot axis, which refers to its manipulator and its joints. Each degree of freedom requires a joint.

Delay time: The time the robot's load device requires to become energized.

Depalletizing: The removal of parts from a uniform stack-up position performed by a pick-and-place robot.

Design for manufacturability: The act of designing parts or products to be assembled by automation and/or robots.

Detector: The part of a sensing system used to detect the presence of an object.

Diagnostic routine: A test program used to detect hardware malfunctions in a computer and/or its peripheral equipment.

Die casting: The part made from hot metal poured into a closed cavity to be formed.

Differential positioning: The difference in positions obtained by providing pulses of compressed air to the air motor in opposite directions, resulting in more-accurate positioning.

Digit: One of the n symbols of integral value ranging from 0 to $n-1$, inclusive, in a scale of numbering of base n. One of the 10 decimal digits, 0–9.

Digital control: The use of a digital computer to perform processing and control tasks in a manner that is more easily changed than an analog control system.

Digital electronics: The discrete data of electrons that controls a robot's peripheral devices and other automated systems.

Digital image: A numerical representation of a picture seen by a TV camera, which is used as a robot's "eye" to analyze digital images to enable the robot to recognize an object.

Digital image analysis: A multistage process that leads to a computer's "understanding" a digital image.

Digital system: The system utilizing mechanical parts, logical elements, and functional units for reading, writing, storing, and manipulating information.

Digital-to-analog converter: Also referred to as D/A, a device that transforms digital data into analog data.

Digitizing electronics: Converting an analog electronic value into numerical value.

Direct current (DC): Electrical energy produced when electrons flow in only one direction.

Direct-drive electric motor: A high-torque motor that drives a robot arm directly without the use of gear reduction.

Direct numerical control (DNC): A system in which a digital computer is directly connected to one or more numerically controlled machine tools and controls their operations.

Discrete variable: A variable that takes on only a finite number of values.

Documentation: A collection of written descriptions and procedures that provide information and guidance about a source of events.

Eddy current proximity sensor: A device that induces electric current to detect an object through a nonuniform magnetic field.

Edit: A computer mode that allows the creation or alteration of a program.

Electric drive: Any actuator driven by an electric motor.

Electric motor: A device that converts electrical energy into mechanical energy.

Electromagnet gripper: An end effector that uses the magnetic field created by an electromagnet or permanent magnet to pick up an object.

Electromechanical system: A system that transfers energy from one point to another through electromechanical means.

Encoder: A transducer used to convert position data into electrical signals.

End-effector: The tool attached to the end of a robot wrist that actually performs the work; also called a gripper, process tooling, or end of arm tooling.

EOAT: Abbreviation for end-of-arm-tooling. The device or tool residing at the end of a robotic arm that allows it to perform its task.

Error: The difference in value between a given response and the desired response in the performance of a controlled machine, system, or process.

Executive control program: A main system program designed to establish priorities and to process and control other programs.

Expert system: A computer program that uses artificial intelligence and a knowledge base acquired from expert data to solve problems and make decisions.

Fail safe: Failure of a device without danger to personnel or damage to product or plant facilities.

Feedback: The return of a portion of the output of a closed-loop system or device to its input for controlling its process.

Fixed automation: The process in which the sequence of operations of a system or production equipment is fixed by the system or equipment configuration.

Fixed-sequence robot: A robot that is programmed to perform successive steps of a predetermined sequence of operations that cannot be easily changed.

Fixture: A device needed to hold a workpiece in a proper position for work performance.

Flexible automation: The process by which machines can be programmed to perform different tasks.

Flexible manufacturing systems (FMS): The type of manufacturing operation in which machines are able to make different products without retooling or changes.

Flowchart: A graphic representation of a sequence of steps or operations using symbols that represent the operations.

Fluid power system: A system designed to transfer power using air, oil, or a combination of both.

Force: Any cause that tends to produce or modify motion.

Fuzzy logic: A type of logic used in computers and other electronic devices for processing imprecise or variable data in place of the traditional binary values. Also, fuzzy logic can employ a larger range of values for greater flexibility. It was coined in 1965 by L. A. Zadeh, U.S. computer scientist.

Gantry: A bridgelike frame along which a suspended robot moves. It is generally used to increase the work envelope of the robot.

Geometric processing: The process of taking measurements characteristic of the geometry of certain objects in an image. Examples are area, orientation, perimeter, number, and location of circles.

Global: Overall, as in a global measurement of an object.

Gripper: An end effector of a robot that is designed to pick up, hold, and/or release an object or part and also move it.

Haptic interfaces: Devices that enable manual interaction with virtual environments (VEs) or teleoperated remote systems.

Haptic perception: The ability to sense information from the joints and muscles; also called *kinesthesia*.

Hardware: The physical, tangible, and permanent components of a computer, data-processing system, or robot.

Hierarchical control: A distributed control technique in which the controlling processes are arranged in a hierarchy or according to task, class, or priority level.

Hierarchy: A relationship of elements in a structure divided into levels, with those at high levels having priority or precedence over those at lower levels.

High-level language: A computer programming language that generates machine codes from problem- or function-oriented statements.

Histogram: The relative frequency of gray-level values in digital image.

Home: A reference point from which all the robot's movements are measured.

Hydraulic system: A system composed of an electric pump, a reservoir tank with fluid, and other auxiliary components that function to generate, transmit, control, and utilize hydraulic energy.

Hysteresis: The lag in an instrument's or process's response when a force acting on it is abruptly changed.

IC: Abbreviation for integrated circuit; a solid-state, microcircuit contained in a chip of semiconductor material.

Image: A spatial array of information corresponding to the reflection of an illuminated object formed by the light rays that traverse an optical system.

Image acquisition: The process of illuminating a workpiece and digitally scanning its image.

Image analysis: The process that involves the analysis of information from an image that will enable the robot to perform a task.

Image interpretation: The process that uses algorithms to describe and measure the features of a digital image.

Image processing: The process of changing an analog signal into an equivalent digital signal to interpret the values of an image.

Incremental optical encoder: A device attached to a shaft to determine position information with respect to a known starting location.

Indicator: A device that displays readings indicating operating conditions of a system.

Inductive proximity sensor: A device used to detect signals based on conductive (usually metallic) materials.

Inductive transducer: A sensory device that has a stationary coil and a movable core to measure object movements.

Industrial robot: A programmable, multifunction manipulator designed to move materials, parts, tools, or special devices through programmed motions for the performance of a variety of tasks.

Initialize: Return a program, system, or hardware device to its original state.

Input: The data supplied to a computer for processing; also, the device used to accomplish this transfer of data.

Input port: The connection through which a robot receives information in the form of digital data.

Instruction: A set of bits that causes a computer to perform a prescribed operation. A computer instruction consists of an operation code that specifies the operation to be performed, one or more operands, and one or more modifiers that modify the operand or its addressee.

Instruction decoder: The application that examines a coded word and decides which operation is to be performed by the arithmetic logic unit.

Integrated circuit (IC): A tiny silicon wafer containing thousands of transistors, resistors, and diodes.

Intelligent robot: A robot that chooses between actions according to the way it senses its environment.

Interface: The concept of interconnecting two or more pieces of equipment with different functions.

Interfacing: The process that allows a robot to communicate and interact with other pieces of equipment.

International Standards Organization (ISO): International organization for standardization, in Geneva, Switzerland.

Interrupt: A break in the normal flow of a system or program that allows an action to be performed and the flow to be resumed from that point at a later time.

Interval timing: Timing that occurs after the robot's load has been energized.

I/O: Abbreviation for input/output.

Joint: A single degree of arm rotation or translation.

Joint-interpolated motion: A method of coordinating the movement of joints in such a way that all joints arrive at the desired location simultaneously.

K: An abbreviation for the quantity 1,000.

Knowledge base: The use of artificial intelligence techniques and a base of information used to control systems automatically.

Knowledge engineering: The use of artificial intelligence techniques and a base of information or knowledge about a specific activity used to control systems automatically. This type of system is called a "knowledge-based system."

Label: An ordered set of characters used to identify an instruction, program, quantity, or data area.

Labeling: The process of assigning different numbers to the picture elements (pixels) of different blobs in a binary image.

Language: A defined group of representative characters or symbols combined with specific rules necessary for their interpretation. The rules enable an assembler or compiler to translate the characters into forms meaningful to a machine, system, or process.

Laser: Acronym for *l*ight *a*mplification by *s*timulated *e*mission of *r*adiation. This highly focused beam of light can cut, carry messages, and perform many kinds of work.

Lateral grip: Gripping movement used to grasp larger objects in a sideways motion.

Lead through: Programming a robot by physically guiding the robot through the desired actions.

Light curtain: A programmable safety barrier consisting of photoelectric presence-sensing devices.

Limited sequence: A simple or nonservo type of robot with movement controlled by a series of limit or stop switches; also called a "bang-bang" robot.

Limit switch: A switch actuated by some part or motion of a robot or machine to alter the electrical circuit.

Linkage: A means of communicating information from one routine to another.

Load capacity: The weight that a robot can manipulate with a fully extended arm.

Loader: A program that operates on input devices to transfer information from off-line to on-line memory.

Local operation: Transformers of the gray-scale value of the pixels according to the gray-scale values of the element itself and its neighbors in a given neighborhood.

Log: A record of values and/or an action for a given function.

Loop: The repeated execution of a series of instructions for a variable number of times.

LPM: Lines per minute.

LSI: Large-scale integration. High-density integration of circuits for complex logic functions. LSI circuits can range up to several thousand logic elements on a silicon chip one tenth of an inch square.

LVDT: Linear variable differential transformer. A transducer used in displacement measurements.

Machine intelligence: The design of machines that incorporate the ability to apply knowledge from a database or sensors to permit them to manipulate their environment.

Machine language: A language written in a series of bits that are understandable by a computer; the "first-level" computer language, compared with a "second-level" assembly language or a "third-level" compiler language.

Machine vision: The capability of a robot to "see", conferred on it by a vision system.

Macro: A source language instruction from which many machine-language instructions can be generated.

Magnetic disk storage: A storage device or system consisting of magnetically coated metal disks.

Magnetic gripper: A type of end effector that is magnetized for grasping ability.

Mainframe computer: The principal computer in a system or network of computers.

Manipulator: That part of the robot that performs mechanical movements.

Manual control: A device containing controls that manipulate the robot arm and allow for the recording of locations and programming motion instructions.

Manual programming: The setup by an operator who adjusts the necessary end-stops, switches, cams, or electric wires, to set up the programming sequence of a machine or robot.

MCR: Abbreviation for master control relay. A device used in logic circuits to control power to an entire circuit or to selected rungs in a program.

Mechanical finger: used on the most common type of robot gripper; these fingers are used for grasping objects.

Memory: A device or medium used to store information in such a form that it can be understood by the computer hardware.

Memory capacity: The number of actions that a robot can perform in a program or the number of storage locations available.

Memory cycle time: The minimum time interval between two successive data accesses from a memory.

Memory, random access: Also referred to as RAM, a memory whose information media are organized into discrete locations or sectors, for example, each of which is uniquely identified by an address, so that data may be recalled from the memory by specifying the appropriate address.

Menu: A display of options on a device, such as a CRT, for user prompting and selection.

Message: A group of words that transports an item of information.

Micro array computer: A special-purpose, multiprocessor system designed for high-speed calculations with arrays of data.

Microbot: A tiny robot designed for such tasks as operating inside the human body.

Microcomputer: A small computer built around a single integrated circuit; also called a PC.

Microprocessor: A single integrated circuit containing most of the elements of a computer.

Microsecond: One-millionth of a second.

Modem: A contraction of *mo*dulator-*dem*odulator, which are associated at the same or opposite ends of a circuit, to form a channel.

Modular robot: A robot made by assembling pre-engineered modules.

Monitor: An operating programming system that provides a uniform method for handling the real-time aspects of program timing, such as scheduling and basic I/O functions.

MTBF: Mean time between failures; the average time that a device will operate before failure.

MTTR: Mean time to repair; the average time needed to repair or service a device after failure or malfunction.

Multiplex: The transmission of multiple data bits through a single transmission line by means of a sharing technique.

NC: Normally closed contacts.

NO: Normally open contacts.

Network: The interconnection of a number of devices by data communications facilities.

Noise: An extraneous signal in an electrical circuit that is capable of interfering with the desired signal; or any disturbance that interferes with the normal operation of a device or system; or any unwanted information in a digitized image.

Nonservo: A system without feedback (open loop system).

Nonservo Robot: A nonintelligent robot. These robots are often referred to as limited-sequence robots or pick-and-place, fixed stops, or bang-bang robots.

Numerical control (NC): The control of machine tools that operate by numerical devices.

Occupational Safety and Health Act (OSHA): The federal act regulating health and safety in the workplace.

Off-line programming: A programming technique by means of a computer where any changes made to the program have no effect on the actual operation of hard-wired components. Isolates the processor from the I/O rack.

Offset: The count value of output from an A/D converter resulting from a zero input analog voltage, used to correct nonzero measurements.

On-line programming: Programming of a robot by means of a computer at the robot's console.

Open loop: A system without feedback.

Operating range: The reach capability of a robot or the work envelope of a robot.

Operating system: A group of programming systems operating under the control of a data-processing monitor program.

Operation, serial: The flow of information through a computer in time sequence, usually by bit but sometimes by characters.

Operational speed: The speed at which the robot can accelerate, decelerate, and stop at a given point.

Optical fibers: Fibers made of glass or plastic; optical fibers can transmit light from one point to another. Also referred to as a light pipe.

Optical proximity sensor: A sensory device that measures the amount of light reflected from an object to determine its position; may respond to either visible or infrared light.

Opto-electronic: Term that refers to the combination of optics and electronics.

Orientation: The movement or manipulation of an object consistently into a controlled position in space.

Orientation axes: The robot's wrist joints labeled pitch, yaw, and roll.

Output: Information transferred from the robot controller through module to control external devices, or the device used to accomplish the transfer of data.

Output port: A connection through which the robot controller sends digital data to peripheral equipment.

Palletizing: Placement of parts in a uniform position, performed by pick-and-place robots.

Parallel port: A connecting point through which multiple bits of digital data can be transmitted very quickly.

Parameter: A quantity that may be given different values when the subroutine is used in different main routines or in different parts of one main routine, but that usually remains unchanged throughout any one such use. In a generator, a quantity used to specify I/O devices, to designate subroutines to be included, or to describe the desired routine to be generated.

Part classification: The identification of different parts by a robot, usually by means of a vision system.

Patch: A section of coding inserted into a routine to correct a mistake or to alter the routine.

Payback period: The time required to recover the amount expended for new equipment through savings in labor or material costs.

Payload: The maximum weight that can be carried by a robot in normal and continuous operation.

PC: Abbreviation for programmable controller.

Period: The interval taken for a timing pulse to pass through a complete cycle from the beginning to the end.

Peripheral: Any equipment or device used for input/output operations with the CPU.

Personal computer (PC): A relatively low-cost portable microcomputer built around a single integrated circuit or microprocessor.

Photoconductive device: A device that changes in conductivity according to variations in light.

Photoemissive device: An opto-electronic device that emits electrons in the presence of light. Phototubes are a type of photoemissive device.

Photovoltaic device: An opto-electronic device that converts light energy into electrical energy; also called a solar cell.

Pick-and-place robot: A simple robot with two to four axes of motion and little or no trajectory control.

Piezoelectric effect: The creation of electrical energy by applying pressure to a certain type of crystal; often used in microphones.

Pitch: The up-and-down motion of the robot's wrist.

Pixel: A picture element of the matrix of gray-scale values that is assumed to be a uniform shade of gray in creating the digital image.

Playback robot: A manipulator that can store in memory and reproduce operations originally executed under human control.

Pneumatic system: A power system that uses air as its power source.

Point operation: The transformation of the gray-scale value of the pixels according to a given function.

Point-to-point robot: A servo or nonservo driven robot with a control system for programming a series of points without regard for coordination of axes, in which the intermediate points are not controlled.

Position axes: The three degrees of freedom of the robot's arm, labeled x, y, and z axes. They are also called rotational, radial, and vertical traverses.

Power supply: The system that provides and regulates the energy required by a robot for operation.

Preventive maintenance: The process of regularly checking equipment, cleaning and maintaining it, and replacing worn parts before breakdowns occur.

Process tooling: The instrument or device attached or mounted to the end of a robot arm.

Program: A series of instructions, stored in the controller's memory, that controls robot movement.

Program counter: A memory device that indicates the location in memory of the next instruction to be executed.

Programmable: A term used to describe a robot or other machine that can be given new instructions to meet new requirements.

Programmable automation: The process in which the sequence of operations of a system or production equipment is designed with the capability to change to accommodate different product configurations. The operation sequence is controlled by a program that is a set of instructions coded so that the system can read and interpret them.

Programmable logic controller (PLC): A simple control device designed specifically for industrial machines that performs logical operations compatible with traditional relay logic.

PROM: An acronym for *p*rogrammable *r*ead-*o*nly *m*emory, a memory that can be changed or initialized more than once by the user.

Proximity sensor: A device that senses the absence or presence of an object within a certain region or that provides feedback about distance between it and something else, such as an end effector.

Pseudo code: A group of characters having the same general form as a computer instruction but not executed by the computer as an actual instruction.

Quality Assurance: An inspection system that ensures the production of the highest quality products.

RAM: Acronym for *r*andom *a*ccess *m*emory.

Range: A characterization of a variable or function.

Range sensor: A device used to determine the precise distance from the sensor to an object.

Rated load capacity: The amount of weight a robot is capable of lifting reduced by a factor of safety.

Read: To sense the data stored on a device.

Reader: A device that senses data as on punch cards or tape, and enters it into the computer.

Read-only memory (ROM): Memory in which data is permanently stored in a microcomputer. This data is not lost when the power source is turned off.

Real time: Refers to the actual time a physical process or operation takes.

Rectification: The process of converting alternating current to direct current.

Reed switch: A device that makes or breaks contact when exposed to a magnetic field.

Register: A memory device capable of containing one or more computer bits or words.

Relay contact: The basic programming element when using relay logic. This contact may be normally open (NO) or normally closed (NC).

Relay ladder: A network of elements used in relay logic.

Relay logic: An assembly language used to program programmable logic controllers (PLCs). A ladder diagram is entered into the system by means of a keyboard.

Reliability: The probability that a component part, equipment, or system will satisfactorily perform its intended task.

Remote-center compliance (RCC) device: A device that fits in the wrist of the robot to give it compliance capability.

Repeatability: A measurement of the deviation between a taught location point and the played-back location, under identical conditions of load and velocity.

Resolution: The number of bits by which a digital image is represented; also, the smallest distance increment that can be commanded by a robot controller.

Revolute Configuration: A term used to describe the unevenly shaped work envelope of a robot with rotary joints.

Robot: A mechanical device that can be programmed to perform some task of manipulation or locomotion under automatic control.

Robotic: Pertaining to robots.

Robotic Industries Association (RIA): An organization founded in 1975 to encourage development and use of robotics in America.

Robotics: The study of robots; also, the science of designing, building, and applying robots.

Roll: Circular motion at an axis, a rotation about the link axis of the robot's wrist.

ROM: Read-only memory. A digital memory containing a fixed pattern of bits, generally unalterable by the user.

Rotary actuator: A fluid power device designed to produce a rotary motion in either direction.

Rotor: The rotating armature, shaft, and associated parts of a motor.

Routine: A series of computer instructions that perform a specific, limited task.

RS-232C: A communication standard computer interface link used for connecting peripheral equipment to CPU devices.

Scan: To examine signals or data point by point in a logical sequence.

SCARA: Selective Compliance Assembly Robot Arm. Developed by Professor Makino of Yamanashi University, Japan. Its design allows it to be firmly yielding in horizontal motions and rigid in vertical motions.

Sensing system: A system that responds to various forms of energy, such as light energy and electrical energy, and conveys information to a control unit.

Sensor: A transducer or device that transmits information to a robot's controller.

Sensory feedback: Information about its environment that affects how a robot responds or interacts.

Sequence controller: A device that uses clock inputs to maintain the proper sequence of events required to perform a task.

Sequencing: The control of a load device that registers in each of its stepping positions.

Serial interface: A method of data transmission that permits transmitting a single bit at a time through a single line.

Serial port: Connecting point through which a computer sends or receives digital data using the serial transmission method.

Servo: A system with feedback (closed-loop system).

Servo-controlled: A closed-loop system with continuously controlled path.

Servomechanism: A closed loop control system for a robot in which the computer issues commands, the motor drives the arm, and a sensor measures the motion and signals the amount of the motions back to the computer repeatedly until the arm is repositioned to the point requested.

Servomotor: A device used to achieve a precise degree of rotary position. Two distinct types of servomotors are used today: the synchronous motor and stepping motor. Usually servomotors are electric, but there are hydraulic and pneumatic ones as well.

Servo robot: A robot classified as intelligent or highly intelligent, determined by its level of awareness of its environment.

Servo system: A machine system that changes the position or speed of a mechanical object.

Set point: The required or ideal value of a controlled variable, usually preset in the computer or system controller by an operator.

Signal processing: Complex analysis of waveforms to extract information.

Significant digit: A digit that contributes to the precision of a numeral.

Simulator: A device or computer program that performs simulation.

Size: A specified value of geometrical characteristics directly related to the measurements of an object.

Slew rate: The maximum rate at which a system can follow a commanded motion.

Smart sensor: A sensor whose output depends on internal data or on input from another part of the system.

Software: The programs, routines, or supporting documentation that instruct the operations of a computer.

Source language: The symbolic language comprised of statements and formulas used to specify computer processing.

Speed: The maximum velocity at which the tip of the robot can move through its work envelope.

Spherical coordinates: Spatial coordinates defined by two angles and a distance.

Statement: A meaningful expression or generalized instruction in a source language.

Stepper motor: A bidirectional, DC permanent magnet motor that turns through one angular increment for each pulse applied to it.

Stop: A mechanical constraint on motion.

Subroutine: A series of computer instructions to perform a specific task for many other routines.

Symbolic coding: Any coding system in which symbols rather than actual machine operations and addresses are used.

Synchronous speed: The stator speed of a motor.

Syntax: The rules governing the structure of statements used in a program.

System: A collection of parts or devices that forms and operates as an organized whole through some form of regulated interaction.

Tactile sensor: A sensory device that indicates the presence of an object by physical contact in order to identify its location.

Teach pendant: A device used to record movements into the robot's memory.

Teach-pendant programming: A programming method in which the operator leads the robot through various positions.

Teleoperation: Having sensors and actuators for mobility and/or manipulation, remotely controlled by a human operator; usually used in hazardous environments.

Tool: A term used loosely to define the device attached or mounted to the end of a robot arm to perform work. For example, welding torch, painting gun, and similar tools.

Tool center point (TCP): The origin of the tool coordinate system or the point of action of the tool attached to the robot arm.

Tool-mounting plate: The plate where the tool is attached at the end of the robot's arm.

Trajectory: The resulting path of a robot's end effector as programmed in the controller.

Transducer: A device that converts energy from one form to another.

Transfer vector: A transfer table used to communicate between two or more programs.

Transmission path: The part of a system that provides a path for the transfer of energy.

Troubleshooting: A method of finding out why something doesn't work properly.

TTY: Abbreviation of teletypewriter; also, trademark of Teletype Corporation. A special electric typewriter that produces coded electric signals corresponding to manual typed characters, and automatically types messages when fed with similar coded signals produced by another machine.

Universal motor: A motor powered by either AC or DC sources.

Vacuum gripper: An end effector consisting of one or more suction cups made of rubber; used to hold an object.

VAL: A high-level robot application programming language developed for Unimation's Unimate and Puma lines of robots.

Variable: A quantity that can assume any of a given set of values.

Velocity: The measurement of the rate of change in motion.

Vision system: A device that can collect data and form an image that can be interpreted by a robot computer to determine an appropriate position or to "see" an object.

VLSI: Abbreviation for very large scale integration. The amount of integration on a chip comprising 100,000 or more gates designed for use in microprocessors. Compare to SSI, MSI, and LSI.

Voice recognition: A technology in which a robot is programmed to respond only to the vocal commands of an operator whose voice frequency has been recorded in memory.

Workcell: A manufacturing unit consisting of one or more workstations.

Work envelope: The outline surface of a robot's work volume, or the extreme point that it can reach.

Workstation: A manufacturing unit consisting of one robot and the machine tools, conveyors, and other equipment with which it interacts.

World coordinate system: A fixed coordinate system that serves as an absolute frame of reference, sometimes also called world frame system.

Wrist: The manipulator device located between the robot arm and end-effector. A general-purpose wrist has three degrees of freedom.

Yaw: Side-to-side motion of the robot's wrist about its vertical axis.

Zero point: The origin of a coordinate system.

Zero position: The configuration at which all principal joint displacements equal zero.

Robot Manufacturers

ABB Flexible Automation Inc.
2487 South Commerce Drive
New Berlin, WI 53151
Phone: (414) 785-3400 (8532)
Fax: (414) 785-0342

Adept Technology
150 Rose Orchard Way
San Jose, CA 95134
Phone: (408) 432-6235
Fax: (408) 432-8707

Advance Research & Robotics, Inc.
341 Christian Street
Oxford, CT 06478
Phone: (203) 264-3333
1-800-827-6268
Fax: (203) 264-4693

**Burns Machinery, Inc.
(BMI Robotics)**
2580 S. Brannon Stand Road
Dothan, AL 36301-7042
Phone: (334) 793-7086
Fax: (334) 671-0310

C&D Robotics, Inc.
690 Hazel Avenue
Beaumont, TX 77701
Phone: (409) 832-4991
Fax: (409) 833-0045

Columbia/Okura, L.L.C.
301 Grove Street, Suite A
Vancouver, WA 98661
Phone: (360) 735-1952
Fax: (360) 905-1707

**Comau North America
Robot Center**
4475 Purks Drive
Auburn Hills, MI 48326
Phone: (248) 475-3240
Fax: (248) 377-3662

Creative Automation, Inc.
4843 Runway Boulevard
Ann Arbor, MI 48108
Phone: (734) 930-0050
Fax: (734) 930-0058

CRS Robotics Corporation
5344 John Lucas Drive
Burlington, ON L7L 6A6
Canada
Phone: (800) 365-7587
Fax: (905) 332-1114

CYBO Robots, Inc.
2040 Production Drive
Indianapolis, IN 46241
Phone: (317) 484-2926
Fax: (317) 241-2727

Denso International America, Inc.
24777 Denso Drive
P.O. Box 5133
Southfield, MI 48086-5133
Phone: (248) 213-2387
Fax: (248) 213-2476

Denso Sales California
3900 Via Oro Avenue
Long Beach, CA 90810
Phone: (310) 513-8550
Fax: (310) 513-7379

Eclipse Automation Corp.
1240-2 Ken Pratt Boulevard
Longmont, CO 80501
Phone: (303) 415-0949
Fax: (303) 682-8822

FANUC Robotics North America, Inc.
2000 South Adams Road
Auburn Hills, MI 48326-2800
Phone: (810) 377-7000
1-800-474-6268
Fax: (810) 377-7366

Flow Robotics
1635 Production Road
Jeffersonville, IN 47130
Phone: (812) 283-7888
Fax: (812) 284-3281

GROB Systems, Inc.
1070 Navajo Drive
Bluffton, OH 45817
Phone: (419) 358-9015
Fax: (419) 358-6719

Gudel Lineartec, Inc.
4881 Runway Boulevard
Ann Arbor, MI 48108
Phone: (734) 214-0000
Fax: (734) 219-9000

IAI America, Inc.
2360 W. 205 St.
Torrance, CA 90501
Phone: (800) 736-1712
Fax; (310) 320-4553

Kawasaki Robotics (USA), Inc.
28059 Center Oaks Court
Wixom, MI 48393
Phone: (810) 305-7610
Fax: (810) 305-7618

KUKA Welding Systems & Robot Corp.
6600 Center Drive
Sterling Heights, MI 48312
Phone: (810) 795-2000
Fax: (810) 978-0429

Motoman, Inc.
805 Liberty Lane
West Carrollton, OH 45449
Phone: (937) 847-3300
Fax: (937) 847-3288

OTC Daihen, Inc.
Dynamic Robotics Division
761 Crossroads Court
Vandalia, OH 45377
Phone: (937) 454-9660
Fax: (937) 454-9661

Panasonic Factory Automation Co.
9377 W. Grand Avenue
Franklin Park, IL 60131
Phone: (847) 288-4490
Fax: (847) 288-4564

PaR Systems, Inc.
899 Highway 96 West
Shoreview, MN 55126
Phone: (612) 484-7261
Fax: (612) 483-2689

Reis Machines, Inc.
1320 Holmes Road
Elgin, IL 60123
Phone: (847) 741-9500
Fax: (847) 888-2762

RMT Engineering Ltd.
623 South Service Road
Grimsby, ON L3M 4E8
Canada
Phone: (905) 643-9700
Fax: (905) 643-9666

The Robot Shop, Inc.
3442 North Shore Drive
Wayzata, MN 55391-9359
Phone: (612) 471-7273
Fax: (612) 471-7290

Sankyo Robotics
1001-D Broken Sound Parkway NW
Boca Raton, FL 33487
Phone: (561) 998-9775
Fax: (561) 998-9778

Seiko Instruments USA, Inc.
2990 W. Lomita Boulevard
Torrance, CA 90505
Phone: (310) 517-7850
Fax: (310) 517-8158

ShinMaywa (America), Ltd.
1603 Barclay Boulevard
Buffalo Grove, IL 60089
Phone: (847) 808-1281
Fax: (847) 808-1286

SMAC
5807 Van Allen Way
Carlsbad, CA 92008
Phone: (760) 929-7575
Fax: (760) 929-7588

Sony Electronics
Factory Automation Division
560 Route 303
Orangeburg, NY 10962
Phone: (914) 365-6000
Fax: (914) 365-6087

Staubli Unimation Inc.
201 Parkway West
Duncan, SC 29334
Phone: (864) 433-1980
Fax: (864) 486-9906

Yamaha Robotics
888 Sussex Boulevard
P.O. Box 190
Broomall, PA 19008
Phone: (800) 829-2624
Fax: (610) 543-8113

Yushin America, Inc.
35 Kenney Drive
Crayston, RI 02920
Phone: (401) 463-1800
Fax: (401) 463-1810

Index

A

Acceleration, 22, 381
Accuracy, defined, 22, 98, 381
Active compliance, 138
Actuators, 34, 381
    electric, 30, 66, 387
    hydraulic, 30, 69, 389
    pneumatic, 30, 67, 396
Adaptive control, 273, 351, 381
Adhesive gripper, 22, 131
Advanced robots, applications of, 315, 321–325
    concepts and procedures, 325
    future developments, 327
    impact on employment, 328
Analog-to-digital converters, 153, 382
Applications, robot, 352
    future, 15, 364–365
    present, 15, 352–363
Arm geometry, 51
    cylindrical, 54
    jointed-arm (vertical), 59
    jointed-arm (horizontal), 59
    rectangular, 51
    spherical, 56
Arm motion, robot, 72
Artificial intelligence, 80, 205, 309
    architecture, 316
    elements of, 309
    systems, 309
Asimov, Isaac, 4
    "laws of robotics," 4
Assembly operations, 359
Automatic guided vehicle, 321, 365, 383
Automatic tool changer, 134, 383

Automation and robots, 1
Automation in manufacturing, 351
Automation, types of, 1, 383
    fixed, 1, 388
    flexible, 2, 388
    programmable, 2, 396

B

Basic components, robot, 23–33
Brief history, 3

C

Cameras, 200
    solid-state, 200
    vidicon, 200
Case studies, 336, 368
Capek, Karel, 3
Challenge for the future, 365
Circular interpolation motion, 72, 292, 384
Closed-loop systems, 39, 83, 384
    advantages, 41
    disadvantages, 41
Collet gripper, 131, 384
Compliance, 103, 136–147, 384
    active, 138
    passive, 141
    instrumented remote center (IRCC), 146
    remote center (RCC), 141
Compliance systems, 138
Computer integrated manufacturing, 287, 385
Computer numerical control, 248
Continuous-path control, 77, 385

Control resolution, 96–98
Control systems, 221
    analysis, 83
    closed-loop, 39, 77, 248
    computer numerical control, 248
    correlation, 221
    interfacing, 254
    microprocessor, 249
    open-loop, 33, 77, 248
    PLC programming terminals, 239
    programmable logic controller, 224
    proportional-integral-derivative, 246
    requirements, 222
    universal robot controller, 254
    workcell control, 256
Controlled-path control, 75, 385
Controller, 23, 31, 385
Cycle, 22, 291, 385
Cylindrical configuration, 42, 54, 385

D

Deceleration, 22, 386
Definitions for safety, 344
Degrees of Freedom, 24, 64, 386
Depalletizing, 354, 386
Diode-array camera systems, 208
Dynamic properties of robots, 94

E

Economic and social issues, 12
Electric power source, 66
Electromechanicl power source, 72
Elements of artificial intelligence, 309

405

advanced robot, 315, 319–325
expert systems, 310, 388
machine learning, 314
machine vision, 194, 312, 392
natural language process, 312
neural networks, 186, 312
Encoders, 40, 176, 387
    absolute optical, 176, 381
    incremental optical, 176, 389
    tachometer, 176
End effector, 22, 25, 387
End effector types, 111
    grippers, 22, 113–134, 388
    process tooling, 134, 396
    special purpose grippers, 131
End-of-arm-tooling (EOAT), 22, 387
Expandable gripper, 131
Expert systems, 310, 388

F

Factory of the future, 324, 352, 365
Feature weighing, 205
Feedback, 33, 246, 388
    derivative, 247
Fixed stops, 37
    integral, 246
    on-off, 246
    proportional, 246
Fixture, 22, 114, 148, 388
Flexible manufacturing systems, 351, 388
Flowchart programming language, 238, 388
Fundamentals, robot, 21
Future developments, 327
Fuzzy logic control, 325, 388

G

Gripper, 111–134, 388
    actuators, 34–37
    adhesive, 131
    characteristics, 113
    force analysis, 117
    magnetic, 128
    mechanical, 114
    selection and design, 133
    special purpose, 131
    vacuum, 124

Gripper classification, 111–119
    angular, 116
    double, 113
    external, 113, 116
    internal, 113, 116
    multiple, 113
    parallel, 116
    single, 113
    2-finger, 116
    3-finger, 116
    4-finger, 116

H

Hierarchical control structure, 90
    hierarchical control, 90, 389
    hierarchy, 90, 389
High-technology intelligence, 78, 256
History, brief, 3, 8
Hook gripper, 131
Human factor issues, 337
Hydraulic power source, 69

I

Image, 193, 389
    acquisition, 194, 389
    analysis, 204, 389
    digitization, 198
    interpretation, 205, 389
    processing, 203, 389
Impulse limit switches, 168
Industrial applications, 271, 282, 351
    assembly operations, 359
    automation in manufacturing, 351
    case studies, 368
    challenge for the future, 365
    future applications, 364
    innovations, 367
    inspection operations, 362
    material handling, 353
    processing operations, 357
Industrial robot, 2, 8, 12, 21, 41 245, 390
    applications, 352
Inflatable gripper, 131
Innovations, 367
    haptic perception robots, 368
    how to make a robot smile, 367
    medical robots, 368
    mouth robot, 368
    telepresence robot, 368
Input, defined, 31, 114, 222, 390
    module, 222
Inspection operations, 362
Intelligent control robots, 79–80, 84, 317
    artificial intelligence, 80, 205, 309, 382
    high-technology, 79
    low-technology, 78
Intelligent systems, 309
Interfacing, 254, 390
    input signals, 254
    output signals, 254
International Standards Organization, 4, 349, 390

J

Joint-interpolated motion, 72, 390
Jointed-arm configuration, 43, 59
    horizontal, 59
    vertical, 59
Joints, 24, 41, 64, 390

K

Knowledge base, 247, 390
Knowledge engineering, 309, 391

L

Ladder logic, 228–230
Ladder logic diagram, 229, 236
Levels of robot programming, 280
Limit switches, 164–168, 391
Limited-sequence control, 73, 391
Line tracking, 93
Linear array, 193
Low-technology intelligence, 78

M

Machine loading/unloading, 354
Machine vision, 194
    linear array, 193
    matrix arrays, 193
Machine vision applications, 206
    inspection, 206

part identification, 207
part location, 207
part orientation, 207
safety monitoring, 207, 258, 338
visual guidance and control, 207
Machine vision system, 194
image acquisition, 194, 389
image analysis, 204, 389
image digitization, 198
image interpretation, 205, 389
image processing, 203, 389
Magnetic gripper, 128, 392
advantages, 128
disadvantages, 128
electromagnets, 129
permanent magnets, 129
Manipulator, 24, 392
Manual switches, 163
Material handling, 353
depalletizing, 354, 386
machine loading/unloading, 354
palletizing, 354, 395
part placement, 353
stacking and insertion operations, 357
Matrix array, 193
Mechanical grippers, 114
Methods defining positions in space, 290
joint movements, 290
tool coordinate motions, 291
x-y-z coordinate motions, 291
Microprocessor, 249, 393
accumulator, 251, 381
address register, 251, 381
arithmetic-logic, 250, 382
buses, 252, 383
data register, 251, 386
instruction decoder, 251, 390
program counter, 251, 396
sequence controller, 252, 398
Microprocessor architecture, 253
Motorola, 253
typical, 253
Microswitches, 160
contact arrangement, 163
inductive, 162
mechanical, 162
resistive, 162

Microswitches, types of, 163
impulse, 168
limit, 164
manual, 163
pressure, 169
reed, 169
Modular robot components, 104
Motion interpolation, 292
Motion, types of, 72
circular-interpolation, 72
joint-interpolation, 72
slew, 72
straight-line interpolation, 72

N

NEMA switch enclosures, 167
Nonservo robot operation, 84
Nonservo system, 33, 394
Numerical Control, 222, 248, 394

O

Off-line programming, 9, 44, 89, 272, 394
advantages, 273
Open-loop systems, 33, 83, 248
actuators, 34
advantages, 38
disadvantages, 38
fixed stops, 37
stepper motors, 37
variable stops, 37
Operator interface, 258
Optical sensing methods, 207
diode-array camera systems, 208
optical triangulation techniques, 212
scanning laser systems, 207
Orientation axes, 25, 394
pitch, 25, 65, 395
roll, 25, 65, 398
yaw, 25, 65, 401
Output, defined, 31, 394
module, 222

P

Palletizing, 354, 395
Part placement, 353

Passive compliance, 141
Path control, 73–78
continuous-path, 77, 385
controlled-path, 75, 385
limited-sequence, 73, 391
point-to-point, 74, 396
Payload, 22, 29, 30, 395
Photoelectric sensors, 173
diffuse scanner, 175
light source and receiver, 173
reflect scanner, 175
retroreflector, 175
Pitch, 25, 65, 395
Pneumatic power source, 67
Point-to-point control, 74–78, 270, 357
Position axes, 25, 396
radial traverse, 65
rotational traverse, 65
vertical traverse, 65
Potentiometers, 181
Power sources, 29, 65, 396
electric, 29, 66
electromechanical, 72
hydraulic, 29, 69, 389
load-carrying capacity, 66, 67, 79
pneumatic, 29, 67, 396
Pressure switches, 169, 170
Processing operations, 351, 352, 357
adhesives & sealants dispensing, 353, 357
continuous arc welding, 357
machining, 353, 357
metal cutting and deburring, 353, 357
rotating and spindling, 357
spot welding, 357
spray painting, 358
Process tooling, 27, 113, 114, 134
multiple (with tool changers), 134
single, 134
Program statements, 294–296
sample programs, 296–298
Programmable logic controllers, 44, 224, 396
advantages, 226
components, 226, 227
I/O interfaces, 227
languages, 228
memory and CPU, 227

power supply, 227
program, 227
programming devices, 227
Programming languages, 274–280
    AL (application), 275
    common, 274
    high-level, 274
    KAREL (application), 277
    low-level, 274
    machine, 274
Programming methods, 270
    computer terminal, 272
    manual, 270
    off-line, 9, 44, 89, 272, 394
    teach pendant, 270
    walk-through, 270
Programming terminals, 239
    computer-based, 243
    dedicated, 239
    hand-held, 245
Programming software, 243
    data collection and analysis, 244
    documentation, 243
    essential, 243
    host computer, 244
    real-time, 244
    simulation, 244
Proportional-integral-derivative, 246
    derivative control, 247
    integral control, 246
    on-off feedback, 246
    proportional control, 246
Proximity probes, 173
Proximity sensors, 172, 396
    capacitive-type, 173
    hall-effect, 173
    inductive-type, 172
    ring-type, 173
    ultrasonic, 173

Q

Quality
    assurance, 328, 397
    control inspection, 184, 193
    standards, 373

R

Rectangular configuration, 42, 51–54

Reed switches, 169
Repeatability, 23, 46, 88, 100, 397
Resolution, 23, 54, 96, 398
Robot, 1, 3–15, 398
    anatomy, 41
    arm, 22, 24–34
    basic components, 23
    characteristics, 21
    chronological developments, 8
    classification, 51
    configurations, 7, 41
    definition (RIA), 2
    description (ISO), 4
    fundamentals, 21
    general characteristics, 21
    generations, 43
    impact on employment, 328
    modular components, 104
    process knowledge, 44
    properties, 94
    safety, 46, 183, 258, 333
    selection, 45
    sensors, 43, 153, 398
    size, 23
    technology, 21
    wrist, 24, 41, 65, 141, 288, 401
Robot applications, 352
    assembly operations, 359
    evaluating the potential, 363
    important characteristics, 352
    inspection operations, 362
    material handling, 353
    processing operations, 357
Robot dynamic properties, 94
    accuracy, 22, 98
    compliance, 103, 136–147
    control resolution, 96
    repeatability, 100
    spatial resolution, 97
    stability, 94
Robot operation, 83
    nonservo, 84
    servo, 87
Robot programming, 8, 44, 267, 275
    data processing, 269
    decision-making, 268
    manipulation, 268
    sensing, 268
Robot programming, basic, 279

common application, 280
development, 280
operating system, 280
special application, 280
Robot programming, levels of, 280
    characteristics of level 4, 282
    example of level 2, 283
    level 1, 280
    level 2, 280
    level 3, 281
    level 4, 281
    path modification, 284
    process control, 284
    program statements, 285
Robot safety, 46, 183, 258, 333
    definitions, 344
    guidelines, 343
    human factor issues, 337
    safeguarding, 340
    sensors and monitoring, 338
    standards, 334
    system reliability, 336
    training, 342
Robota, 3, 8
Robotic Industries Association, 2, 334, 349, 398
Robotics, 1, 4, 398
Roll, 25, 398
Rotary position sensors, 176
    encoders, 176
    potentiometers, 181
    resolvers, 180
    synchros, 179

S

Safeguarding, 340
Safety, 46, 183, 258, 333
    definitions, 344
    sensors and monitoring, 338
    standards, 334
    training, 342
Safety guidelines, 343
Safety monitoring, 258
Sample programs, 296
SCARA, 9, 43, 60, 398
Scoop and ladle gripper, 131
Sensors, 7, 43, 153, 398
    acquisition, 154

classification, 154
complex, 154, 160
contact, 159
discrete, 159
electrical, 154
external, 157
interlock, 159
internal, 154
magnetic, 154
manipulation, 154
mechanical, 154
noncontact, 159
photoelectric, 171, 173
physical activation, 159
process, 160
proximity, 172
rotary position, 176
smart, 43, 117, 399
thermal, 154
Sensors, usage and selection of, 183
control interlocking, 183
position and related information, 184
quality control inspection, 184
safety monitoring, 183
Sensors and control integration, 186
Servo-controlled system, 33, 40, 79
advantages, 41
disadvantages, 41
Servomechanism, 39, 399
servo robot operation, 87
Sequence control, 257
Sequence controller, 252, 398
Signal processing, 184, 313, 399
controller, 31–41, 184
decision-making, 184
sensors, 184
Size, 23
Slew rate, 22, 399
Slew motion, 49, 72
Solid-state switches, 170
Solid-state cameras, 170

Spatial resolution, 97
Space position programming, 288
methods defining positions, 290
reason for defining points, 291
speed control, 292
Special-purpose grippers, 114, 131
collet, 131
expandable, 131
hook, 131
inflatable, 131
scoop and ladle, 131
Speed, 15, 22, 399
acceleration, 22, 381
deceleration, 22, 386
slew rate, 22, 399
Spherical configuration, 42, 56, 396
Stability, 94
Stacking and insertion operations, 357
Stepper motor, 34, 37, 38, 400
Straight-line interpolation motion, 72
Strain gauges, 136, 157
Switches, 8, 44, 73, 160–172
electromechanical, 159, 160
photoelectric, 173
proximity, 172
rotary, 176
solid state, 170
Synchro system, 179
System architecture, 316
System reliability, 336

T

Technology of robots, 7
Template matching, 205
Tool, defined, 8, 25, 27, 400
Tool center point, 27, 400
Tool mounting plate, 22, 25, 111, 400
Training, 342

U

Universal robot controller, 11, 254

interface, 254
specifications, 255

V

Vacuum, 124
Vacuum surface, 127
Vacuum cups, 27, 124–128
Vacuum gripper, 124, 401
advantages, 127
capacity, 124
multiple-cup, 124
Variable stop, 37
Velocity, 22, 401
Venturi devices, 127
Vidicon cameras, 200
Vision, 8, 158, 193
basic lighting devices, 201
commercial types, 205
other optical methods, 207
system, 193, 401
Visual sensing, 193

W

Work envelope, 22, 25, 52, 346, 401
Workcell, 6, 401
Workcell control, 256
criteria, 256
operator interface, 258
safety monitoring, 258
sequence control, 257
World coordinate system, 25, 291, 401

Y

Yaw, 25, 65, 401

Z

Zero point, 401
Zero position, 401

Robot Manufacturers

ABB Flexible Automation Inc.
2487 South Commerce Drive
New Berlin, WI 53151
Phone: (414) 785-3400 (8532)
Fax: (414) 785-0342

Adept Technology
150 Rose Orchard Way
San Jose, CA 95134
Phone: (408) 432-6235
Fax: (408) 432-8707

Advance Research & Robotics, Inc.
341 Christian Street
Oxford, CT 06478
Phone: (203) 264-3333
1-800-827-6268
Fax: (203) 264-4693

Burns Machinery, Inc. (BMI Robotics)
2580 S. Brannon Stand Road
Dothan, AL 36301-7042
Phone: (334) 793-7086
Fax: (334) 671-0310

C & D Robotics, Inc.
690 Hazel Avenue
Beaumont, TX 77701
Phone: (409) 832-4991
Fax: (409) 833-0045

Robot Manufacturers

Columbia/Okura, L.L.C.
301 Grove Street, Suite A
Vancouver, WA 98661
Phone: (360) 735-1952
Fax: (360) 905-1707

Comau North America Robot Center
4475 Purks Drive
Auburn Hills, MI 48326
Phone: (248) 475-3240
Fax: (248) 377-3662

Creative Automation, Inc.
4843 Runway Boulevard
Ann Arbor, MI 48108
Phone: (734) 930-0050
Fax: (734) 930-0058

CRS Robotics Corporation
5344 John Lucas Drive
Burlington, ON L7L 6A6
Canada
Phone: (800) 365-7587
Fax: (905) 332-1114

CYBO Robots, Inc.
2040 Production Drive
Indianapolis, IN 46241
Phone: (317) 484-2926
Fax: (317) 241-2727

Denso International America, Inc.
24777 Denso Drive
P.O. Box 5133
Southfield, MI 48086-5133
Phone: (248) 213-2387
Fax: (248) 213-2476

Denso Sales California
3900 Via Oro Avenue
Long Beach, CA 90810
Phone: (310) 513-8550
Fax: (310) 513-7379

Eclipse Automation Corp.
1240-2 Ken Pratt Boulevard
Longmont, CO 80501
Phone: (303) 415-0949
Fax: (303) 682-8822

FANUC Robotics North America, Inc.
3900 W. Hamlin Road
Rochester Hills, MI 48309-3253
Phone: (248) 377-7570
Fax: (248) 377-7365

Flow Robotics
1635 Production Road
Jeffersonville, IN 47130
Phone: (812) 283-7888
Fax: (812) 284-3281

GROB Systems, Inc.
1070 Navajo Drive
Bluffton, OH 45817
Phone: (419) 358-9015
Fax: (419) 358-6719

Gudel Lineartec, Inc.
4881 Runway Boulevard
Ann Arbor, MI 48108
Phone: (734) 214-0000
Fax: (734) 219-9000

IAI America, Inc.
2360 W. 205 St.
Torrance, CA 90501
Phone: (800) 736-1712
Fax: (310) 320-4553

Kawaski Robotics (USA) , Inc.
28059 Center Oaks Court
Wixom, MI 48393
Phone: (810) 305-7610
Fax: (810) 305-7618

KUKA Welding Systems & Robot Corp.
6600 Center Drive
Sterling Heights, MI 48312
Phone: (810) 795-2000
Fax: (810) 978-0429

Motoman, Inc.
805 Liberty Lane
West Carrolton, OH 45449
Phone: (937) 847-3300
Fax: (937) 847-3288

OTC Daihen, Inc.
Dynamic Robotics Division
761 Crossroads Court
Vandalia, OH 45377
Phone: (937) 454-9660
Fax: (937) 454-9661

Panasonic Factory Automation Co.
9377 W. Grand Avenue
Franklin Park, IL 60131
Phone: (847) 288-4490
Fax: (847) 288-4564

PaR Systems, Inc.
899 Highway 96 West
Shoreview, MN 55126
Phone: (612) 484-7261
Fax: (612) 483-2689

Reis Machines, Inc.
1320 Holmes Road
Elgin, IL 60123
Phone: (847) 741-9500
Fax: (847) 888-2762

RMT Engineering Ltd.
623 South Service Road
Grimsby, ON L3M 4E8
Canada
Phone: (905) 643-9700
Fax: (905) 643-9666

The Robot Shop, Inc.
3442 North Shore Drive
Wayzata, MN 55391-9359
Phone: (612) 471-7273
Fax: (612) 471-7290

Sankyo Robotics
1001-D Broken Sound Parkway NW
Boca Raton, FL 33487
Phone: (561) 998-9775
Fax: (561) 998-9778

Seiko Instruments USA, Inc.
2990 W. Lomita Boulevard
Torrance, CA 90505
Phone: (310) 517-7850
Fax: (310) 517-8158

ShinMaywa (America), Ltd.
1603 Barclay Boulevard
Buffalo Grove, IL 60089
Phone: (847) 808-1281
Fax: (847) 808-1286

SMAC
5807 Van Allen Way
Carlsbad, CA 92008
Phone: (760) 929-7575
Fax: (760) 929-7588

Sony Electronics
Factory Automation Division
560 Route 303
Orangeburg, NY 10962
Phone: (914) 365-6000
Fax: (914) 365-6087

Staubli Unimation Inc.
201 Parkway West
Duncan, SC 29334
Phone: (864) 433-1980
Fax: (864) 486-9906

Yamaha Robotics
888 Sussex Boulevard
P.O. Box 190
Broomall, PA 19008
Phone: (800) 829-2624
Fax: (610) 543-8113

Yushin America, Inc.
35 Kenney Drive
Crayston, RI 02920
Phone: (401) 463-1800
Fax: (401) 463-1810